PERFORMANCE AND DURABILITY ASSESSMENT:
Optical Materials for Solar Thermal Systems

Elsevier Internet Homepage - http://www.elsevier.com
Consult the Elsevier homepage for full catalogue information on all books, journals and electronic products and services.

Related Journals:
A sample journal issue is available online by visiting the homepage of the journal (homepage details at the top of this page). Free specimen copy gladly sent on request. Elsevier Ltd, The Boulevard, Langford Lane, Kidlington, Oxford, OX5 1GB, UK

Optical Materials
Renewable Energy
Renewable and Sustainable Energy Reviews
Solar Energy
Solar Energy Materials and Solar Cells
Thin Solid Films

To contact the Publisher
Elsevier welcomes enquiries concerning publishing proposals: books, journal special issues, conference proceedings, etc. All formats and media can be considered. Should you have a publishing proposal you wish to discuss, please contact, without obligation, the Publishing Editor responsible for Elsevier's Energy programme:

Tony Roche
Senior Publishing Editor
Elsevier Ltd
The Boulevard, Langford Lane   Phone: +44 1865 843887
Kidlington, Oxford   Fax: +44 1865 843931
OX5 1GB, UK   E.mail: t.roche@elsevier.com

General enquiries, including placing orders, should be directed to Elsevier's Regional Sales Offices – please access the Elsevier homepage for full contact details (homepage details at the top of this page).

PERFORMANCE AND DURABILITY ASSESSMENT:

Optical Materials for Solar Thermal Systems

Edited by

Michael Köhl

Fraunhofer-Institut für Solare Energiesysteme, ISE, Freiburg, Germany

Bo Carlsson

SP Swedish National Testing and Research Institute, Borås, Sweden

Gary Jorgensen

Center for Performance Engineering and Reliability,
Golden, CO 80401-3393 USA

A.W. Czanderna

Czanderna Consulting, Denver, CO, USA

2004

ELSEVIER

Amsterdam – Boston – Heidelberg – London – New York – Oxford
Paris – San Diego – San Francisco – Singapore – Sydney – Tokyo

ELSEVIER B.V.
Sara Burgerhartstraat 25
P.O.Box 211, 1000 AE
Amsterdam,
The Netherlands

ELSEVIER Inc.
525 B Street, Suite 1900
San Diego,
CA 92101-4495
USA

ELSEVIER Ltd
The Boulevard
Langford Lane
Kidlington
Oxford OX5 1GB, UK

ELSEVIER Ltd
84 Theobalds Road
London WC1X 8RR
UK

First edition 2004
Library of Congress Cataloging in Publication Data
A catalog record is available from the Library of Congress.

British Library Cataloguing in Publication Data
A catalogue record is available from the British Library

ISBN 0-08-044401-6

∞ The paper used in this publication meets the requirements of ANSI/NISO Z39.48-1992 (Permanence of Paper).
Printed in the United Kingdom.

Working together to grow
libraries in developing countries

www.elsevier.com | www.bookaid.org | www.sabre.org

ELSEVIER   BOOK AID
            International   Sabre Foundation

Preface

The purposes of this Preface are (1) to describe how the expertise from six countries was brought together in the technical areas required for measuring the performance and assessing the durability of selected solar thermal components and materials, (2) to provide a perspective of why the work is crucially important for advancing a large number of solar technologies with spin off to other industrial products, (3) to provide an overview of why the service lifetime prediction (SLP) methodology is also an approach that is generically applicable to a wide variety of materials, components, and systems, and (4) to summarize the past and present projects of the Solar Heating and Cooling Program of the International Energy Agency (IEA), which provided the framework for carrying out part of the work described in this book. This book is divided into six parts. Important definitions are given throughout the book, but commonly used terms such as performance criteria, evaluation criteria, testing methods, degradation stresses, mechanisms, modes, and effects have been concisely collected.

For (1), the collaborative research was begun in 1985 as an IEA Task 10, Solar Materials R&D project and concluded in 1991, during which time an assessment was made of the durability of solar absorber coatings. The principal members of the research group at that time were Bo Carlsson, Ueli Frei, Michael Köhl, and Kenneth Möller, all of whom are authors in this volume. The scientific background of the participating experts permitted the development of a generic methodology for a service lifetime prediction of selective solar absorber coatings used in flat-plate collectors. An IEA Working Group on Materials for Solar Thermal Collectors (MSTC) functioned from 1994 through 1999 for the express purpose of validating the methodology by comparing the degradation in performance of absorber coatings during in-service use with the results from accelerated life testing procedures. Beginning in 1994, the remaining authors contributing to this volume, i.e., Stefan Brunold, Markus Heck, Ole Holck, Gary Jorgensen, Henk Oversloot, Arne Roos, and Svend Svendsen joined the group. Most of the results presented in this volume were obtained by the joint efforts made by the experts in the expanded group. The work has been published, in part, in more detail in other publications and reports, which are referenced in a number of the chapters. The book was prepared as a shared effort, i.e. the principle authors are acknowledged for the different chapters in the book simply because they assumed responsibility for the content of the chapter.

For (2), applying a SLP methodology to other solar technologies is crucially needed. Beginning in 2000, the principal expertise of the MSTC group has been applied for studying other advanced materials in solar thermal systems, i.e., glazing materials including electrochromic and gasochromic windows, reflector materials, solar facades, and advanced building facade materials under IEA Task 27. Many solar devices studied consist of multilayer stacks of materials. As such, they are thermodynamically unstable, so the kinetics of degradative processes control the usable service life of the devices. Therefore, the SLP methodology described in this book also has many other applications where complex multilayer materials are used, e.g., in photovoltaic (PV), biomedical, aerospace, electronic, and coatings technologies.

For (3), using a SLP methodology is especially essential for materials, components, and systems, which require lifetimes exceeding 30 years. Technological advances cannot wait for 30 years of testing to determine the potential for a product. The SLP methodology depends on comparing the results of degradation using accelerated life testing (ALT) or aging with those from only a few years of real time testing, and ideally not more than two years. From materials analyses of the ALT test specimens, degradation mechanisms must be deduced and compared with the mechanisms determined after various times of exposure of replicate specimens in an in-service, multi-stress environment but without acceleration. Once the degradation mechanisms are known, new or existing products can be prepared or improved by using different materials or constructions that will slow the rate of degradation for next generation devices.

For (4), the International Energy Agency (IEA) was established in 1974 as an autonomous agency within the framework of the Economic Cooperation and Development (OECD) to carry out a comprehensive program of energy cooperation among its 25 member countries and the Commission of the European Communities. An important part of the IEA programs involves collaborative research, development and demonstration of new energy technologies to reduce excessive reliance on imported oil, increase long-term energy security, and reduce greenhouse gas emissions. Their R&D activities are headed by the Committee on Energy Research and Technology (CERT) and supported by a small Secretariat staff, headquartered in Paris. In addition, three Working Parties are charged with monitoring the various collaborative energy agreements, identifying new areas for cooperation and advising the CERT on policy matters.

Collaborative programs in the various energy technology areas are conducted under Implementing Agreements, which are signed by the contracting parties (government agencies or entities designated by them). There are currently over 40 Implementing Agreements covering renewable energy technologies, efficient energy end-use technologies, fossil fuel technologies, nuclear fusion science and technology, and energy technology information centers.

The Solar Heating and Cooling Programme was one of the first IEA Implementing Agreements to be established. Since 1977, 20 member countries have been collaborating to advance active solar, passive solar, and photovoltaic technologies and their application in buildings. They are:

Australia	Finland	Norway
Austria	France	Spain
Belgium	Italy	Sweden
Canada	Japan	Switzerland
Denmark	Mexico	United Kingdom
European Commission	Netherlands	United States
Germany	New Zealand	Portugal

More than 30 Tasks have been initiated, 20 of which have been completed. An Operating Agent from one of the participating countries manages each Task. Overall control of the program rests with an Executive Committee comprised of one representative from each contracting party to the Implementing Agreement. In addition, a number of special ad hoc activities – working groups, conferences, and workshops – have been organized.

The completed and current Tasks (with Tasks of particular relevance to this book underlined) of the IEA Solar Heating and Cooling Programme are:

Completed Tasks and Topics

Task 1   *Investigation of the Performance of Solar Heating and Cooling Systems*
Task 2   *Coordination of Solar Heating and Cooling R&D*
Task 3   *Performance Testing of Solar Collectors*
Task 4   *Development of an Insolation Handbook and Instrument Package*
Task 5   *Use of Existing Meteorological Information for Solar Energy Application*
Task 6   *Performance of Solar Systems Using Evacuated Collectors*
Task 7   *Central Solar Heating Plants with Seasonal Storage*
Task 8   *Passive and Hybrid Solar Low Energy Buildings*
Task 9   *Solar Radiation and Pyranometry Studies*
Task 10 *Solar Materials R&D*
Task 11 *Passive and Hybrid Solar Commercial Buildings*
Task 12 *Building Energy Analysis and Design Tools for Solar Applications*
Task 13 *Advanced Solar Low Energy Buildings*
Task 14 *Advanced Active Solar Energy Systems*
Task 16 *Photovoltaics in Buildings*
Task 17 *Measuring and Modeling Spectral Radiation*
Task 18 *Advanced Glazing and Associated Materials for Solar and Building Applications*
Task 19 *Solar Air Systems*
Task 20 *Solar Energy in Building Renovation*

Task 21 *Daylighting in Buildings*
Task 22 *Building Energy Analysis Tools*
Task 23 *Optimization of Solar Energy Use in Large Buildings*
Task 24 *Solar Procurement*

Current Tasks and Topics
Task 25 *Solar Assisted Air Conditioning of Buildings*
Task 26 *Solar Combisystems*
Task 27 *Performance of Solar Facade Components*
Task 28 *Solar Sustainable Housing*
Task 29 *Solar Crop Drying*
Task 31 *Daylighting Buildings in the 21st Century*
Task 32 *Advanced Storage Concepts for Solar Thermal Systems*
Task 33 *Solar Heat for Industrial Processes*
Task 34 *Testing and Validation of Building Energy Simulation Tools*

To receive a publications catalogue or learn more about the IEA Solar Heating and Cooling Programme please visit the Internet site at http://www.iea-shc.org or contact the SHC Executive Secretary, Pamela Murphy, Morse Associates Inc., 1808 Corcoran Street, NW, Washington, DC 20009, USA, E-mail: pmurphy@Morse AssociatesInc.com. Tel: +1 (202) 483-2393, Fax: +1 (202) 265-2248.

The work which is described in this book was financially supported by the Danish Energy Agency, the German Federal Ministry of Education and Research under Contract No. 0329600 and the German Federal Ministry of Economics and Labour under Contract No. 0327276A, the Netherlands Agency for Energy and Environment under contract no. Novem 143.300-529.0.LB, the Swedish Council for Building Research (Now Swedish Council for Environment, Agriculture Sciences and Spatial Planning), Swedish Energy Agency, and Swedish National Testing and Research Institute, the US Department of Energy under Contract No. DE-AC36-99GO10337 and the Swiss Federal Office for Energy.

The authors are very grateful to Stefanie Brambach and Susanne Ehling for their tireless support in preparing the manuscript.

Michael Köhl. Freiburg, Germany
Bo Carlsson, Borås, Sweden
Gary Jorgensen, Golden, CO, U.S.A
A.W. Czanderna, Lakewood, CO, U.S.A
June 2004

List of Contributors

Stefan Brunold, Institut für Solartechnik SPF, Hochschule für Technik Rapperswil HSR, Oberseestr. 10, CH-8640 Rapperswil, Switzerland

Bo Carlsson, SP Swedish National Testing and Research Institute, P.O. Box 857, S-501 15 Borås, Sweden

Ueli Frei, Institut für Solartechnik SPF, Hochschule für Technik Rapperswil HSR, Oberseestr. 10, CH-8640 Rapperswil, Switzerland

Markus Heck, Fraunhofer-Institut für Solare Energiesysteme, ISE, Heidenhofstr. 2, D-79110, Freiburg, Germany

Ole Holck, Department of Civil Engineering, Building 118, Technical University of Denmark, DK-2800 Lyngby, Denmark

Gary Jorgensen, Center for Performance Engineering and Reliability, MS-3321, National Renewable Energy Laboratory, 1617 Cole Boulevard, Golden, CO 80401-3393 USA

Michael Köhl, Fraunhofer-Institut für Solare Energiesysteme, ISE, Heidenhofstr. 2, D-79110, Freiburg, Germany

Volker Kübler, Fraunhofer-Institut für Solare Energiesysteme, ISE, Heidenhofstr. 2, D-79110, Freiburg, Germany

Kenneth Möller, SP Swedish National Testing and Research Institute, P.O. Box 857, S-501 15 Borås, Sweden

Henk Oversloot, TNO Building and Construction Research, Department of Sustainable Energy and Buildings, P.O. Box 49, 2600 AA – Delft, The Netherlands

Arne Roos, The Angstrom Laboratory, Uppsala University, Box 534, S-751 21 Uppsala, Sweden

Svend Svendsen, Department of Civil Engineering, Building 118, Technical University of Denmark, DK-2800 Lyngby, Denmark

Contents

Part 1

Introduction

In this introductory part, an overview is provided about the essential concepts for the performance and durability assessment of the specific solar collector systems that are featured in this volume. The essential roles of durability assessment, accelerated life testing, and service lifetime prediction are discussed followed by a presentation of the need for solar products to meet three important criteria, i.e., minimum cost, adequate performance, and demonstrated durability for achieving successful and sustainable commercialisation. Descriptions are given of protective transparent covers, e.g., glass and polymer covers, and several types of multilayer designs of metalized polymers for solar reflector applications. A flate-plate collector, how a selective absorber functions, and typical multilayer solar absorber coatings are then described. The specific solar collector systems described have been studied for more than ten years, and representative results of these studies are presented in this volume.

Chapter 1.1

Introduction to the Performance and Durability Assessment of Optical Materials for Solar Thermal Systems

B. Carlsson,[1] G. Jorgensen,[2] and M. Köhl[3]

[1]*Swedish National Testing and Research Institute, P.O. Box 857, S-501 15 Borås, Sweden*
[2]*National Renewable Energy Laboratory, 1617 Cole Blvd, Golden, CO 80401, USA*
[3]*Fraunhofer-Institut für Solare Energiesysteme, Heidenhofstr. 2, D-79110, Freiburg, Germany*

Abstract: The purposes of this chapter are to provide an overview of the essential concepts about the performance and durability assessment of solar thermal components and materials and to describe the specific systems that are featured in this volume. The essential roles of durability assessment, accelerated life testing and service lifetime prediction are discussed first. The need for solar products to meet three important criteria, i.e., minimum cost, adequate performance, and demonstrable durability, for achieving successful and sustainable commercialization is then presented. Descriptions are given of protective transparent covers, e.g., glass and polymer covers, and several types of multilayer designs of metalized polymers for solar reflector applications. A flat-plate collector, how a selective absorber functions, and typical multilayer solar absorber coatings are then described. The specific solar thermal components and materials described have been studied for more than ten years, and representative results of these studies are presented in this volume, especially to illustrate the essential concepts.

Keywords: Accelerated life testing, Solar system cost, Durability, Performance, Service lifetime predictions, Solar reflectors, Flat-plate collectors, Solar thermal absorbers

1.1.1 INTRODUCTION TO DURABILITY, ACCELERATED LIFE TESTING, AND SERVICE LIFETIME PREDICTION

To achieve successful and sustainable commercialization, solar products must meet three important criteria, namely minimum cost, adequate performance, and demonstrable durability. Durability assessment directly addresses all three segments

3

of this triad. First, it permits analysis of life-cycle costs by serving as a first step toward a service lifetime prediction, operating and maintenance costs, and realistic warranties. Understanding how performance parameters are affected by environmental stresses (for example by failure analysis) allows improved products to be designed. Finally, mitigation of known causes of failure directly results in increased product longevity. Thus, the accurate assessment of durability is of paramount importance for assuring the success of solar thermal and building products.

To select the most cost-effective material, component or product requires durability to be expressed in quantitative terms in the form of an expected service life for the material, component or product in the application considered. Durability in this case is the ability of a material or a product to withstand deterioration caused by all external factors in the environment, which may influence the performance of the material, component or product under service conditions. The time aspect of durability is related to the concept of "performance over time," which is the variation in time of specific properties important for the performance of the material, component or product. "Service life" is defined as the period of time after installation during which these properties – the performance – meet or exceed minimum acceptable values. Consequently, the service life of a material, component or product is not only dependent on its properties, but also on its performance requirements in the application considered, and on the external environmental factors that influence performance under service conditions.

Data on the durability of materials and service life of components or parts of components can be obtained or estimated from different kinds of sources, for example, feedback from practice, results of long-term tests under in-use conditions, and estimates based on accelerated life testing. In the case of components with traditional materials in conventional applications, feedback data from practice are often available in the form of actual service lifetimes of materials and parts of components, and such data can serve directly as an aid for selecting the material in component design. A problem that may arise, however, is that environmental factors contributing to material degradation vary with time and location. This makes it difficult, as a rule, to transfer material life data from one climate to another. Long-term tests under in-use conditions are one way of generating service life data for components or parts of components with new or substitute materials, or for components or parts of components with traditional materials in new applications. However, it may take a long time to obtain results from in-service tests, unless changes in the properties of the materials can be detected at an early stage of degradation.

One alternative to long-term testing under in-service conditions is accelerated life testing in which the stress levels of one or more degradation factors are kept higher relative to in-use conditions. As a result, the testing time can be much shorter than the expected service life. For the development of components with new materials, accelerated life testing is, as a rule, the only realistic way for estimating the likely long-term performance of the components prior to their market introduction in new products. Early efforts for developing accelerated testing to age materials were often aimed at attempting to speed up the degradation of the materials, under laboratory conditions, to as great an extent as possible. In many cases, the exposure conditions employed in the laboratory tests were altogether too extreme, resulting in damage to the materials of a type that would not be observed under normal conditions of use. The suitability of these laboratory tests varied from one type of material to another.

For developing the present-day accelerated methods of aging, R&D efforts have become more mechanism-related. When developing a new method, it is essential that the same type of aging mechanisms occur in the materials during the accelerated testing as during normal in-service conditions. A general methodology, which allows for all the factors associated with the work to be considered, and a quantitative approach are necessary in to be able to predict the expected service life of a component and the limitations in the service life set by the durability of its materials.

Several research activities were carried out within the IEA Solar Heating and Cooling Programme (Task 10 "Solar Materials Research and Development," Working Group on "Materials of Solar Thermal Collectors" and Task 27 "Performance of Solar Facade Components," which was begun in early 2000) with the objectives of developing and applying appropriate methods for the assessment of durability, reliability, and environmental impact of advanced materials and components for solar thermal applications.

There are two main objectives for the work on durability described in this volume. The first is to develop a general protocol for durability testing procedures and service lifetime prediction (SLP) methods that are applicable to a wide variety of advanced optical materials and components used in energy efficient solar thermal and buildings applications. The second is to apply the appropriate durability testing procedures, which allow prediction of the service lifetime on specific materials or components, and to generate proposals for international standards.

1.1.2 PERFORMANCE AND DURABILITY, RISKS AND COSTS

As mentioned above, solar building components must fulfil three important criteria to achieve successful and sustainable commercialization, namely: (1) maximum performance, (2) demonstrable durability, and (3) minimum cost. This triad of requirements is shown schematically in Figure 1-1. *Performance* is a measure of a product's ability to function in a useful or desired manner. For example, the merit of a glazing material may be characterized by its solar-weighted hemispherical transmittance; a solar energy collection system may be quantified by its overall

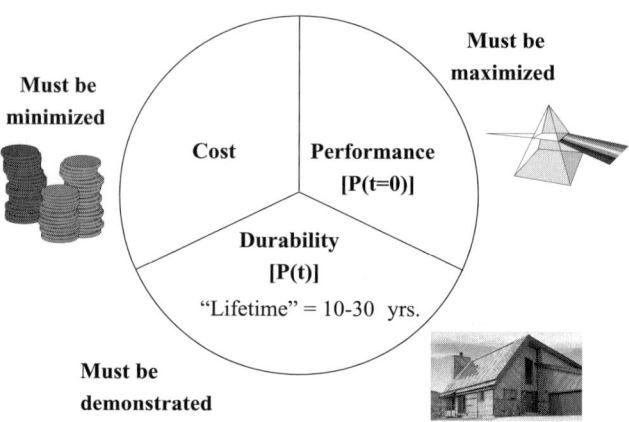

Figure 1-1.   The product requirement triad.

efficiency. *Durability* is the ability to retain performance (avoid degradation) during in-service use throughout a products lifetime. *Costs* include the initial capital expenditure as well as cost of upkeep. Durability assessment directly addresses all three segments of this triad. First, it permits analysis of life-cycle costs by providing estimates of service lifetime, O&M costs, and realistic warranties.

Because performance goals are generally specified as the ability to maintain an average value above some critical level (P_{crit}) for a specified length of time, and where $P(t')$ is an instantaneous performance parameter at time t'. It is convenient to represent performance as a time-averaged quantity (Jorgensen *et al.*, 1996) defined as:

$$\overline{P}(t) = \frac{\int_0^t P(t')\,dt'}{t} \tag{1}$$

In addition, the time-weighted performance can represent an associated series of measured values with a single quantity. For example, Figure 1-2 shows two sets of actual performance data, P_1 (unfilled diamonds) and P_2 (unfilled squares). Often, the "time to failure" is defined as the time, t_f, at which the performance drops below some predefined value of P_{crit}:

$$P_{crit} \equiv \overline{P}(t_f) = \frac{\int_0^{t_f} P(t')\,dt'}{t_f} \tag{2}$$

A time-averaged value also contains more information and is more meaningful in terms of determining lifetime distributions from observed failure times.

A useful example is shown in Figure 1-2. Both P_1 and P_2 drop below $P_{crit} \equiv 90\%$ at $t = 2$ years. However, the corresponding time-averaged performance measures, P_1 (filled diamonds) and P_2 (filled squares) "fail" at $t = 3$ years and $t = 4$ years,

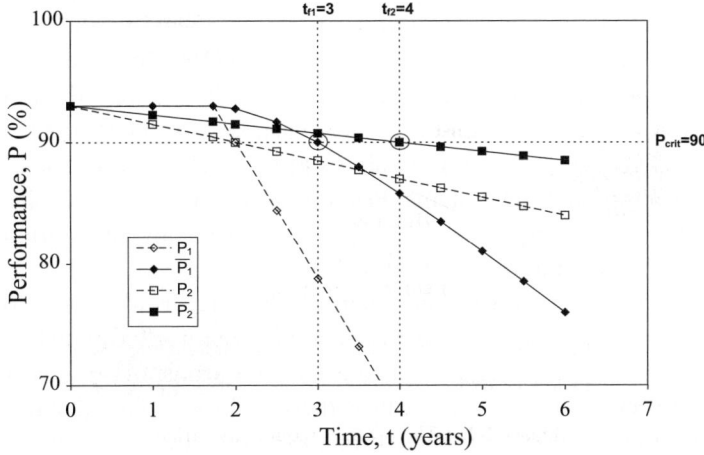

Figure 1-2. Relationship between performance and time-averaged performance.

respectively. Thus, a material or component exhibiting performance P_2 (linear degradation with time) can provide a substantially longer service lifetime (in a time-averaged sense) than a material that does not degrade at all for some induction period, but then has a rapid loss in performance (such as P_1).

A detailed approach for the financial analysis of the costs, $C(t)$, associated with energy efficiency and renewable energy technologies and projects has been provided (Short *et al.*, 1995). Quantities such as cash flow, inflation rates, discount rates, cost of capital, present value, taxes, depreciation, fixed charge rates, financing, and risk and uncertainty are included in the cost analysis. In general, it is desired to minimize cost and maximize performance for a product lifetime (t). Thus, the cost-per-performance ratio (CPR) must be minimized:

$$\mathrm{CPR}(t) = \frac{C(t)}{P(t)} = \text{Minimum} \tag{3}$$

1.1.3   SOLAR THERMAL MATERIALS AND COMPONENTS

1.1.3.1 Introduction

The oil crisis in the early 1970s revealed the drastic consequences of energy-dependent countries with insufficient fossil fuel resources. Studies of alternative energy sources revealed that the principle options are to use nuclear energy, renewable energy, and energy-saving (conservation) technologies. Energy saving and renewable energies have the additional advantage of sustainability and low-environmental impact. The disadvantages of renewable energies are the daily and seasonal fluctuations, the potential need for some storage, the low solar power density, i.e., $1\,kW/m^2$ as a maximum for the solar resource, and their incompatibility with traditional power grids. However, their exploitation offers the possibility of a truly sustainable energy supply in the future. This book is about some of the aspects of solar thermal energy conversion materials at low temperatures used for domestic hot water, heating or cooling systems. Because of the low power density of the solar irradiation, the solar energy has to be collected from an appropriate area by a collector shown in the diagram of Figure 1-3.

This device looks very simple at first glance. The solar irradiation is absorbed and converted into internal energy by the absorber coating on a flat-plate heat exchanger, where the energy gained is transferred to a fluid for transport to a thermal storage unit. The absorber is insulated to reduce thermal losses. The glazing, acting as a convection barrier, is used for insulating the front aperture. Solar collectors in temperate climates must perform well to be competitive with conventional heating sources. The materials properties, particularly the optical properties during the interaction with solar radiation, contribute essentially to the efficiency of the components and the overall performance of a solar system.

Solar radiation has to interact with materials to be converted into thermal or electrical energy. These materials reflect, transmit, or absorb the irradiation according to the physical laws formulated by Maxwell and Fresnel (see Figure 1-4.).

Energy conservation requires the incoming radiation (normalized to unity) to be the sum of the reflected, absorbed, and transmitted radiation: $A + R + T = 1$. The ratio among the different components is determined by the materials properties.

1.1.3.2 *Reflectors*

Reflectors are used to redirect solar radiation and to concentrate it to increase the power density on an absorber and thus achieve higher temperatures. Because the reflectors are essentially opaque, the losses upon reflection usually result from absorption in the materials. Solar mirrors must be low cost and maintain high

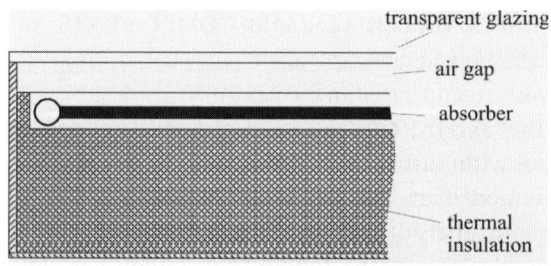

Figure 1-3. Schematic drawing of a flat-plate solar collector.

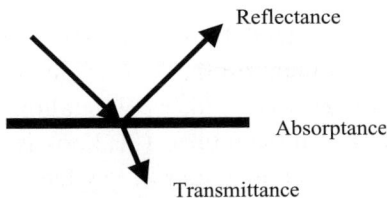

Figure 1-4. Radiation and optical properties of materials.

reflectance for extended lifetimes, e.g., typically at least 10 years, during harsh outdoor service conditions. Additional requirements for solar reflector materials include low weight, mechanical robustness (nonfragile), shape flexibility, and formability. Manufacturability in terms of available width and lamination characteristics is also critical. Mirrors, which are abrasion resistant, cleanable, replaceable, and easily handled during manufacture, transport, and field installation, are desirable as well.

Appropriate reflector materials for solar applications are silver and aluminum. Silver has the highest reflectance, but has to be protected against weathering, which is expensive. Silver thin films are formed on glass or polymeric substrates with additional corrosion-protection and reflectance enhancing coatings (see Chapters 4.7 and Section 6, especially). Aluminum is suitable for lower concentration factors or lower efficiency. Low-end products are simply anodized for corrosion protection. For corrosion protection, high-end reflectors are coated with a reflecting metal layer (silver or aluminum) and additional dielectric coatings are formed using physical deposition techniques such as reactive sputtering or evaporation and applying and transparent polymeric corrosion protection (see Figure 1-5).

Candidate solar reflector material constructions include metalized thick and thin glass, metalized polymeric sheets and films, protected front surface mirrors, and multilayer dielectric designs (Jorgensen, 1993). Accelerated and outdoor exposure test results for a large number of these types of samples have been reported (Jorgensen *et al.*, 2000). The most common solar reflectors are metalized "thick" glass mirrors that are produced by applying a silver reflective layer and an adhesion-promoting layer (usually copper) onto glass (typically > 1-mm thick) using wet chemistry processes, followed by applying a protective backside paint. Because the reflective layers are corrosion resistant, thick mirrors have excellent durability; they are readily available, have the confidence of the solar manufacturing industry, but are heavy and fragile. Curvature is difficult and requires using slumped glass that is expensive, but thick mirrors have been commercially deployed. Depending upon how the structural support of the mirrors is provided, high winds can cause significant breakage. Their unweathered solar-weighted hemispherical reflectance is ~88 to 92% and their cost is ~\$15 to 40 m^{-2}.

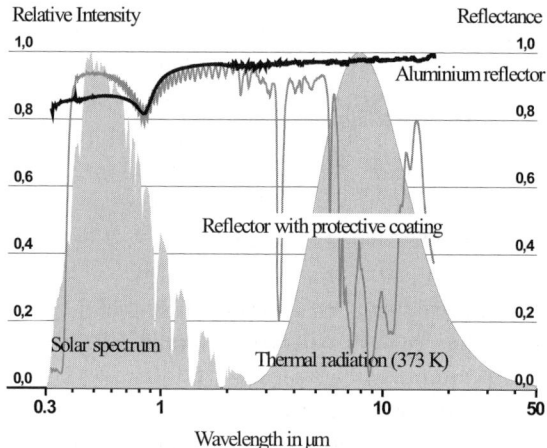

Figure 1-5.   Spectra of an aluminium reflector, an optimized reflector with protection coating, the solar irradiation, and the Planck function for 373 K.

Thin glass mirrors have a durability, which is comparable to that for thick glass mirrors because they are made using the same wet silvering processes, but use a relatively lightweight glass that is ≤1-mm thick; they also have greater material costs, are more difficult to handle, and have higher associated labor costs (25–40%) than other advanced reflector technologies. Their unweathered solar-weighted hemispherical reflectance is ~93 to 96% and their cost is ~$15 to 40 m^{-2}. The choice of adhesive, which is used to bond the mirror to a structural substrate, can adversely affect the performance of weathered thin glass mirrors; corrosion has been detected in thin glass mirrors deployed in the field.

Polymer reflector constructions are attractive for solar applications because they minimize weight and cost and allow for ease of handling during manufacture, transport, and field installation. However, because polymers are significantly more permeable than glass to water, oxygen, and pollutants, corrosion of the metal reflective layer and subsequent loss in reflectance is a difficult problem that must be overcome (Schissel and Czanderna, 1980). A number of advanced metalized polymer constructions have been developed (Jorgensen and Schissel, 1989; Schissel *et al.*, 1994); these may have lifetimes of 5 to 10 years depending on where they are used. Their unweathered solar-weighted hemispherical reflectance is ~92 to 94%. Uncertainties regarding commercial availability exist but cost estimates range from $15 to 45 m^{-2}.

One of the most promising ways to reduce the cost of solar mirrors is to metallize an appropriate, inexpensive substrate material, such as polyethylene terephthalate (PET) that is a thermoplastic co-polyester film, and then overcoat the reflector with an abrasion-resistant, durable, protective top layer. Some reflectors with this generic construction have demonstrated encouraging results in accelerated durability tests. In particular, a construction being developed, which uses a dense alumina overcoat: Al_2O_3/Silver/PET, has demonstrated excellent optical durability and potential for low cost (Kennedy *et al.*, 1997). In an attractive variation of this construction, silver is "directly deposited" onto appropriately levelized stainless steel substrate to allow obtaining a high specular reflectance and electrochemical isolation between the dissimilar reflective and substrate metal layers.

Front surface aluminized reflectors are typically prepared by polishing an aluminum substratem, electrochemically depositing a reflective layer of aluminum, and allowing a protective oxidized topcoat (alumina) to form. This type of construction exhibits adequate optical durability in nonindustrial/urban environments; however, they corrode very rapidly in the presence of atmospheric pollutants. A recent formulation (Fend *et al.*, 2000), in which the aluminum reflective layer is deposited by physical vapor deposition and a thin protective polymeric overcoat is used to protect the alumina layer, has survived for more than three years outdoor exposure in Köln, Germany. Traditional front surface aluminized reflectors experience significant degradation in less than six months at this highly industrialized location. The unweathered solar-weighted hemispherical reflectance of these aluminized reflectors is ~89 to 90%; the product is commercially available for $< \$20\,m^{-2}$.

For the multilayer dielectric design concept (Fink *et al.*, 1998), an all-polymeric version has been developed that uses thousands of alternating coextruded layers of low-cost commercially available transparent thermoplastics to achieve high reflectance. These mirrors do not corrode because no metals are used. Both thin film and thermoformed sheet reflectors have been fabricated; thermoforming is used to provide the desired shape and curvature. Presently, multilayer dielectric mirrors only reflect over a limited spectral range. However, a broadband, i.e., across the terrestrial solar spectrum, reflector material is being developed. Further improvement is being sought by the development of UV-screening skin layers to impart adequate weatherability.

1.1.3.3 *Transparent Covers*

Transparent covers usually have to prevent convective heat losses and protect against weathering in the same way as windows. A low-iron glass is typically used, because its absorptivity is lower than that of ordinary float glass, yielding a higher solar transmittance. The other transmittance losses result from reflection from both sides of the pane, which is typically about 8% per glass pane. These losses can be reduced by using antireflective (AR) coatings, such as the well-known examples for ophthalmic glasses. However, this type of multilayer coating is not suitable for solar energy applications, because the AR properties are only for the visible part of the spectrum and also enhance the reflectance in the near-infrared range. The best AR coatings for solar applications need to have a refractive index n of about 1.2, which corresponds to the square root of the refractive index of glass, over the entire solar spectral range. Some fluoride materials could be considered, but they are not durable in humid environments.

Current research is focused on developing porous coatings that allow the integrated solar transmittance to be increased from 0.90 to 0.96 (Gombert *et al.*, 2000). When the pores are much less than the wavelength of light, i.e., smaller than 300 nm, the coating can be considered as an "effective medium" (Maxwell Garnett 1906; Bruggemann, 1935). The optical properties are then a kind of average of those of the component materials, depending on their volume fraction. The effective refractive index of a porous silica layer, for example, depends on the volume fraction of the pores ($n = 1$). Porous silica layers can be deposited by sol–gel processes (Nostell *et al.*, 1999). Porous layers of glass are obtained by etching the glass pane with hydrofluoric acid. Durability testing protocols for candidate AR coated glazing materials, and initial results, have been discussed (Jorgensen *et al.*, 1999).

1.1.3.4 Solar Absorber Coatings
The most important material in collectors is the absorber coating. The absorber should be an optical coating that absorbs as much as possible of the incident solar radiation, i.e., in the spectral wavelength range from 0.3 to 3 µm (see Figure 1-6).

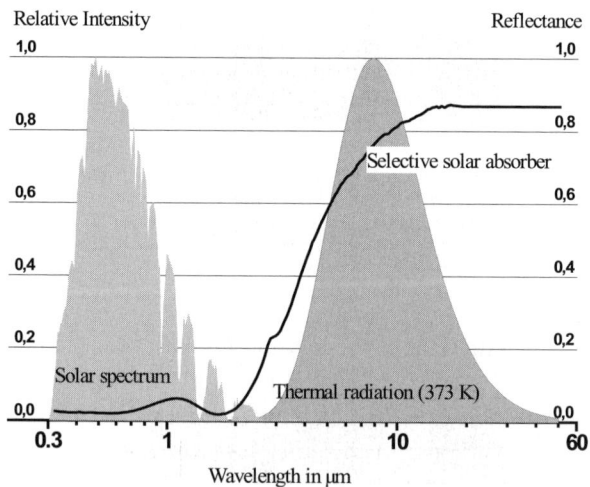

Figure 1-6. Spectra of a selective solar absorber coating, the solar irradiation, and the Planck function for 373 K.

Moreover, it should not lose energy by emitting thermal radiation in the infrared according to Planck's law. These radiative losses depend on the fourth power of the surface temperature and the spectrally dependent emittance in the infrared (IR) from 3 to 50 μm of the coating system. An optimized coating, showing a high absorptance in the solar range and a low emittance in the thermal range, is called a spectrally selective solar absorber coating.

Solar absorbers are typically prepared by depositing a coating, which absorbs the solar irradiation and transmits the thermal radiation, to an IR-reflecting metal substrate or metal coating. Therefore, the absorber coating has to be relatively thin, e.g., between 0.15 and 1 μm, and usually has also to serve as corrosion protection for the metal substrate. Because the ideal combination of these properties cannot be provided by one single material, composite materials and multilayer coatings are used (see Figure 1-7).

An electroplated black chrome coating, which was the most popular selective absorber coating in the 1970s is composed of single particles of metallic chromium and chromium oxide, forming an effective medium and providing a porous anti-reflective surface, as can be seen in Figure 1-8.

It is obvious that such a porous layer requires an additional corrosion barrier between the metal substrate and the coating. An electroplated nickel coating, which is about 5 μm thick, is used for this purpose. The IR emittance is limited to more than 7% by the properties of nickel. A similar coating system based on nickel and

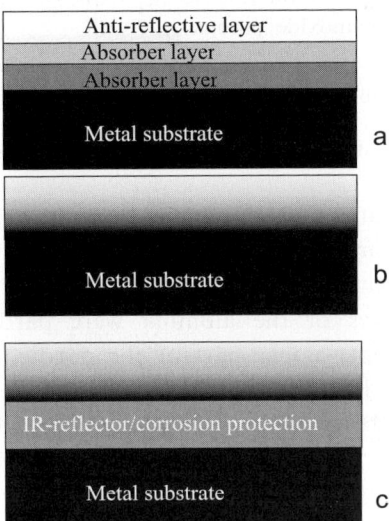

Figure 1-7. Design possibilities of solar absorber coatings: (a) multilayer coating; (b) graded index coating; and (c) graded index coating on a protective layer against corrosion.

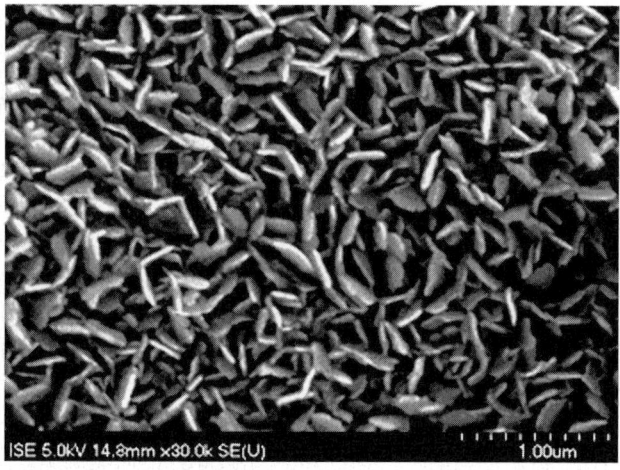

Figure 1-8.   Scanning electron micrograph of an electroplated black chrome coating.

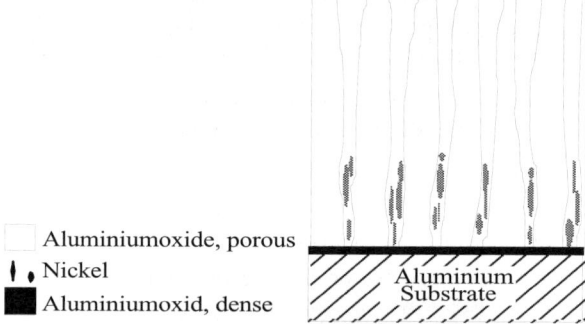

Figure 1-9.   Schematic drawing of an anodized nickel-pigmented aluminum coating.

nickel oxide composite materials is known as black nickel. Currently, this kind of coating is not available commercially.

Porous anodized aluminum has also been used as a substrate material for selective solar absorbers. The pores of the alumina were partly filled with nickel by electrodeposition. The empty upper part of the porous alumina forms the anti-reflective layer (see Figure 1-9). The thickness of the different layers of the stack has to be optimized. The durability of the porous aluminum absorber is poor in humid environments.

A wide variety of coatings have been developed in the last few decades and are described in numerous publications. Novel coating technologies facilitate the exploitation of new coatings. Coatings produced with modern vapor deposition technologies are becoming commercially available (Graf *et al.*, 1995). These are

Figure 1-10. Sputter coil coating plant for depositing solar absorber coatings on copper (source: Interpane Solarbeschichtungs GmbH).

usually of the design shown in Figure 1-7 (a or b) because sputtering or vapor deposition results in dense layers that prevent oxygen diffusion. Figure 1-10 shows a coil coating plant for preparing sputtered selective solar absorber coatings on metal coils with a width of up to 1.1 m.

The coated metal needs to be in contact with a heat-transfer medium, such as water, a water–glycol mixture, or air. Usually tubes are connected to the absorber sheets by welding, soldering, or roll forming (see 2.2.2.3). The durability of the coating could be reduced because of the type of processing used. Special attention has to be paid to areas where the protective coatings have been partially removed.

An overview of the present state of the art of solar energy including aspects from materials to applications can be found in Gordon (2001).

REFERENCES

Bruggemann, D.A.G. (1935) *Ann. Phys.*, **24**, 635.

Fend, T., Jorgensen, G. & Küster, H. (2000) Applicability of Highly Reflective Aluminum Coil for Solar Concentrators, *Solar Energy*, **68**, 361–370.

Fink, Y. *et al.* (1998) A Dielectric Omnidirectional Reflector, *Science*, **282**, 1679–1682.

Gombert, A., Glaubitt, W., Rose, K., Dreibholz, J., Bläsi, B., Heinzel, A., Sporn, D., Döll, W. & Wittwer, V. (2000) Antireflective Transparent Covers for Solar Devices, *Solar Energy*, **68:4**, 357–360.

Gordon, J., (Ed.) (2001) *Solar Energy: The State of the Art*, James & James, London.

Graf, W., Brucker, F., Köhl, M., Tröscher, T., Wittwer, V., Blessing, R. & Herlitze, L. (1995) Sputtered Solar Absorber Coatings with High Spectral Selectivity and Good Durability, *International Symposium on Optical Science, Engineering and Instrumentation, Proceeding SPIE 2531*, p. 27.

Jorgensen, G. & Schissel, P. (1989) Optical Performance and Durability of Silvered Polymer Mirrors, In: Mittal, K.L. & Susko, J.R. (Eds.), *Metallized Plastics 1, Fundamental and Applied Aspects*, pp. 79–92, Plenum Press, NY.

Jorgensen, G. (1993) Reflective Coatings for Solar Applications, *Society of Vacuum Coaters 36th Annual Technical Conference Proceedings*, pp. 23–31, Society of Vacuum Coaters, Albuquerque, NM.

Jorgensen, G., Brunold, S., Köhl, M., Nostell, P., Oversloot, H., & Roos, A. (1999) Durability Testing of Antireflection Coatings for Solar Applications, In: Lampert, C.M. & Granqvist, C.-G. (Eds.), *Solar Optical Materials XVI, Proceedings of SPIE 3789*, pp. 66–76, International Society for Optical Engineering, Bellingham, WA.

Jorgensen, G., Kennedy, C., King, D. & Terwilliger, K. (2000) *Optical Durability Testing of Candidate Solar Mirrors*, NREL/TP-520-28110, National Renewable Energy Laboratory, Golden, CO.

Jorgensen, G.J., Kim, H.M. & Wendelin, T.J. (1996) Durability Studies of Solar Reflector Materials Exposed to Environmental Stresses, In: Herling, Robert J. (Ed.), *Durability Testing of Non-Metallic Materials, ASTM STP 1294*, pp. 121–135, American Society for Testing and Materials, Philadelphia.

Kennedy, C.E., Smilgys, R.V., Kirkpatrick, D.A. & Ross, J.S. (1997) Optical Performance and Durability of Solar Reflectors Protected by an Alumina Coating, *Thin Solid Films*, **304**, 303–309.

Maxwell Garnett, J.C. (1906) *Philos. Trans. R. Soc.*, London A, **205**, 237.

Nostell, P., Roos, A. & Karlsson, B. (1999) Optical and Mechanical Properties of Sol-gel Antireflective Films for Solar Energy Applications, *Thin Solid Films*, **351**, 170–175.

Schissel, P. & Czanderna, A.W. (1980) Reactions at the Silver/Polymer Interface: A Review, *Solar Energy Materials*, **3**, 225–245.

Schissel, P., Jorgensen, G., Kennedy, C. & Goggin, R. (1994) Silvered-PMMA Reflectors, *Solar Energy Materials and Solar Cells*, **33**, 183–197.

Short, W., Packey, D.J. & Holt, T. (1995) *A Manual for the Economic Evaluation of Energy Efficiency and Renewable Energy Technologies*, NREL/TP-462-5173, National Renewable Energy Laboratory, Golden, CO.

Part 2
Materials Performance and System Performance

Part 2 consists of five chapters about optical materials and system performance. Service life prediction needs a clear definition of what loss in performance is acceptable to define the end of service life. In the first chapter, an overview is presented about how to measure optical transmittance and reflectance of typical materials and surfaces used in solar energy applications. In the second chapter, different designs of solar collectors are compared and contrasted in terms of their material properties and optical and thermal characteristics by using different models for calculations. In the third chapter, objective criteria are defined for comparing system performance and sophisticated simulation programs are used to estimate the annual solar gain of a particular system. In the fourth chapter, a summary is given about the concepts of system performance, performance criteria, loss in performance, and that system performance usually depends on a few important properties of the materials used. In the last chapter, various ways are outlined for quantifying changes in performance. The choice of an appropriate and sensitive measure of performance degradation is critical.

Chapter 2.1

Optical Properties and Measurements

A. Roos

The Angstrom Laboratory, Uppsala University, Box 534, S-751 21 Uppsala, Sweden

Abstract: This chapter provides an overview of how to measure optical transmittance and reflectance of typical materials and surfaces used in solar energy applications. The presentation is based on using a standard spectrophotometer for the measurements, a type of instrument readily available in many laboratories. Both double-beam and single-beam measurements are described. Apart from describing the basic measurements, descriptions of how to use various accessories are provided. Special attention is paid on how to avoid systematic errors when measuring at oblique angles of incidence and when using an integrating sphere. Most of the results are valid independently of wavelength and the problems and concepts presented are true for both the solar and infrared wavelength intervals. It is the intention that the reader will become aware of various sources of errors and learn how to take the necessary steps to avoid them.

Keywords: Optical measurements, Reflectance, Transmittance, Integrating spheres, Spectrophotometry

LIST OF ACRONYMS AND ABBREVIATIONS

ADOPT	Angular Dependent Optical Properties of Coated Glass and Glazing Products: Measurement Procedures and Validation of Associated Predictive Method
ALTSET	Angular Dependent Light and Total Solar Energy Transmittance for Complex Glazings
UV	Ultraviolet
VIS	Visible
NIR	Near Infrared
IR	Infrared

LIST OF SYMBOLS

A	amplification factor
B	high angle fraction of scattered radiation
F	fraction of scattered light contained in sphere
f	port area fraction
R	reflectance
S	detected signal
T	transmittance
V	detected signal
I	intensity

2.1.1 INTRODUCTION AND BACKGROUND

In all solar energy related research, it is of vital importance to know the optical properties of the studied materials. The optical reflectance and transmittance are the key properties that determine how the device is going to function during operation. There are several ways of measuring the optical properties and all these measurements must be compared to in-service tests, simulations and aging tests. Different ways of measuring the same property must be compared and evaluated for correctness and accuracy. Considering the diversity of materials used in solar energy applications, this is not a simple task and many investigations and research projects have been conducted to evaluate the optical properties of solar energy materials. Within the International Energy Agency – Solar Heating and Cooling Programme (IEA-SHCP), the optical properties of solar absorber materials were investigated in Task 10 and a solar absorber-working group. Glazing materials were studied in Task 18. The performance of building envelope materials is the subject of Task 27.

The optical properties of interest are sometimes the spectral properties of the materials. However, in solar energy applications, it is often the integrated properties over the visible or solar spectral range that are of importance. This means that the measured spectra have to be weighted with a suitable solar spectrum to obtain the final property. In materials research optical properties can be studied to understand the mechanisms of importance for the materials properties, and optical spectroscopy is used as a materials analysis tool. To predict the function of a device the optical properties of the complete device must be measured or calculated from the component properties. Thus, optical measurements must be performed both at the materials level and the component level. This means making measurements on both small laboratory-sized samples and large full-scale devices.

Materials used in solar energy applications range from materials for the absorption and reflection of solar irradiation to materials with a certain level of transmittance. The transmittance should sometimes be as high as possible and sometimes as low as possible without being totally opaque. All these materials can be perfectly smooth and nonscattering or completely diffuse or translucent. For surfaces that are not smooth, the surface texture or irregularities can be microscopic or macroscopic, and the resulting scattering can be isotropic or anisotropic. For a thermal solar absorber, which is the main subject of this book, the absorbing surface is usually made from rolled metal sheets that are never perfectly smooth. In fact, a certain surface roughness is usually used to enhance the absorptance. The collector cover can be a sheet of clear float glass without any scattering or some kind of textured glass or plastic cover, which scatters the solar radiation and reduces glare from the absorber surface. Antireflection coatings can be used to enhance the transmittance and improve the efficiency of the collector. Recent research has resulted in product development of antireflection coatings that are likely to become common on solar collectors in the future. Solar collectors can be combined advantageously with focusing or booster reflectors to enhance the efficiency of the system. These reflectors must be durable and as highly reflecting as possible. They cannot be completely diffuse like white paint, but they do not need to be perfect mirrors. A certain level of light scattering is acceptable and maybe even favorable.

Most solar devices are used in a stationary mode without tracking the sun. This means that the solar irradiation reaches the surfaces at different angles of incidence during different times of the day and during different seasons of the year. Therefore, the optical properties must be measured at different angles of incidence, and this results in several problems. In general, these measurements are difficult to perform and many laboratories are not adequately equipped to perform such measurements.

The diversity of materials and requirements for the optical properties make the optical measurements complex. In this chapter, standard spectrophotometric measurements are described and procedures for the measurements and interpretation of the results are presented. The presentation is in no way complete and is mostly focused on making spectrophotometric measurements on a laboratory scale. Some of the methods presented are general and do not depend on the sample size, but several problems encountered when using large integrating spheres will not be discussed. Some of the problems that have to be taken into account will be treated, so that the operator becomes aware of such problems and will avoid making the more common systematic errors when measuring the optical properties of solar energy materials.

2.1.2 OPTICAL PROPERTY MEASUREMENTS

2.1.2.1 *Basic Definitions*
When electromagnetic radiation, such as an electromagnetic monochromatic plane wave, reaches an interface between two media with different refractive indices, it is partly reflected and partly transmitted. The optical reflectance (R) and transmittance (T) are thus defined as the ratio between the reflected (I_r) or transmitted intensity (I_t) and the incident intensity (I_o):

$$R = I_r/I_o \tag{2.1}$$

$$T = I_t/I_o \tag{2.2}$$

Because the quantities are defined as ratios, the units, in which the intensity is measured, are not important. The reflectance and transmittance are dimensionless and are always in the range $0 < R, T < 1$. Both can be readily derived from the amplitude reflectance or transmittance as defined by the Fresnel equations. The magnitude and phase shift of the Fresnel coefficients depend on the complex refractive indices of the two media. The Fresnel formalism is a consequence of Maxwell's equations and will not be dealt with here. In this chapter, we are only concerned with the problems of measuring R and T as defined above. Throughout this chapter it is also assumed that R and T are independent of the incident intensity I_o. This is equivalent to saying that the electric field of the incident light is small compared to the local electric fields of the material. This is generally true for ordinary daylight and artificial illumination. In highly concentrated solar irradiation, the limit may be approached when this is no longer the case.

In principle, measuring R and T as defined above is very simple. All we have to do is to take the ratio of the measured reflected or transmitted intensity and the incident intensity. We need a light detector to measure the intensity. The eyes of the animals and humans are extremely refined light detectors with a tremendous dynamic range. We humans can detect very low levels of light intensities, see quite well in bright sunshine, and detect colours, i.e., different frequencies of the light. To achieve this dynamic range, the sensitivity of the eye has to be logarithmic, and therefore we have problems quantifying different intensities. We can see that something is brighter or darker than something else, but we cannot usually put a number to it. To be able to use equations (2.1) and (2.2), we need not just a detector, but a detector that can quantify the detected intensities. This is in practice achieved by transforming the light intensity to a voltage or current that can be measured accurately. Photomultiplier tubes, lead sulphide cells, and silicon detectors are examples of such detectors. Modern silicon semiconducting detectors with appropriate amplifier circuits have a close to linear response over as much as eight orders of magnitude of light intensity. The detector problem will not be further considered in this paper and, in the following, it is always assumed that the detector response is perfectly linear. For high-precision measurements, however, the linearity of the detector response should be validated, and this is especially important in the near infrared range.

Returning to equations (2.1) and (2.2), we assume an appropriate detector is available. If we focus a beam of light, with intensity I_o, on this detector, we measure a signal, S_o, which can be denoted as $S_o = A I_o$, in which the parameter A is an amplification factor that depends on the detector efficiency and the electronic amplifier used. If we insert a flat transparent sample in front of the detector, we believe that we are measuring the light transmitted through the sample. The detected signal should be $T_s S_o$, where T_s is the transmittance of the sample. The ratio $T_s S_o / S_o$ is obviously equal to T_s, or is it? What if S_o has changed while we inserted the sample in front of the detector? What if some light reaches the detector without passing through the sample, such as stray light or light reflected by the sample? These problems are illustrated in Figure 2.1-1.

We can solve the problem with stray light by using a light chopper and phase sensitive detection. Then only light with the chopper frequency will contribute to the detected signal. By assuring that the chopped light is not scattered around the room in an uncontrolled way, our detector signal will only result from I_o in Figure 2.1-1. If I_o is not constant with time, we can solve this problem by measuring I_o separately at the same time as we measure $T_s I_o$. We then have a configuration for our measurement as shown in Figure 2.1-2. The beam-splitter divides the incident beam into two beams, which are simultaneously detected and fed into the amplifier. The output reference signal, S_r, is always the ratio between the signals from detectors 1 and 2. The output signal for the reading without the sample is then

$$S_r = T_b I_o A_1 / R_b I_o A_2 = T_b A_1 / R_b A_2 \tag{2.3}$$

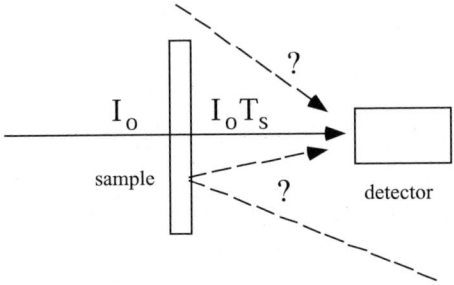

Figure 2.1-1. Sample and detector configuration with possible sources of stray light.

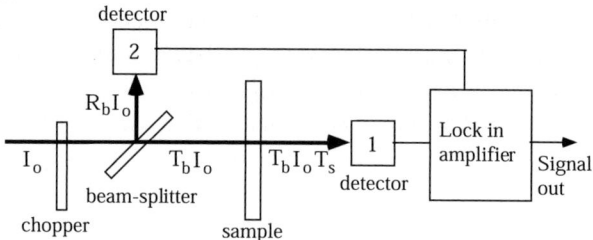

Figure 2.1-2. Configuration for correct measurement of transmittance.

T_b and R_b are the transmittance and reflectance values of the beam-splitter and A_1 and A_2 the amplification factors for the two detectors. It can be seen that S_r is now independent of I_o. The output signal, S_s, when the sample is inserted becomes

$$S_s = T_b A_1 T_s / R_b A_2 \qquad (2.4)$$

We now obtain T_s, the transmittance of our sample, from the ratio S_s/S_r. If no other light with the chopper frequency reaches the detector without passing through the sample, the correct value of the sample transmittance is obtained from S_s/S_r.

Measuring the reflectance is more complicated. The light path is then changed and the detector needs to be moved to the position where the reflected beam is to be found. A more common way is to use a series of mirrors, usually three or four, to deflect the beam to the sample and then focus the reflected beam onto the detector. This will be described later.

2.1.2.2 Spectrophotometers

We have now arrived at the basic construction of a spectrophotometer. The beam I_o is supplied by a monochromator, usually a grating monochromator, and everything is confined in a neat and compact package with a control panel or an external computer with the control software and user interface. The radiation source is usually a deuterium lamp for ultraviolet (UV) radiation, a tungsten lamp for visible and near infrared radiation (NIR), and a glow bar for infrared (IR) radiation. Thus, commercial spectrophotometers are available covering the wavelength range from 0.2 to 50 µm. There are instruments for the UV-VIS range, the UV-VIS-NIR range and for the IR range. In the IR range, Fourier transform instruments have taken over the market and the dispersive type of instruments, using a grating for the wavelength resolution, are no longer for sale. Thus, the effective wavelength range is in practice reduced to about 2 to 25 µm. This is unfortunate for two reasons. First, the loss of the longer wavelength range makes it necessary to extrapolate data to wavelengths $> 25\,\mu m$ when the thermal emittance of a surface is calculated. According to existing standards, this requires measured values up to $50\,\mu m$. Second, the accuracy in the photometric intensity is lower with the Fourier transform instruments than with the dispersive instruments. The issue of obtaining wavelength resolution using the dispersive or the Fourier transform techniques is not discussed further in this chapter.

Most commercial spectrophotometers are double-beam instruments. This means that the light beam emerging from the monochromator is divided into two beams i.e., a sample beam and a reference beam. The beam splitting is usually obtained with a rotating mirror instead of with a beam-splitter, and the same detector is used for both beams (see Figure 2.1-3). With this instrument, the intensity of these two beams is compared at the detector, and the signal output that is shown by the display or the recorder is always the ratio between the intensities of the sample and reference beams. With no sample, the instrument reading becomes:

$$S_o = (I_s/I_r)A \qquad (2.5)$$

where I_s and I_r are the intensities of the sample and reference beams respectively, and A is the instrument amplification factor. This is identical to equation (2.3) with $I_s = T_b I_o$, $I_r = R_b I_o$, and $A = A_1/A_2$. Equation (2.5) is just a more compact version of equation (2.3).

In Fourier transform instruments, the monochromator is replaced by a Michelson interferometer. The entire spectrum is detected simultaneously and thus the Fourier transform instruments are faster and especially suitable for dynamic studies. The wavelength resolution and accuracy is very good for these instruments but the photometric accuracy, as mentioned above, is not as good as for dispersive instruments. Fourier transform instruments also have a smaller spectral range and do not allow measurements in the UV and far IR ranges, at least not with the currently available technology. Two separate beam-splitters are required to cover the spectral range of interest. Fourier transform instruments are usually single-beam

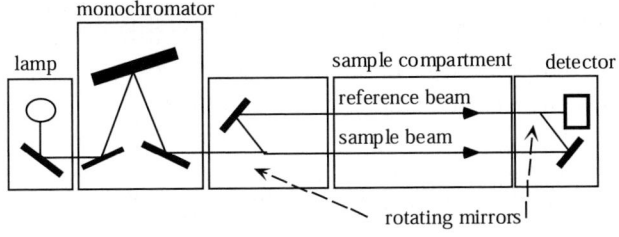

Figure 2.1-3. Schematic plan of a typical spectrophotometer.

instruments. The technique applies a special principle for the signal detection and the interested reader is referred to the literature for further details. The principles of how to measure reflectance and transmittance in this chapter are general and are valid for both dispersive and Fourier transform instruments.

2.1.3   SPECULAR OPTICAL MEASUREMENTS

2.1.3.1   *Specular Transmittance*
When a sample with transmittance T_s blocks the sample beam, the intensity of this beam is reduced and the instrument reading, S_T, is:

$$S_T = (I_s T_s / I_r) A \qquad (2.6)$$

Ideally the intensities I_r and I_s are equal and the factor A can be set to one, i.e. $T_s = S_T$. In actual practice, this is only approximately true and, in particular, the quantities are not independent of wavelength. If we assume that the deviation from the ideal situation is constant from one scan to another and take the ratio between the two instrument readings, we find that

$$S_T / S_o = T_s \qquad (2.7)$$

This is an exact relation, which is independent of I_r, I_s, and the amplification A. In other words, to measure transmittance accurately two scans are required.

2.1.3.2   *Specular Reflectance*
To measure reflectance, the light beam must be deflected, hit the sample surface, and then be transferred back to its original light path. The most common configuration used is shown in Figure 2.1-4, where both beams are reflected by three aluminum mirrors. Ideally these are identical, but this is not necessary.

We can assume that the reflectance values of these mirrors are R_1, R_2, R_3, R_4, R_5, and R_6. Again, we have to perform two scans. The first is the reference scan giving the instrument reading, S_o,

$$S_o = A(I_s R_4 R_5 R_6)/(I_r R_1 R_2 R_3) \qquad (2.8)$$

Replacing mirror 5 by the sample yields a reflectance signal, S_R

$$S_R = A(I_s R_4 R_s R_6)/(I_r R_1 R_2 R_3) \qquad (2.9)$$

and taking the ratio between the two readings now gives

$$S_R/S_o = R_s/R_5 \tag{2.10}$$

and the reflectance of our sample, R_s, is known if R_5 is known. In other words, this is a relative measurement so we need a reference mirror of known reflectance. There is another more complicated configuration that gives an absolute measurement. This type of configuration is called V-W for reasons that become obvious from the configuration shown in Figure 2.1-5. In the "V" configuration, the detected signal is

$$S_o = A(I_s R_4 R_5 R_6)/(I_r R_1 R_2 R_3) \tag{2.11}$$

which is identical to equation (2.8). Mirror number 5 is now rotated and the sample inserted as shown in Figure 2.1-5. The detected signal is now

$$S_s = A(I_s R_4 R_s R_5 R_s R_6)/(I_r R_1 R_2 R_3) \tag{2.12}$$

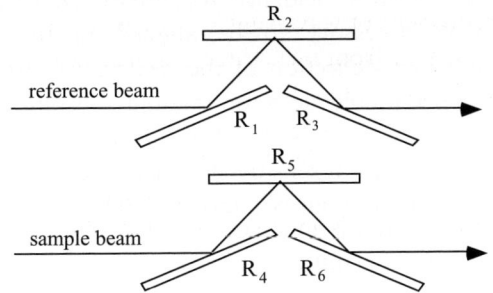

Figure 2.1-4. Schematic drawing of a reflectance accessory for relative reflectance measurements.

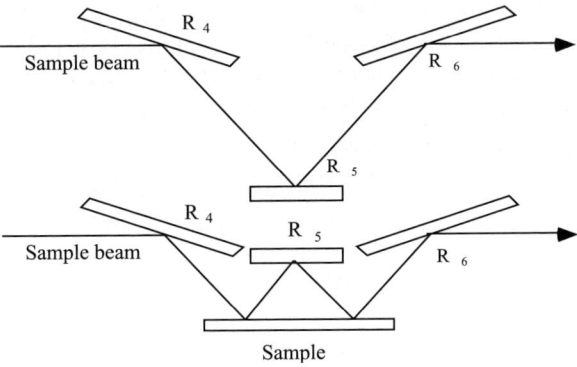

Figure 2.1-5. Schematic drawing of a V-W reflectance accessory for absolute reflectance measurements. Only the sample path is shown, with and without the sample. The reference beam remains unchanged.

and the signal ratio becomes

$$S_R/S_o = R_s^2 \qquad (2.13)$$

Thus, we obtain the sample reflectance as the square root of the signal ratio, S_R/S_o. As can be seen in equation (2.13), the reflectance value thus obtained does not depend on any known reference sample. This method is an absolute method.

In both these cases, the mirrors in the reference beam are not necessary, but it is common practice to deflect the reference beam in the same way as the sample beam to maintain the signal balance between the two beams.

For both these methods of measuring specular reflectance, it can be seen that the angle of incidence has to be a few degrees off normal. In practice the angle of incidence for spectrophotometer accessories is usually in the range of 10 to 20°. This is why the term "near normal" is often used. For very accurate measurements, the effect of the angle of incidence must be considered. At 20°, the reflectance value is a fraction higher than at normal incidence for unpolarized light. It is also very important for the V-W accessory to be aligned in the best possible way. Misalignment will result in a deflection of the beam away from its original path, which may cause an error in the detector reading (absolute is not the same as accurate!). For V-W accessories, it is therefore always best to use an integrating sphere detector. This detector is not sensitive to small shifts in the beam position, but will result in a loss of sensitivity (van Nijinatten, 2002).

2.1.3.3 *Oblique Incidence*

Complete characterization of optical surfaces includes determining the reflectance and transmittance at oblique angles of incidence. For the determination of optical constants, an angle of incidence of about 60° for s- or p-polarized light is useful. This results in further complications, because we have to take polarization effects into account. For transmitting samples, we have to correct for the parallax shift of the beam and back surface reflections. The detectors of the instrument may be more or less sensitive to lateral shifts of the beam reaching the detector. This depends on how the beam is focused on the detector, and the operator does usually not know this. For all measurements as described in the manuals provided, it is assumed that the beam is not shifted. This is not the case at oblique angles of incidence as indicated in Figure 2.1-6. Furthermore, the beam in any spectrophotometer is polarized in an uncontrolled way because of reflections from mirrors and in particular by the grating. It is therefore necessary to use polarizers, so that purely s- or p-polarized light can be obtained. A depolarizer can also be used to obtain approximately unpolarized light, but this can never replace polarizers for precision

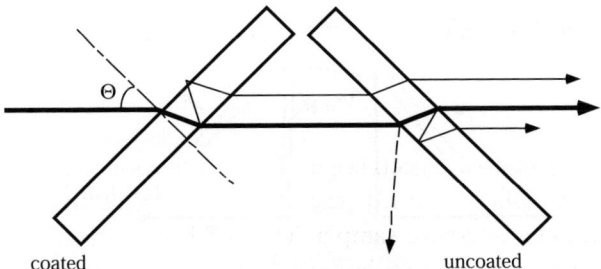

coated   uncoated

Figure 2.1-6. Schematic representation of a variable angle transmittance accessory with parallax compensation.

measurements. Polarizers are preferred because quite often the optical properties for different polarization states are desired and the polarization state is more precisely controlled. The value for unpolarized light can always be obtained as the arithmetic mean value of the values for s- and p-polarization. For dispersive instruments using a grating monochromator, a depolarizer is recommended for use in front of the polarizers to yield essentially the same intensity levels in the two states of polarization.

Transmittance measurements at oblique incidence are often performed using a double sample holder with parallax compensation as shown in Figure 2.1-6. A reference scan without samples and then a sample scan have to be performed for both s- and p-polarization. The transmittance value of the sample is taken as

$$T(\text{sample}) = \{S(\text{sample})/S(\text{ref})\}^{1/2} \tag{2.14}$$

where S(sample) and S(ref) are the instrument signals for the two scans, respectively. This is an absolute measurement. No known reference sample is needed.

A disadvantage of this method is that T^2 is measured, and that two identical samples are needed. Transmittance values below 10% will then generate a 1% signal level, and low levels of transmittance cannot be measured very accurately. This can be avoided if a highly transmitting reference sample, which has the same thickness and refractive index as the sample, can be used in one of the two sample positions in Figure 2.1-6 to take care of the parallax compensation. This has to be used when only one sample is available. In Figure 2.1-6, parallel shifted multiple reflections also occur between the surfaces of each sample. These are *not* compensated for by this set up. Furthermore, multiply reflected components between the panes disappear and are not recorded. The true transmittance through a double-glazed unit at oblique incidence *cannot* be measured in this configuration.

The reflectance measurements at oblique angles of incidence can be performed with a variable angle of incidence accessory together with the polarizers. The

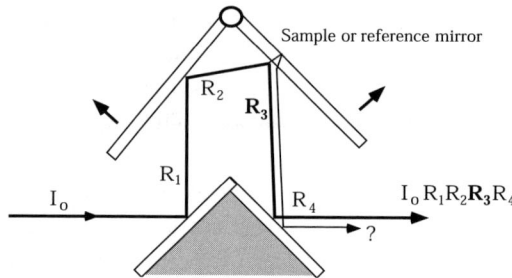

Figure 2.1-7.   Schematic representation of a variable angle reflectance accessory.

accessory usually consists of four mirrors for the deflection of the light beam, one of which is the sample. The mirror configuration and the light path are shown schematically in Figure 2.1-7. The same mirror configuration with four aluminum mirrors can be used in the reference beam, but is not necessary. The intensity of the sample beam leaving the reflectance accessory is proportional to $R_1R_2R_3R_4$. A reference scan with a reference mirror in the R_3-position and a sample scan with the sample in this position have to be performed for each polarization.

The ratio between the signals from these two scans is

$$S(\text{sample})/S(\text{ref}) = R(\text{sample})/R(\text{ref}) \qquad (2.15a)$$

To obtain the correct reflectance value of the sample, the reflectance of the reference mirror needs to be known for the particular angle of incidence and polarization. This is a serious problem because it is difficult to obtain good standards for oblique angles of incidence. The reflectance value for the sample can be calculated from equation (2.15a) as

$$R(\text{sample}) = R(\text{ref})\{S(\text{sample})/S(\text{ref})\} \qquad (2.15b)$$

2.1.3.4   Back Surface Correction for Reflectance Measurements

The optical properties of thin films are commonly determined on transparent substrates. The substrates materials are usually quartz or various qualities of glass. If the optical properties of the uncoated substrate are known, it is possible to compensate for the back surface component of the reflected light. Such a compensation is instrument dependent, although the principle is general, and becomes more important, obviously, when the substrate is very thick. For a transparent substrate, a back surface component (plus several multiply reflected components, but these can usually be neglected) is shifted laterally relative to the front surface component. This shift, y, depends on the angle of incidence as well as on the refractive index and thickness of the substrate. The back surface component reaching the detector appears to have been reflected by a virtual surface that is parallel shifted a distance c relative to the sample front surface. This shift is depicted by the reflected beams R_1 and R_2 in Figure 2.1-8 for a substrate of thickness d and refractive index n. For the majority of transparent samples, the substrate material is ordinary clear float glass, Corning glass, quartz, or something similar. This means that the refractive index n is nearly always about 1.5 and the extinction coefficient, k, is low. A known refractive index makes it possible to calculate the component from an uncoated substrate surface and thus make corrections for the back surface component R_2 in Figure 2.1-8. The correction method is valid for uncoated samples and for samples with one surface coated and the other uncoated.

The first step is to establish how the two parameters c and y in Figure 2.1-8 depend on the substrate thickness, d, and angle of incidence, φ. These functions can be obtained by using standard trigonometric functions. The derivation is straightforward and is not given here. The parallel shift c for the virtual position of the back surface can thus be written

$$c = d\cos\varphi/(n_2 - \sin^2\varphi)^{1/2} \tag{2.16}$$

and the parallel shift y of the beam

$$y = 2c\sin\varphi \tag{2.17}$$

equations (2.16) and (2.17) can be presented in graphical form and in Figure 2.1-9 the function y is shown for $n = 1.5$ and substrate thickness $d = 6\,\text{mm}$. The shape of this curve is perhaps not immediately obvious. One would be tempted to anticipate that the situation becomes gradually worse when the angle of incidence increases.

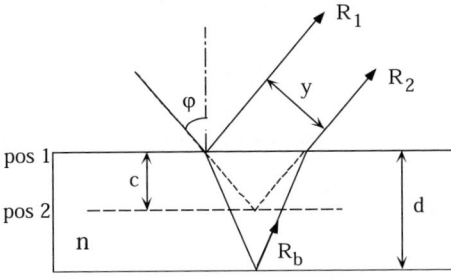

Figure 2.1-8. Schematic picture of reflected components for a transparent substrate. Indicated parameters are defined in the text.

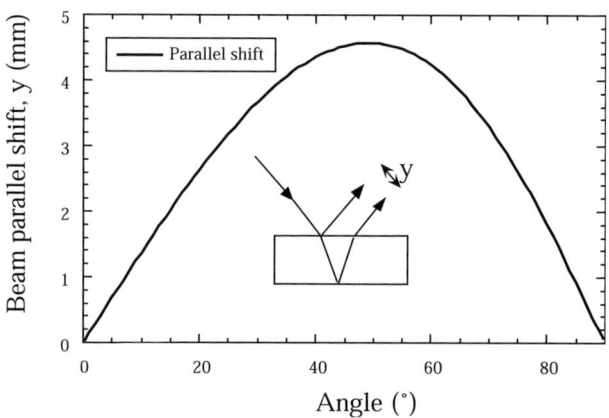

Figure 2.1-9. Parallel shift (y in Figure 2.1-8) of the back surface reflected component relative to the front surface component.

In Figure 2.1-9, it is shown, however, that the parallel shift of the beam is as large at 30° incidence as it is at 70°.

The graph in Figure 2.1-9 offers a way of calibrating an instrument for the back surface reflectance component. The resulting parallel shift of the beam will, in most cases, result in a decrease in the detected signal because the sensitivity of the detector decreases when the light beam is shifted laterally. This change in the detected signal can easily be determined by simply measuring an evaporated opaque aluminum film at the two positions indicated in Figure 2.1-8. The film at the shifted position of the virtual surface generates the beam R_2 and the detected signal S_2. R_1 is then zero because there is no surface at this position. With the sample in the normal position, the beam R_1 is generated and the recorded signal is S_1. R_2 is now zero because the sample is opaque. The reflectance of the sample is obviously identical in the two cases and taking the ratio between S_2 and S_1 gives the function F, which is then a measure of how much the signal from the back surface reflection of a sample is reduced because of the parallel shift in the beam.

$$F = S_2/S_1 \qquad (2.18)$$

After establishing how the instrument responds to a parallel shift of the beam, the next problem is to separate the measured signal into the two components R_1 and R_2 in Figure 2.1-8. In the general case when both surfaces are coated with an unknown coating, this is not possible, but when one of them may be calculated from known optical constants, this separation can be performed and the R_2 component can be corrected using the correction function F. This is the case when only one surface is coated. By defining the front surface reflectance as R_f and the back surface reflectance as R_b, it is immediately clear that $R_1 = R_f$. The back surface contribution R_2 can then be written as

$$R_2 = (1 - R_f)^2 R_b \exp(-2ax) \tag{2.19}$$

where $(1 - R_f) = T_f = (1 - R_1)$ is the transmittance through the front surface and $\exp(-2ax)$ is the extinction of the beam traversing the substrate twice. In the following, it is assumed for simplicity that the factor $\exp(-2ax)$ is included in R_b. The reflectance of the sample can then be written as the sum of the two components

$$R = R_1 + (1 - R_1)^2 R_b \tag{2.20}$$

According to the previous discussion only a fraction F of the back surface component is actually detected by the instrument. The detected signal is therefore proportional to

$$R_{\text{det}} = R_1 + F(1 - R_1)^2 R_b \tag{2.21}$$

It is now possible to calculate the true reflectance R from the detected value R_{det} using equations (2.20) and (2.21). The two cases of samples with a coated front surface or a coated back surface have to be treated separately. The simplest case is when the back surface is coated and the front surface is uncoated. In this case, it is assumed that R_1 can be correctly calculated from the optical constants of the substrate. Equation (2.21) can be written as

$$(1 - R_1)^2 R_b = (R_{\text{det}} - R_1)/F \tag{2.22}$$

and inserted into equation (2.20), this yields

$$R = R_1 + (R_{\text{det}} - R_1)/F \tag{2.23}$$

The true reflectance value is hence a function of the detected value R_{det}, the calculated front surface value $R_f (= R_1)$ and the correction function F.

A little more algebra is needed when the front surface is the coated one. R_1 is then the unknown quantity and equation (2.21) results in a second-order equation in R_1:

$$R_{\text{det}} = R_1 + FR_b - 2R_1 FR_b + R_1^2 FR_b \tag{2.24}$$

Solving equation (2.24) gives R_1 as a function of R_{det}, R_b, and F. Equation (2.20) then gives the correct reflectance value of the sample. Note that the value R_b is not equal to R_2 because the back surface component of the substrate includes absorption but not the transmittance through the front surface.

2.1.3.5   Absolute Instrument for Measuring Directional Reflectance and Transmittance

The problems with the parallel shift of the beam can largely be avoided if an integrating sphere is used as the detector. No corrections are required if the sphere port is large enough to collect all multiply reflected components and also the shifted main beam. This has also been recognized by spectrophotometer manufacturers and many instruments now have an integrating sphere detector. The disadvantage is a loss of sensitivity, although this is not a serious problem with modern detectors that use low-noise amplifiers. A thorough investigation of problems encountered when measuring optical transmittance and reflectance at oblique angles of incidence has been conducted in a European project, ADOPT (Angular Dependent OPTical properties of coated glass and glazing products). Some of the results have already been presented in some published papers, but several of the results are yet to be published (Roos *et al.*, 2000).

An instrument with a movable integrating sphere as the detector can have a design as depicted in Figure 2.1-10. The instrument does not need any reflectance standards and is based on the ideal situation described by equations (2.1) and (2.2). The sample

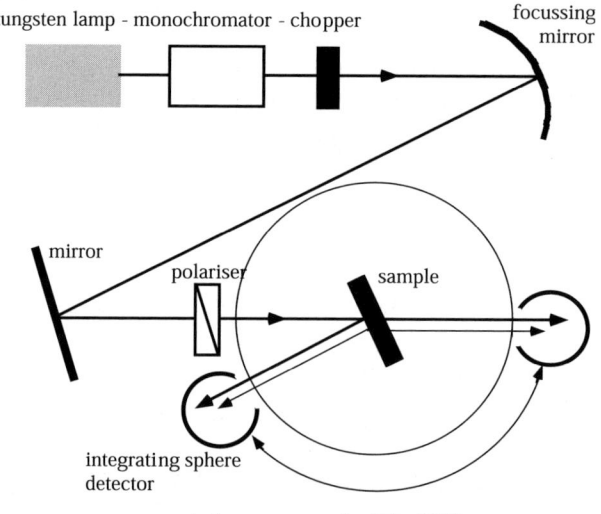

tungsten lamp - monochromator - chopper

focussing mirror

mirror

polariser

sample

integrating sphere detector

spectral range presently 400 - 1000 nm

Figure 2.1-10.   Schematic drawing of the absolute reflectometer.

reflectance or transmittance is the ratio between the sample signal and the reference signal. The instrument is set up in a single-beam mode using an ordinary tungsten lamp combined with a grating as the light source. The light is collected by a focusing mirror and the exit slit of the monochromator is focused on the detector port. The distance between the focusing mirror and the detector is about 260 cm, which means the convergence of the beam is less than 1°. This is an advantage when measurements are made at high angles of incidence. A polarizer is placed immediately in front of the sample. The sample holder is a rotating table that allows the angle between the light beam and the sample to be set to a precision of better than ±0.2°. For commercial instruments, the accessories provided do not, in general, have the required precision in the angular setting. Exact normal incidence can be established by tracing the reflected beam back along the light path of the instrument. A stabilized power supply is used for the tungsten lamp, so no reference reading is needed for the light source. The detected signal is constant to better than 1% over a period of hours. This instrument has been designed and constructed in Uppsala, Sweden, and is described in more detail (Nostell *et al.*, 1999).

The detector is a 5-cm diameter integrating sphere with a tandem Si/PbS photodetector. The entry port has an oblong shape to allow multiply reflected components to be properly detected. The dimension of the entry port was chosen in such a way that all first-order multiply reflected components will be collected at 60° incidence for a triple glazing sample configuration with 4-mm panes and a spacing of 1 mm between the panes. A schematic drawing of the detector sphere is shown in Figure 2.1-11. The internal shield prevents the detector from receiving any reflections from the light spot of the direct beam entering the sphere. Because no scattering sample is used to cover the sphere entry port, no shield is needed to screen this port. As indicated in the top view in Figure 2.1-11, the incident beam can be parallel shifted several centimetres and still be collected by the sphere. Ideally this parallel

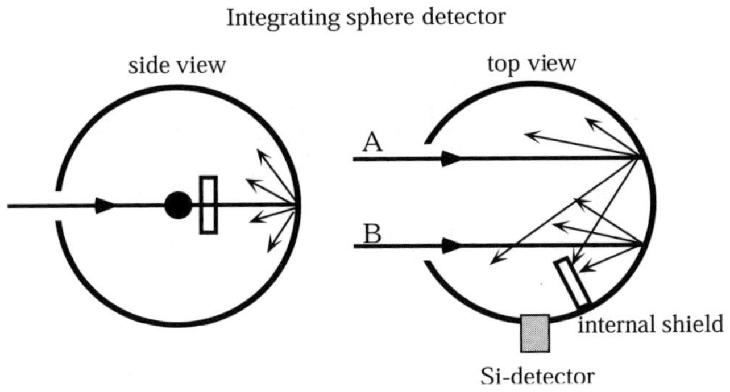

Integrating sphere detector

Figure 2.1-11. Schematic drawing of the integrating sphere detector.

shift does not influence the detected signal. In practice, a slight variation of the reflectance of the wall of the sphere cannot be excluded and this would cause a variation of the detected signal as the light beam is scanned across the entry port. This has been verified experimentally and the variation has been found to be within 0.5%. The sphere was coated with Kodak $BaSO_4$ paint.

To test the instrument, an uncoated 4-mm clear float glass sample was chosen. This type of sample should be simple to measure. However, at high angles of incidence, there is a considerable parallel shift of multiply reflected components, and it is important for the detector to collect all these components. Measured reflectance and transmittance values should always vary with the angle of incidence according to standard Fresnel formalism. Because there is no film with unknown optical constants on these samples the results can be compared with calculated data. There is a slight uncertainty in the optical constants of the glass itself, but this only marginally affects the angular dependence of the optical properties. In Figures 2-12 and 2-13, the experimental and calculated transmittance and reflectance values for s- and p-polarized light versus angle of incidence are shown. The refractive index was $n = 1.53$ in both cases and the extinction coefficient $k = 6 \times 10^{-7}$. As can be seen in Figures 2-12 and 2-13, the agreement between experimental and calculated data is very encouraging.

2.1.4 MEASUREMENT OF LIGHT SCATTERING

In the prior sections, we have only discussed measurements on nonscattering samples, but still encountered several problems. These problems are simple

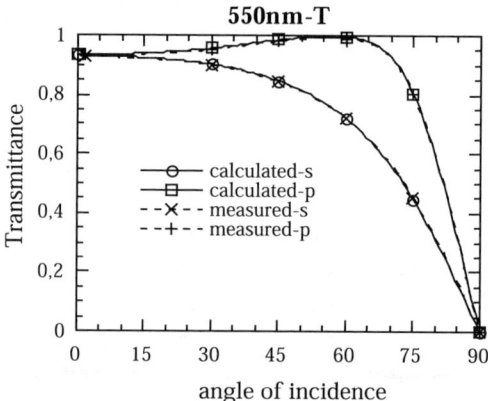

Figure 2.1-12. Experimental and calculated transmittance versus angle of incidence for 4-mm clear float glass. s- and p-polarized light of 550-nm wavelength as indicated.

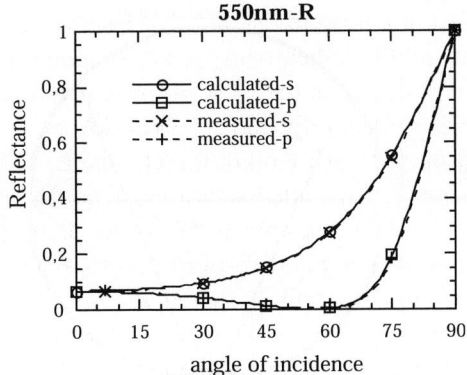

Figure 2.1-13. Experimental and calculated reflectance versus angle of incidence for 4-mm clear float glass. s- and p-polarized light of 550-nm wavelength as indicated.

compared to what happens when the studied surfaces scatter the incident light. The light scattering from surfaces and/or the bulk of the sample range from nearly 100% for white diffuse surfaces, such as white paint, and down to 0.0001% for smooth optical surfaces, such as polished silicon wafers. A perfect Lambertian surface is defined as a hypothetical surface that scatters 100% of incident light evenly distributed into the hemisphere above the surface. There are two kinds of instruments used for the measurement of light scattering. In the first kind, angle-resolved scattering is measured in which all possible angles above the surface are scanned by the detector. With this type of instrument, the angular distribution of the scattered light is obtained. In the second kind, which is the most common, the total scattering from the surface is measured. It is basically simpler to measure the total scattering, often called the total integrating scattering (TIS), or total hemispherical scattering, because less intensity is required to detect all the scattered light than is required to measure the light scattered into a small solid angle.

Total scattering can be measured in two basically different ways. The most common way is to use an integrating sphere in which the inside is coated with a white diffuse coating, usually $BaSO_4$ or a polymeric material, such as Spectralon (Ulbricht, 1900). The scattering from the inside surface is a close approximation to that of a Lambertian surface. An integrating sphere is shown schematically in Figure 2.1-14. The sample is placed against the sample port in the sphere wall, and light reflected (or transmitted) by the sample is trapped inside the sphere. The light intensity inside the sphere is obtained with a detector, which must be screened from the sample so that no light can reach it directly from the sample (Hansen, 1989, 1996; Snail and Hansen, 1989). The second way is to use a focusing mirror that is usually a smooth aluminum mirror. Ideally, this should be elliptical with the sample at one of the foci and the detector in the other. In practice, spherical mirrors are used. The advantage

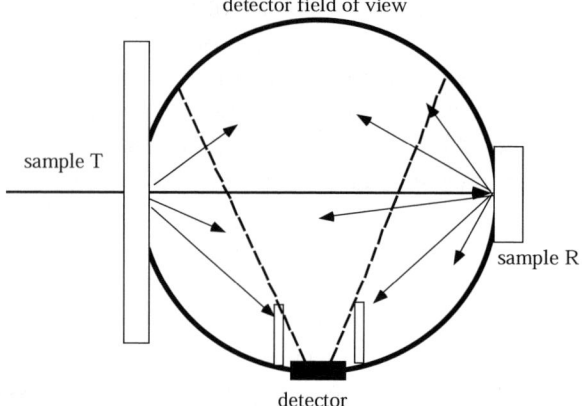

Figure 2.1-14. Schematic picture of an integrating sphere.

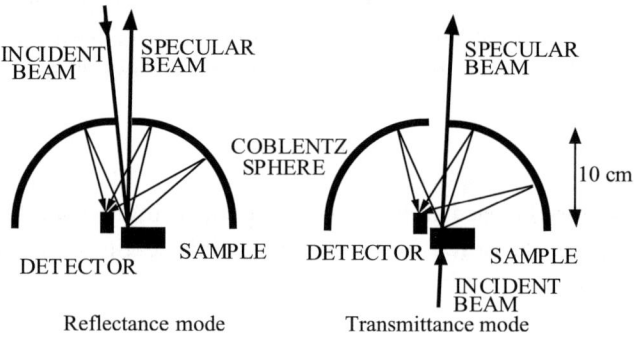

Figure 2.1-15. Schematic picture of a focusing Coblentz sphere.

of this method is that all reflected or transmitted light is focused on the detector, and lower levels of scattering can be measured (Coblentz, 1913). A focusing sphere, also called a Coblentz sphere, is shown schematically in Figure 2.1-15 (Rönnow and Veszelei, 1994). Because integrating spheres are much more commonly used than Coblentz spheres, we will discuss the problems with taking measurements with an integrating sphere in more detail than those associated with Coblentz spheres.

2.1.4.1   Measurements Using an Integrating Sphere

An integrating sphere is in principle a very simple device. Light reflected or transmitted by the sample impinges on the inside wall of the sphere and is reflected again and again until eventually it is absorbed by the detector or the sphere wall, or escapes through one of the entrance ports for the light beams. The detector is usually placed at 90° relative to the plane of incidence of the light beam. No line-of-sight is permitted between the detector and the sample or the specular light spot inside the sphere. Only the average light intensity inside the sphere is supposed to reach the detector. This is usually arranged by the proper positioning of light shields near the detector as illustrated in Figure 2.1-14. A double-beam integrating sphere designed for both reflectance and transmittance measurements can have as many as seven ports, and we shall soon see that this can cause some problems. A single-beam instrument and a sphere dedicated only to reflectance (or transmittance) measurements can be designed with fewer ports. Figure 2.1-14 shows a sphere in single-beam configuration. A double-beam instrument would have a reference beam entry port and a reference port covered by a reference plate, usually in a configuration perpendicular to the plane of the figure. The sphere wall and the reference sample are coated with a white diffusing coating.

The principle of the measurements is the same as for specular samples. With a double-beam instrument, the ratio between the sample beam and the reference beam is always obtained, and two scans are usually required for accurate measurements. The reference reading with diffusing plates at both the sample and reference ports gives

$$S_o = (I_s R_b / I_r R_b) A \tag{2.25}$$

where R_b is the reflectance of the diffusing plate (and the sphere wall). When measuring reflectance, the diffusing plate at the sample port is replaced by the sample and the reading becomes

$$S_R = (I_s R(\text{sample}) / I_r R_b) A \tag{2.26}$$

The reflectance of the sample is obtained from the ratio

$$S_R / S_o = R(\text{sample}) / R_b \tag{2.27}$$

if R_b is known.

It is sometimes convenient to avoid the reference reading by setting $A = R_b$. This is done by setting the reference reading to the value of the reference plate at 550 nm. If $I_s = I_r$, then $R(\text{sample}) = S_R$ from equation (2.27). This approximation is often acceptable in the visible spectral range, but it can result in large errors in the near IR and UV parts of the spectrum depending on the wavelength variation of the sphere coating.

For transmittance measurements the sample is placed at the sample beam entry port and the instrument reading becomes

$$S_T = (I_s T(\text{sample}) R_b / I_r R_b) A \tag{2.28}$$

By taking the ratio between the sample and reference readings, we obtain

$$S_T/S_o = T(\text{sample}) \tag{2.29}$$

which is an exact relation but it is valid only for samples that only have specular reflections.

Equations (2.25)–(2.29) above are approximations and in general cannot be used to obtain satisfactory results. Equation (2.27) yields an acceptable result if R(sample) represents a nearly Lambertian distribution of the reflected light, or is at least similar to R_b. Equation (2.29) is valid only for specular samples, and the idea was to measure scattering samples.

To perform more accurate measurements for any kind of sample, we need a better model for calculating the reflectance and transmittance values from the detected signals. It is beyond the scope of this brief summary to describe these models in detail, and the interested reader is referred to the literature (Roos and Ribbing 1998; Roos, 1991, 1993). Only a short overview of the basic principles is given.

The reflected or transmitted light from a sample may have an angular distribution that varies within wide limits. The first step is to separate the specular and scattered components. This is usually performed by letting the specular component of the reflected or transmitted light escape into a beam dump, but the next step, which is equally important, is more seldom carried out. The scattered light has to be divided into high- and low-angle scattering, as depicted in Figure 2.1-16. The reason for this becomes obvious when we consider what happens inside the sphere. The specular light impinges on the sphere wall and is scattered into the sphere from this position. The high-angle scattering component is scattered into the sphere directly by the sample. This means that the specular beam has to be reflected off the sphere wall one more time than light scattered directly by the sample. This is only strictly true, however, for light scattered in the same way by the sample as by the sphere wall, i.e. in a "Lambertian manner". Many samples scatter light at angles close to the specular beam, and this scattered light is detected by the sphere as if it were specular. This is illustrated in Figure 2.1-17, which shows how light is scattered from three different

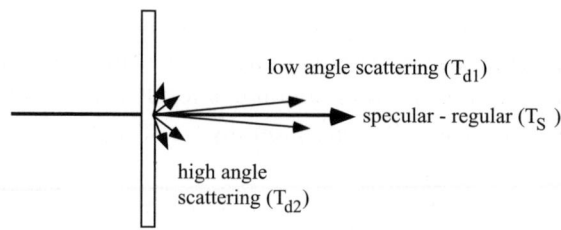

Figure 2.1-16. Schematic representation of how transmitted light is divided into specular, low-angle scattering T_{d1} and high-angle scattering T_{d2}.

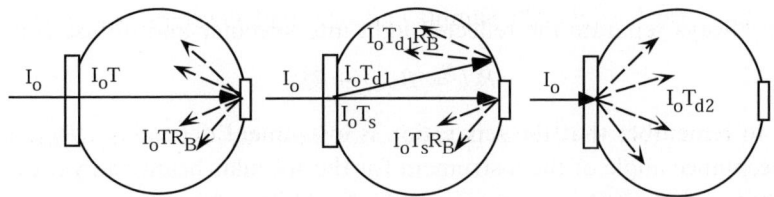

Figure 2.1-17. Schematic pictures illustrating how transmitted light enters the integrating sphere after being transmitted by the sample. *Left*: specular sample; *Centre*: low-angle scattering; *Right*: high-angle scattering.

samples into the sphere. The first sample is perfectly specular and no light is scattered. All light is scattered into the sphere from the closed specular exit port. The total power of this scattered light is proportional to I_oTR_B, where T is the transmittance of the sample. In the second case, light is mainly scattered at small angles with a high ratio of specular light. In this case, the light is also scattered into the sphere from a position opposite the sample, some from the specular exit port, and some from the sphere wall immediately around the exit port. The total power of this scattered light is proportional to $I_oT_sR_B+I_oT_{d1}R_B$, where T_s is the specular transmittance of the sample and T_{d1} is the scattered transmittance. In terms of detection, these two contributions are equivalent. The third sample is a completely diffuse sample. All light is homogeneously scattered into the sphere directly by the sample, not by the sphere wall. The total power of this scattered light is I_oT_{d2}. Note that this term does not contain the factor R_B. The diagrams in Figure 2.1-17 illustrate the concept of high- and low-angle scattering. In the center diagram, light just missing the specular exit port enters the sphere in the same way as the specular light even though it is registered by the sphere as being scattered. "Specular light" is in this case defined as light falling on the exit port. The light scattered at high angles from the diffuse sample on the right hand side of Figure 2.1-17 illuminates the sphere evenly. This is the key to understanding the concept of dividing the scattered light into near-specular and completely diffuse fractions.

2.1.4.2 *Integrating Sphere Correction Factors*
When measuring with an integrating sphere, we have just shown that it is sometimes necessary to divide the scattered radiation into high- and low-angle components. This separation has been extensively discussed in previous publications for reflectance (Roos and Ribbing 1988) and transmittance (Roos, 1991). Our insight also resulted in a suggested modification of the correction factor for sample reflectance when measuring transmittance with single beam integrating spheres (Grandin and Roos, 1994). This is further discussed below.

We can always separate the reflected light into specular and diffuse components:

$$R_{tot} = R_{spec} + R_{diff} \tag{2.30}$$

We have to remember that the separation is instrument dependent. R_{spec} is defined by the acceptance angle of the instrument for the specular beam and varies from one instrument to another. R_{spec}, as measured by two different instruments, can therefore be different. R_{tot}, on the other hand, is not instrument dependent and the same result should always be obtained for R_{tot} if measured correctly. According to Figure 2.1-17 we should divide the scattered light further and we have to modify equation (2.30)

$$R_{tot} = R_{spec} + R_{d1} + R_{d2} \tag{2.31a}$$

It is convenient to define a separation parameter, B, according to

$$R_{tot} = R_{spec} + (1 - B)R_{diff} + BR_{diff} \tag{2.31b}$$

where $B \leq 1$. B is in other words defined as the high-angle fraction of the scattered radiation.

Light losses through ports can result in severe problems when measuring scattering samples because light scattered by the sample may escape through ports on the opposite side of the sphere. The specular beam, on the other hand, is reflected back into the sphere by the sphere wall close to these ports and is completely contained in the sphere. This is also the case for the fraction of the low-angle scattering components, which do not escape through the beam entry ports. For a correct interpretation of the detected signals, it is therefore necessary to introduce one more parameter, F_s; the fraction of the low-angle scattering that is contained in the sphere. The fraction $(1 - F_s)$ escapes through the entry ports opposite the sample. The factor F_s is different for different sphere designs and it also depends on the angular distribution of the scattered light from the sample. For the sphere wall coating, which is assumed to be Lambertian, $F_s = F_B = 1 - x$, where x is the fraction of the total wall area occupied by the entry ports.

Because the specular and the diffuse signals have different throughputs, it is necessary to measure them separately. Together with the reference signal, this means three scans are needed to measure the sample reflectance. Additional scans may be required if the zero level is not correctly adjusted. This is of importance when very low levels of scattering are being measured (Rönnow and Roos, 1994). In the following, it is assumed that the zero level is correct. The following signals need to be recorded: $V_{ref} = BaSO_4$ reference reflectance (or whatever reference is being used), $V_{tot} = $ total sample reflectance with the specular exit port closed, and, finally, $V_{diff} = $ diffuse sample reflectance with the specular exit port open. It is assumed in the following that the reference reflectance standard used is identical to that from the coating on the sphere wall. The total and diffuse reflectance values can now be obtained from the detected signals according to

$$R_{\text{diff}} = \frac{F_B R_B V_{\text{diff}}}{[F_B B + F_s(1 - B)R_B]V_{\text{ref}}} \tag{2.32}$$

$$R_{\text{spec}} = \frac{F_B(V_{\text{tot}} - V_{\text{diff}})}{V_{\text{ref}}} \tag{2.33}$$

R_B is the reflectance of the diffuse reflectance standard. The total reflectance R_{tot} of the sample is given by equation (2.30).

A similar approach is needed for transmittance measurements. Light entering the sphere through a sample can be specular (direct) or diffuse (scattered). The diffuse light can in turn be separated into a low-angle and a high-angle component. The low-angle component becomes diffuse after the first reflection of the sphere wall; the high-angle component can be treated as if it were diffused by the sample. In accordance with equations (2.31a) and (2.31b) we write for the transmittance

$$T_{\text{tot}} = T_{\text{spec}} + T_{d1} + T_{d2} \tag{2.34}$$

and

$$T_{\text{tot}} = T_{\text{spec}} + (1 - B)T_{\text{diff}} + BT_{\text{diff}} \tag{2.35}$$

Three scans are required for samples that scatter the transmitted light. The detected signals are V_{ref}, V_{tot}, and V_{diff} in complete analogy with the reflectance case. The corresponding diffuse and specular transmittance values are

$$T_{\text{diff}} = \frac{V_{\text{diff}}}{V_{\text{ref}}[1 - B + (B/F_B R_B)]} \tag{2.36}$$

$$T_{\text{spec}} = \frac{V_{\text{tot}} - V_{\text{diff}}}{V_{\text{ref}}} \tag{2.37}$$

Another mode of operation is possible for transmittance measurements. A specular mirror can be used instead of covering the reflectance sample port with a diffuse reflectance standard. In this mode, the specular exit port for the reflectance mode is used for the separation between specular and diffuse components. By combining the two modes of operation for transmittance measurements, it is possible to obtain information about the angular distribution of the scattered light (Roos, 1991).

It can be seen from equations (2.30)–(2.37) that the diffuse components cannot be calculated as simply as is generally believed. In the limit $B \rightarrow 0$ (no high-angle scattering), the expressions for R_{diff} and T_{diff} become simple ratios between recorded signals. For most real surfaces $0 < B < 1$ and $F_s < 1$. Under normal conditions it is difficult to determine exact values of the factors B and F_s. As long as the diffuse component is small, not knowing it is not a serious problem, but for highly scattering samples it is sometimes not possible to obtain reliable results because the *B* and F_s factors are not known.

2.1.4.3   Light Trapping in the Sample

In all the above, it has been assumed that all the scattered light actually enters the sphere through its port. For diffusing samples, however, this is not always the case. When incident light is scattered at high scattering angles, either by one of the surfaces or by inhomogeneities inside the sample, some of this light is temporarily trapped by total internal reflection. The problem is illustrated in Figure 2.1-18, which shows how light can be transmitted through the sample but miss the sphere entry port, so the detected signal is reduced. Light may then exit the sample at a position far away from the incident light beam or even through the edge of the sample. This is a difficult problem to deal with and no simple solution is available. The corresponding situation holds for reflectance measurements. The level of light trapping increases with sample thickness.

When calculating the absorptance as $A = 1 - R - T$, the value of *A* may become too large. In fact, it is always a good idea to calculate the absorptance to check the measurements. If $A < 0$, which indeed happens, then something in the measurements is wrong. If *A* is considerably greater than zero when measuring a dielectric sample with very low internal absorptance, then again something is wrong.

A possible way of minimizing the problem with light trapping is to use an aperture immediately in front of the sample to "stop down" or focus the beam in the transmittance mode as illustrated in Figure 2.1-19. The idea is that the illuminating beam must be considerably smaller than the sample port. This allows for some of (hopefully most of) the trapped light to enter the sphere and become detected. In the

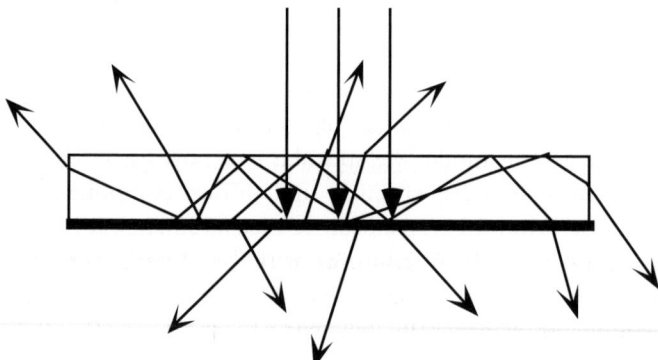

Figure 2.1-18.   Schematic picture showing how scattered light is trapped into the sample and escapes at positions far away from the illuminating beam.

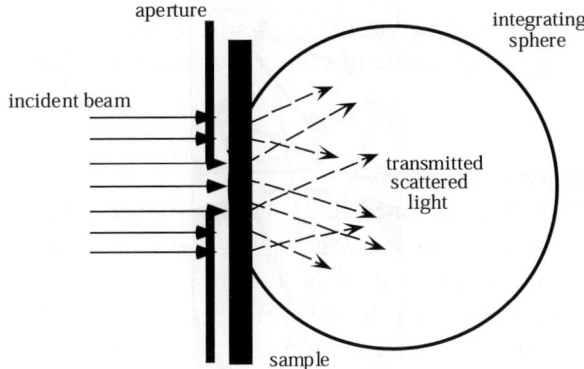

Figure 2.1-19. Picture showing how an aperture in front of the sample can reduce the problem of light coupled into the sample.

reflectance mode, an aperture cannot be used immediately in front of the sample, but the beam can always be focused or stopped down to a smaller size also in the reflectance mode.

It is a coincidence that the systematic errors caused by the light trapping effect result in values that are too low, and errors caused by neglecting the angular distribution of the scattering result in values that are too high. To some extent these systematic errors may cancel, but obviously this is not necessarily the case.

The problem with light trapping in the sample is related to a problem that appears when measuring large samples with structures or irregularities of a dimension of several centimeters or more. This is common for many solar control devices, such as blinds, and transparent insulating materials, such as honeycomb structures. In this case the illuminating beam must be much larger than the typical dimension of the sample irregularity to obtain a true average. Large integrating spheres have been specially designed for this purpose and it is common practice to use an illuminating beam that is much larger than the aperture of the sphere (Milburn and Hollands, 1994). Several other papers have been published on this subject and in a recent European project (ALTSET); this problem has been addressed for a number of diffusing, light scattering, and light deflecting glazing materials.

2.1.4.4 *Measuring Low Levels of Scattering with Integrating Spheres*
Most double-beam integrating spheres are commercial instruments with a compact design. The optical system guiding the light beams into the sphere consists of several mirrors confined to a small housing as illustrated in Figure 2.1-20. It is inevitable that some scattering from all these mirrors will enter the integrating sphere and result in a nonzero level of stray light. This causes problems when low levels of scattering

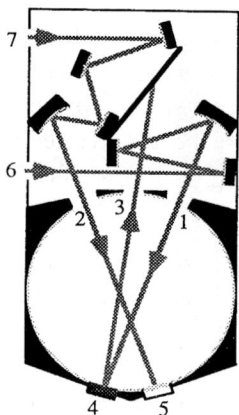

Figure 2.1-20. Example of an integrating sphere showing the optical paths for the sample and reference beams. From reference Rönnow and Roos (1994).

are being measured. Low levels for the integrating sphere mean scattering from a few percent to the detection limit, which is a few tenths of a percent. Another source of error is the beam dump that is used in some measurements. This is often a black cone or wedge. The reflectance of a beam dump is not identically zero but is about a few tenths of a percent. If the exact zero level of the detection system is established, i.e., a zero reading is obtained when it is made absolutely certain that no light can reach the detector, it is possible to measure all levels of stray light and to take these into account when the sample scattering is calculated. Expressions such as those in equations (2.38) and (2.39) are needed to accurately calculate the scattering from the samples.

$$R_{\text{diff}} = \left[\frac{\left(F_B\left(R_B - R_{\text{dump}}\right)(S_7 - S_8)\right)/(S_1 - S_4) - CR_s - F_B R_{\text{dump}}\left(T_{\text{tot}}^2 - 1\right)}{(1 - B)R_B + F_B B} \right] \qquad (2.38)$$

$$T_{\text{diff}} = \frac{S_3 - S_5 - T_{sb}(S_4 - S_5)}{S_4 - S_5 + (S_1 - S_4)\left[1 + (B(1 - F_B B_B)/F_B(R_B - R_{\text{dump}}))\right]} \qquad (2.39)$$

The measured signals S_1 to S_8 represent measurements of reference and sample signals, and stray light levels entering the different ports of the sphere. These equations have been derived for a Beckman 5240 spectrophotometer equipped with the Beckman integrating sphere accessory (Rönnow and Roos, 1994). Other integrating spheres with different geometries will have different but similar expressions. This means that up to seven scans are needed to determine accurately the scattering from a sample. This would of course make the measurements very tedious, but fortunately many of these scans for stray light can be performed once and stored for future use. It is reasonable to assume that the stray light levels do not change much with time. The importance of correctly taking stray light into account is demonstrated in Figure 2.1-21. The uncorrected values are 2 to 3 times too large, which means an error of the order of 100 to 200%.

2.1.4.5 *Integrating Spheres used with Single Beam Spectrophotometers*
Integrating spheres are commonly used on instruments in many laboratories and range in size from only a few centimeters up to as much as 3 m. Many spheres made for special purposes are used with single-beam instruments. The measurement technique is the same as for spheres on double-beam instruments. Reference and sample readings are needed and, in principle, the result is the ratio between these readings. In practice, however, a more careful analysis is necessary to obtain correct results. When transmittance is measured in a single-beam instrument, the reference reading is taken with the entrance port open. The sample reading is taken with the sample covering this port. The problem then is that the presence of the sample changes the geometry of the sphere. Light is reflected back into the sphere not only by the sphere wall, but also by the sample itself. The throughput of the integrating sphere is not the same for the sample reading as for the reference reading. If the sample reflectance is known, this can be taken into account when the true

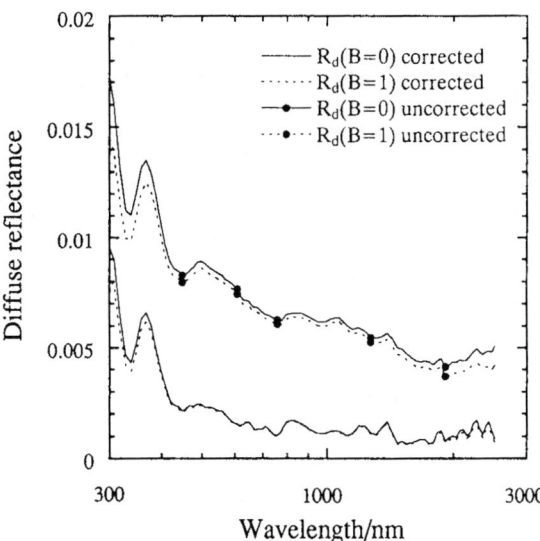

Figure 2.1-21. Diffuse reflectance spectra of a low scattering tin oxide film on glass showing the difference between corrected and uncorrected spectra. From reference Rönnow and Roos (1994).

transmittance of the sample is calculated, but the angular distribution of the transmitted light becomes an additional problem. The correction factor for light scattered by the sample is not the same as the factor for specularly transmitted light. Instead of the simple ratio between the sample and reference readings, the following formula gives the transmittance of the sample (Grandin and Roos, 1994)

$$T_m = \frac{S_s}{S_r} = \left\{ \frac{1 - R_w(1 - f_s)}{1 - [R_w(1 - f_s) + f_s R_s]} \right\} \left\{ \frac{T_{\text{diff}} + T_{\text{spec}}[R_w(1 - x f_s) + f_s R_s]}{R_w(1 - x f_s)} \right\} \quad (2.40)$$

R_w is the reflectance by the sphere wall, R_s is the sample reflectance, and f_s the entry port area fraction of the total wall area. T_m is the measured transmittance value and the true values are T_{spec} and T_{diff}. As can be seen in equation (2.40), the correction factors for the diffuse and the specular components of the transmitted signal are different. The difference is quite significant and not recognizing this fact can result in large errors in the measured transmittance values. The measured ratio between the sample and reference signal is always too large. The correction depends not only on the fraction of the port area, f_s, but also on the reflectance of the sample and is therefore wavelength dependent. For a sample with $R_s = 0.5$ and $f_s = 0.0037$, which is a very small fraction, the correction factor is 0.056 if the sample is specular and 0.093 if the sample is diffuse. A highly transmitting sample with $R_s = 0.04$ gives 0.004 and 0.038 respectively.

It must be remembered that equation (2.40) is not an exact solution. If the sphere has no specular exit port, the separation into diffuse and specular components is difficult. In a real case the transmitted light can be scattered into the sphere by the sample in such a way that neither of the correction factors is exactly correct. If the scattering is mostly low-angle scattering close to the specular direction, it is more correct to treat the scattered component as being specular.

Similar considerations must be made for reflectance measurements in a single beam instrument, since the sphere throughput is different for the reference and sample scans. It is also strongly recommended that the reference level is recalibrated at regular intervals to avoid problems with drift and fluctuations in the light source and detector systems.

2.1.4.6 *Integrating sphere with center mounted sample*

The Edwards type is a special design of a reflectance sphere, where the sample is mounted on a sample holder in the center of the sphere, as shown schematically in Figure 2.1-22. This is necessary for measurements of the reflectance as a function of the angle of incidence. As in the case for measuring the transmittance, the total reflectance has to be separated into its specular and diffuse components. The fraction of reflected light that escapes through the entrance port is a special complication

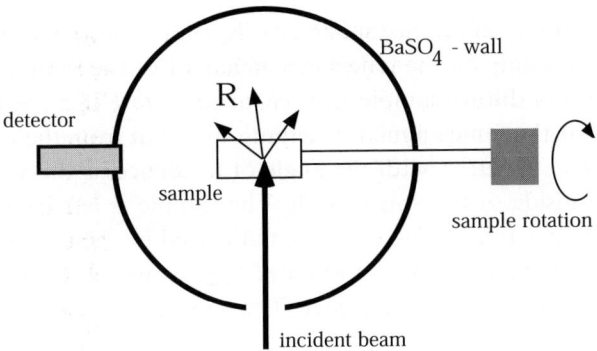

Figure 2.1-22. Schematic drawing of an integrating sphere with center-mounted sample.

because it depends on the incidence angle. The signals measured are related to the optical properties by the following equations:

$$S(\text{sample}) = AI_o R_{\text{spec}}(\theta) R_{\text{wall}} F_s + AI_o R_{\text{diff}}(\theta) F_d \tag{2.41}$$

For the reference signal we have

$$S(\text{ref}) = AI_o R_{\text{wall}} F. \tag{2.42}$$

By taking the ratio between equations (2.41) and (2.42), and assuming $F_d = F$, we obtain

$$S(\text{sample})/S(\text{ref}) = R_{\text{spec}}(\theta) F_s/F + R_{\text{diff}}(q)/R_{\text{wall}} \tag{2.43}$$

In most cases the ratio, F_s/F is ≥ 1 and $R_{\text{wall}} < 1$. A is the instrument amplification and the various F-factors are defined below. It can be seen that measuring a specular sample is an absolute measurement and measuring a diffuse sample is a relative measurement. The problem is we often do not know the ratio F_s/F and the reflectance of the reference plate R_{wall}. An example of how R_{wall} can be determined is given by Grandin and Roos (1994). In most real cases, the reflectance has specular and diffuse components and the signal has to be separated into these components. For the case of near normal incidence, the specular beam can be directed out through the entrance port and thus only the diffuse component is measured. For any other angle of incidence, this is a difficult problem, and there is no general way of solving equation (2.43). More detailed discussions of the importance of the diffuse and specular components, which was implied in connection with equations (2.30)–(2.37), are available (Nostell *et al.*, 1999).

The reference signal is measured with the sample holder rotated by 180° so the incident light strikes the back surface of the sample holder, which is coated with the same $BaSO_4$ paint as the sphere wall. As for the transmittance sphere, the fractions

of light contained in the reflectance sphere, F_s, F_d, and F, are not the same. Here, the
F-factors depend not only on the reflectance behavior of the sample, but also on the
incidence angle. For a diffuse sample F_d is *about equal to* F if the scattered radiation
from the sample has the same angular distribution as that from the reference. For the
specular component, F_s varies with the angle of incidence as the specular light spot
moves across the inside of the sphere wall. The sample is left in the sphere during
the reference reading, so the diffused light is influenced by the presence of the sample
in both readings. Furthermore, the amount of light escaping through the entry port
is different for different angles of incidence. The factors $F_{s,d}$ should in other words be
written as

$$F_{s,d} = F_{s,d}(R_{\text{sample } \theta}) \tag{2.44}$$

The angular dependence of $F_{s,d}$ results from the asymmetrical design of the sphere
in combination with the hemispherical reflectance of the sample for different angles
of incidence. To be able to use the sphere for various types of samples with different
reflectance characteristics, we have to evaluate $F_{s,d}$ in equation (2.41). There is no
possibility of calculating this factor in advance because we are interested in the
properties of an "unknown" sample. As examples, we can think of the two extreme
cases of totally specular and totally diffuse characteristics. For the first case, F_s
gradually decreases as the incidence angle increases. In the second case, F_d is a
constant and independent of the angle of incidence.

The exact response of the sphere when the sample is turned inside the sphere
depends on the design and there is no general way for making the corrections. The
detector field of view becomes extremely important and the configuration used for
the instrument designed in Uppsala is to be preferred. The detector is not positioned
at the sphere wall but at the end of the sample holder. In this way, the detector never
has a line-of-sight with the sample and cannot detect any change of the signal caused
directly by the rotating sample. In another design, a white screen is positioned in
front of the detector, between the sample and the detector. The detector can then
only respond to light scattered from the back of this screen. It is inevitable that the
sphere throughput changes when the sample is turned and this has to be taken into
account. Nostell *et al.* (1999) have investigated these effects and suggested a number
of test measurements that can be performed to establish which corrections need to be
made and the impact on the final result.

2.1.4.7 *Measurement of Samples with Azimuthal Anisotropy*
There is a special class of samples that are particularly difficult to measure. These are
samples with azimuthal anisotropy in the light scattering distribution, such as in an
ordinary rolled metal sheet. These samples usually have a surface texture or fine

grooves or scratches along the surface, which are somewhat like grooves in a grating (Roos *et al.*, 1988). Such samples tend to scatter the light into a plane perpendicular to the direction of the grooves. Because no instrument is rotationally symmetric around the incident beam, great care must be taken when measuring such samples. Port losses and the detector response may vary considerably when the sample is rotated around an axis parallel to the incident light. To make things worse, such samples usually also scatter a large fraction of the light into small scattering angles so that the formalism outlined in the previous sections must be applied. In a recent study, measurements with the Coblentz sphere also had to be corrected for the angular distribution of the scattered radiation (Rönnow and Roos, 1995; Lindström and Roos, 2000).

There is also another effect that requires attention. Grooved or linearly textured surfaces may have a reflectance (or transmittance if transparent) that depends on the polarization state of the incident light at near normal incidence. Light polarized along the grooves has a different reflectance from light polarized perpendicular to the grooves. Because the light from the monochromator can be strongly polarized with a polarization that is strongly wavelength dependent, some caution must be taken when measuring such samples.

2.1.4.8 *Measurement of Foils and Thin Flexible Materials*
The transmittance and reflectance of thin films of plastic or metal foils including some flexible materials are difficult to measure accurately. In these materials the sample surface is not completely flat, but the surface can be smooth on a microscopic scale and does not scatter light in the same way as discussed previously. The surface can be slightly curved, as for instance a section of a cylinder or a sphere, or have surface irregularities in the form of small dints and hillocks. A deflection of the light beam from the specular direction results from such a reflecting surface or, in the case of transmitting materials, a deviation from the original light beam results from refraction of the light beam. A smoothly curved or bent surface acts as a focusing or defocusing lens that may alter the focal distance of the optical path. A typical bathroom window and a hammered metal surface are examples of smooth and macroscopically flat but light deflecting surfaces. Solar collector covers and the mirrors used as booster mirrors for solar absorbers often belong in this category of samples. The principles of how light deflecting surfaces result in errors are illustrated in Figure 2.1-23. The typical error caused by this kind of surface irregularity is that a large fraction of the reflected radiation can be deflected out through the integrating sphere entry port. For transmittance measurements, the transmitted beam is smeared out across the opposite side of the sphere and if there is a specular exit port in this position some of the transmitted light will strike the edge around this port. The edge

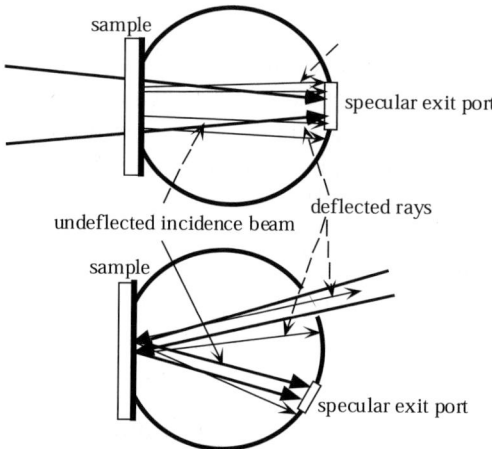

Figure 2.1-23.   Illustration of how light deflecting surfaces can result in a signal error in an integrating sphere.

tends to absorb more radiation than the surface of the sphere and thus the measured signal is reduced.

It is difficult to correct the errors resulting from light deflecting as shown in Figure 2.1-23. There is, in general, no way of measuring exactly how much light is lost or how much the signal is reduced because of port edge absorption. The best advice is simply to be aware of the problem and try to minimize it, for example, by making the foil as flat and smooth as possible. For reflectance measurements of opaque flexible materials, it may help to attach the sample to a more solid backing material, such as a piece of ordinary float glass.

2.1.5   CALCULATION OF INTEGRATED SOLAR PARAMETERS

In most solar energy applications, some integrated property over the entire or part of the solar spectrum is usually of primary interest. Such a property can be the light transmittance calculated for the visible part of the solar spectrum, or the total solar reflectance or transmittance calculated over the entire solar spectrum or even the chromaticity coordinates as defined in one of the color systems. To perform these calculations, an appropriate solar spectrum is needed for the integration procedure, a standard luminous light source, and the photopic sensitivity of the human eye. In this chapter, a short overview is given of how to perform these calculations. The reader is referred to the literature for more details (Duffie and Beckman, 1991).

For solar thermal absorbers, the parameters of interest are the total solar energy absorptance and the total hemispherical emittance. The absorptance of a solar

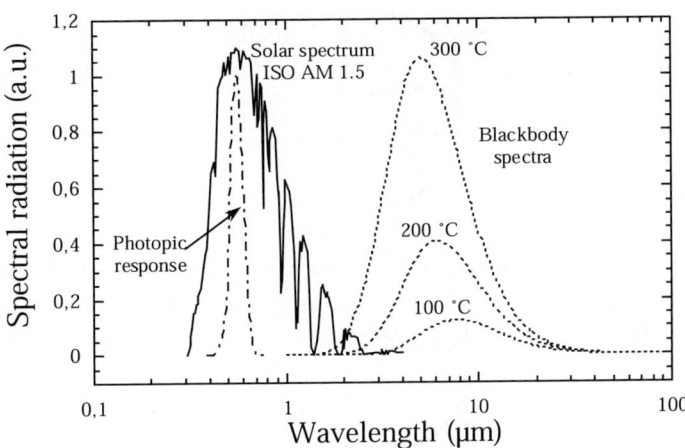

Figure 2.1-24. Solar spectrum at air mass 1.5, blackbody radiation spectra and photopic response of the human eye.

thermal absorber should be as high as possible for all solar irradiation and, obviously, a surface with 100% absorptance for all wavelengths does not exist. Thus, less than perfect absorbers are used in practice. The intensity of the solar spectrum varies with wavelength as shown in Figure 2.1-24, where the solar spectrum $S(\lambda)$ is shown for air mass 1.5 according to the ISO 9845 standard. Air mass 1.5 means that the solar radiation has an angle of incidence relative to a horizontal surface such that the distance through the atmosphere corresponds to 1.5 times the distance at exactly normal incidence. Also shown in Figure 2.1-24 are the blackbody radiation spectra, $P(\lambda)$, for surfaces with temperatures of 100, 200, and 300°C and the photopic response of the human eye, $E(\lambda)$.

Ideally, the experimental spectra measured for any solar device should be weighted with the spectra in Figure 2.1-24. This means taking the integral of the recorded spectrum multiplied by one of the spectra in Figure 2.1-24 and then normalizing by dividing by the weighting spectrum. In practice, this is extremely difficult mathematically so an approximation in the form of a summation is usually performed. Thus the spectra are divided in m intervals and a summation is performed according to

$$\alpha = \frac{\sum_{j=1,m}(1 - R_j)S_j}{\sum_{j=1,m}S_j} \tag{2.45}$$

The parameter α is the solar absorptance of the surface, R_j is the average reflectance in the wavelength interval j, and S_j is the solar intensity in this wavelength interval. By dividing by the intensity $\sum S_j$, the α value represents the absorbed fraction of the incident solar power. This procedure means that the recorded spectrum must be divided into the same intervals as the available solar spectral distribution. The average reflectance from this spectrum should be used but, in practice, the midpoint value is frequently used.

2.1.6 SUMMARY

This chapter provides a summary of different techniques used for the measurements of optical reflectance and transmittance. In some cases, a more detailed description is presented and, in other cases, only the main concept is described. We primarily discussed measurements with commercially available spectrophotometers and accessories that are used in many laboratories. For many types of measurements, a better result can be obtained if a custom designed instrument is used, for example one with a large integrating sphere. Such instruments are, however, not readily available in most laboratories, and where they are found, the user is assumed to have the necessary knowledge about how to use the instrument. The interested reader can find more detailed presentations of the ideas together with more complete experimental results in the papers in the reference list. It is our hope that the presented concepts and ideas will result in improvements in future experimental optical results performed by the readers of this chapter. The list of references provides a comprehensive source for further reading for anyone who would like to probe more deeply into the concepts of optical measurements.

REFERENCES

Coblentz, W.W. (1913) The Diffuse Reflecting Power of Various Substances, *National Bureau of Standards Bulletine* (*USA*), **9**, 283–325.
Duffie, J.A. & Beckman, W.A. (1991) *Solar Engineering of Thermal Processes*, Wiley, New York.
Grandin, K. & Roos, A. (1994) Evaluation of Correction Factors for Transmittance Measurements in Single Beam Integrating Spheres, *Appl. Opt.*, **33**, 6098–6104.
Hansen, L.M. (1989) Effects of Restricting the Detector Field of View when Using Integrating Spheres, *Appl. Opt.*, **28**, 2097–2103.
Hansen, L.M. (1996) Effects of Non-Lambertian Surfaces on Integrating Sphere Measurements, *Appl. Opt.*, **35**, 3597–3606.
Lindström, T. & Roos, A. (2000) Reflectance and Transmittance Measurements of Anisotropically Scattering Samples in Focussing Coblentz Spheres, *Rev. Sci. Instrum.*, **71**, 2270–2278.
Milburn, D.I. & Hollands, K.G.T. (1994) Solar Transmittance Measurements Using an Integrating Sphere with Broad Area Irradiation, *Solar Energy*, **52**, 497–507.
Nostell, P., Roos, A. & Rönnow, D. (1999) Single Beam Integrating Sphere Spectrophotometer for R and T Measurements Versus Angle of Incidence in the Solar Wavelength Range on Diffuse and Specular Samples, *Rev. Sci. Instrum.*, **70**, 2481–2494.

Rönnow, D. & Roos, A. (1994) Stray Light Corrections in Integrating Sphere Measurements of Low Scattering Samples, *Appl. Opt.*, **33**, 6092–6097.

Rönnow, D. & Roos, A. (1995) Correction Factors for Reflectance and Transmittance Measurements of Scattering Samples in Focussing Coblentz Spheres and Integrating Spheres, *Rev. Sci. Instrum.*, **66**, 2411–22.

Rönnow, D. & Veszelei, E. (1994) Design Review of an Instrument for Spectroscopic Total Integrated Light Scattering Measurements in the Visible Wavelength Region, *Rev. Sci. Instrum.*, **65**, 327–334.

Roos, A. & Ribbing, C.-G. (1988) Interpretation of Integrating Sphere Signal Output for Non-Lambertian Samples, *Appl. Opt.*, **27**, 3833–3837.

Roos, A. (1991) Interpretation of Integrating Sphere Signal Output for Non Ideal Transmitting Samples, *Appl. Opt.*, **30**, 468–474.

Roos, A. (1993) Use of an Integrating Sphere in Solar Energy Research, *Solar Energy Materials and Solar Cells*, **30**, 77–94.

Roos, A., Ribbing, C.-G. & Bergkvist, M. (1988) Anomalies in Integrating Sphere Measurements on Structured Samples, *Appl. Opt.*, **27**, 3828–3832.

Roos, A., van Nijnatten, P.A., Hutchins, M.G., Polato, P., Olive, F. & Anderson, C. (2000) Angular Dependent Optical Properties of Low-e and Solar Control Windows – Simulations Versus Measurements, *Solar Energy*, **69 (suppl)**, 15–26.

Snail, K.A. & Hansen, L.M. (1989) Integrating Sphere Designs with Isotropic Throughput, *Appl. Opt.*, **28**, 1793–1799.

Ulbricht, R. (1900) Die Bestimmung der mittleren räumlichen Lichtintensität durch nur eine Messung, *Electrotech. Z.*, **21**, 595–597.

van Nijnatten, P.A. (2002) A Spectrophotometer Accessory For Directional Reflectance and Transmittance of Coated Glazing, *Solar Energy*, **73**, 137–149.

Chapter 2.2

Performance Models of Solar Collectors and Systems

U. Frei

Institut für Solartechnik SPF, Hochschule für Technik HSR, Oberseestr. 10, CH-8640 Rapperswil, Switzerland

Abstract: The purpose of this chapter is to compare and contrast different designs of solar collectors in terms of their material properties and optical and thermal characteristics. Solar thermal collectors are used world wide in both domestic and industrial applications. Depending on their design, they can deliver low-temperature heat for domestic hot water or produce overheated steam for thermal electric power plants. Different designs are discussed on the basis of material properties, including optical and thermal characteristics. Collector performance is described using models that can be implemented in system simulation programs. Finally, both steady-state and dynamic approaches for testing collectors are presented.

Keywords: Solar thermal collector, Flat plate collector, Tubular collector, Collector efficiency, Collector efficiency factor, Absorptance-transmittance product, Compound parabolic concentrators, Collector models, Steady-state testing, Dynamic collector testing, Thermal capacitance

ACRONYMS, ABBREVIATIONS, AND NOMENCLATURE

CPC	Compound parabolic concentrator
IAM	Incident angle modifier
PVD	Physical vapour deposition
ISO	International Standard Organization
CEN	Commission European de la Normalisation
EN	Europäische Norm
A_a [m^2]	absorber area
A_e [m^2]	aperture area
A_b [m^2]	gross area
C_C [J/(m^2 K)]	collector capacity

E_L [W/m^2]   longwave irradiance ($\lambda > 3\,\mu$m)
G [W/m^2]   hemispherical irradiance in the collector plane
G_b [W/m^2]   beam solar irradiance
G_d [W/m^2]   diffuse solar irradiance
K_θ incident angle modifier
$K_{\theta b}$ incident angle modifier for beam solar irradiance
$K_{\theta d}$ incident angle modifier for diffuse solar irradiance
T_a [K]   ambient temperature
T_m [K]   collector mean temperature $T_m = (T_i + T_o)/2$
T_i [K]   collector inlet temperature
T_o [K]   collector outlet temperature
F' [–]   collector efficiency factor
$(\tau\alpha)_e$ [–]   product of the effective absorptance and transmittance
α [–]   absorptance of the solar absorber
τ [–]   transmittance of the collector cover
η [–]   collector efficiency
η_0 [–]   maximal collector efficiency for $T_m = T_a$ for
                          hemispherical irradiance (for the part of the
                          beam that is perpendicular to the collector plane)
η_{0b} [–]   maximal collector efficiency for $T_m = T_a$ and
                          perpendicular beam solar irradiance
η_{0d} [–]   maximal collector efficiency for $T_m = T_a$ and
                          diffuse solar irradiance
U_L [W/(m^2 K)]   heat transfer coefficient plate temperature
U_O [W/(m^2 K)]   heat transfer coefficient fluid temperature
a_1, c_1 [W/(m^2 K)]   algebraic constant, in reference to x
a_2, c_2 [W/(m^2 K^2)]   algebraic constant, in reference to x
x [m^2 K/W]   reduced temperature difference
t_k [s]   collector time constant
σ [W/(m^2 K^4)]   Stefan–Boltzman constant

2.2.1 INTRODUCTION

Detailed models are needed to assess the performance of systems. The system itself may consist of numerous components that are also described by their respective models. In the case of solar systems, the main components are the solar collectors, storage tanks and, depending on the system type, pumps, valves, controllers and other parts. Detailed models of components using material properties as the input data, e.g., absorptance and emittance of the solar selective coating, may also be used to evaluate the impact of these properties on the performance of the individual components and/or the performance of the whole system. Using detailed simulation methods, the "performance criterion" (PC) of an element, e.g., a material or a fabrication technique, may be derived. The value of the PC is selected, typically to be a 5 or 10% loss of performance. The value of the PC chosen is a measure of the relative importance of the particular performance property of a system.

The variation of the absorptance and/or emittance of a solar selective coating are typical examples of the influence of PCs on the solar system performance. In the case of solar systems, the annual performance under all the relevant boundary conditions may be used. In the following three sections, solar collector and system models are discussed.

2.2.2 OPTICAL AND THERMAL PERFORMANCE OF COLLECTORS

2.2.2.1 Introduction

The absorber of the collector is heated by absorption of solar radiation onto the absorber coating. In the absorber plate, the heat transfer medium (liquid or gaseous) is heated. The medium is used to transport the energy from the collector to the consumer.

The energy flow in the collector may be discussed in terms of optical and thermal characteristics. Optically, the solar radiation has to be transmitted through the cover, and then the transmitted radiation has to be absorbed by the absorber coating. The thermal characteristics include the description of the heat losses after the radiation is transferred into internal energy of the absorber (Figure 2.2-1).

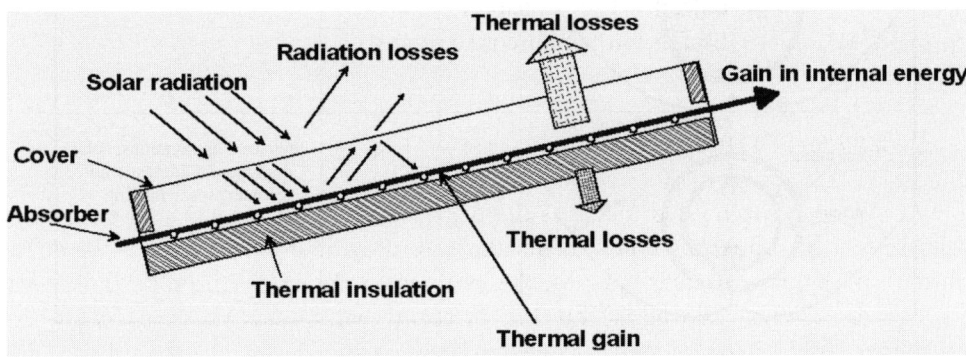

Figure 2.2-1. Energy flow in a thermal collector.

Flat-plate solar collectors are primarily used domestically for heating water or for any other heating application in buildings. Evacuated tubular collectors are able to provide higher temperature thermal energy because thermal losses are reduced. In conjunction with single-axis parabolic mirrors (parabolic trough collectors), evacuated tube collectors are also used for industrial applications, or as a source for producing steam in thermal electric power plants. In Figure 2.2-2, cross sections of

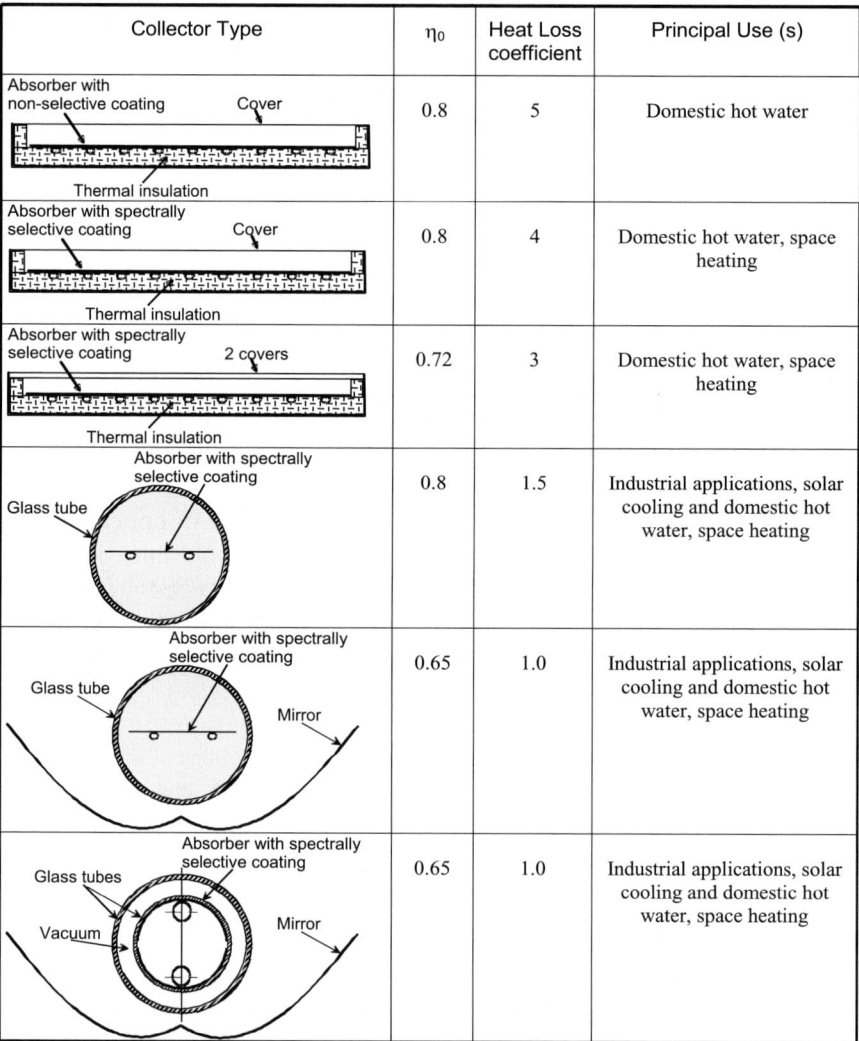

Collector Type	η_0	Heat Loss coefficient	Principal Use (s)
Absorber with non-selective coating — Cover — Thermal insulation	0.8	5	Domestic hot water
Absorber with spectrally selective coating — Cover — Thermal insulation	0.8	4	Domestic hot water, space heating
Absorber with spectrally selective coating — 2 covers — Thermal insulation	0.72	3	Domestic hot water, space heating
Absorber with spectrally selective coating — Glass tube	0.8	1.5	Industrial applications, solar cooling and domestic hot water, space heating
Absorber with spectrally selective coating — Glass tube — Mirror	0.65	1.0	Industrial applications, solar cooling and domestic hot water, space heating
Absorber with spectrally selective coating — Glass tubes — Vacuum — Mirror	0.65	1.0	Industrial applications, solar cooling and domestic hot water, space heating

Figure 2.2-2.   Cross-sectional view of the principal designs for solar thermal collectors and absorber tubes and coatings.

the principal designs for solar thermal collectors and absorber tubes and coatings are illustrated.

2.2.2.2 Optical Characteristics

Solar radiation is transmitted through the cover of the collector, as illustrated in Figure 2.2-1. The incoming radiation intensity is reduced by absorption in the cover material as well as by reflectance at both interfaces, from air to cover material, and from cover material to air. In the case of an evacuated tubular collector, the second interface is between the cover material and the vacuum. This material property is described by the solar transmittance. The transmitted solar radiation is then partially absorbed by the solar absorber. Some solar radiation is reflected by the absorber and redirected to the cover. In addition, multiple reflections between the cover and the absorber are possible. If multiple reflections are considered, the individual properties, solar absorptance and solar transmittance, are combined to give the transmittance–absorptance product ($\tau\alpha$).

This absorptance–transmittance product is not just the product of the two individual properties, but rather a compound property that results from a particular combination of a cover and an absorber. Detailed information about the solar absorptance and the solar transmittance, especially the determination of these properties, was presented in Chapter 2.1 of this volume.

Evacuated tubular collectors are as well as flat-plate collectors commercially available. Their optical characteristics may be similarly described as for flat-plate collectors. As shown in Figure 2.2-2, a combination of mirrors is used to concentrate solar radiation on evacuated tubular collectors. Nonimaging concentrators, or compound parabolic concentrators (CPC), are best suited for use as nontracking collectors. CPCs, within their acceptance angle, reflect incident radiation (specular and diffuse) to the absorber. The area concentration ratio is defined by the acceptance angle of the CPC (see Figure 2.2-3).

For CPC evacuated tubular collectors, it is not as simple to assess the optical performance as it is for flat-plate collectors. Depending on the geometry of the CPC mirror, multiple reflections of the incident radiation take place even before reaching the absorber tube. For any analysis of the optical performance, ray-tracing methods are required in which the optical paths of a large number of incident rays, which represent the solar radiation, are followed (Figure 2.2-4).

The optical properties of absorptance, reflectance, and transmittance depend on the angle of incidence. Standard installations of solar collectors in buildings are in a fixed position, so the "incident angle modifier" (IAM) is of great importance (Figure 2.2-5). For flat-plate collectors, the IAMs for both main axes are identical, whereas

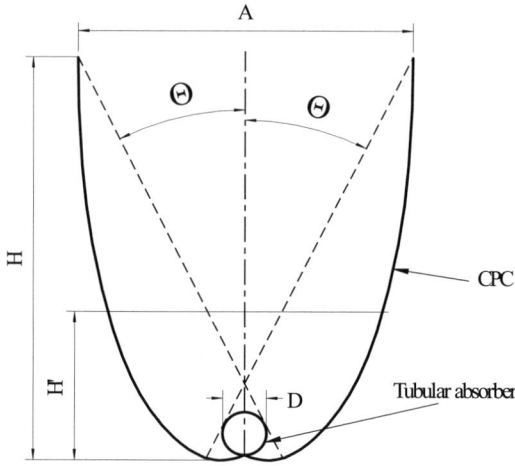

Figure 2.2-3. Acceptance angle (2θ) of a symmetrical CPC with tubular absorber of diameter D. The CPC may be truncated, for example, from H to H'.

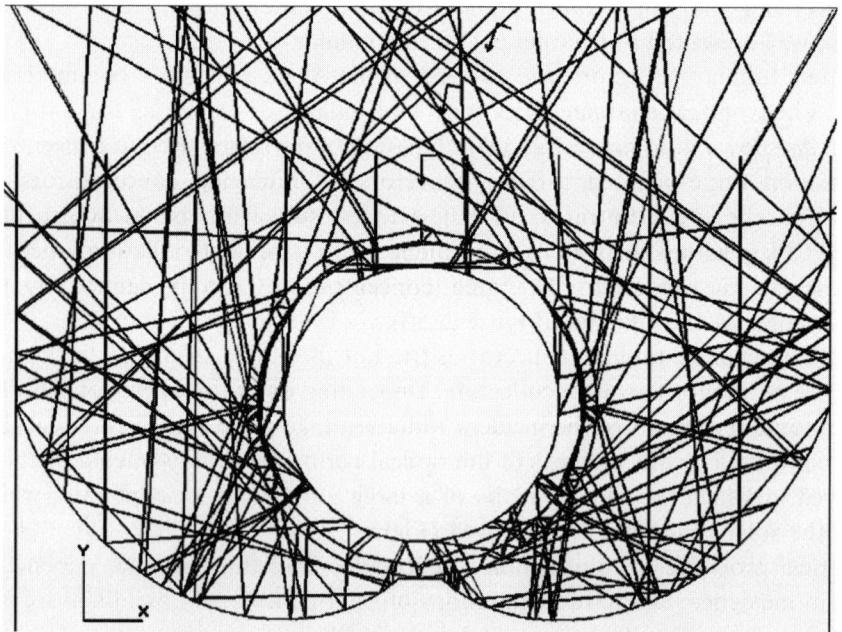

Figure 2.2-4. Ray tracing using standard personal computer software.

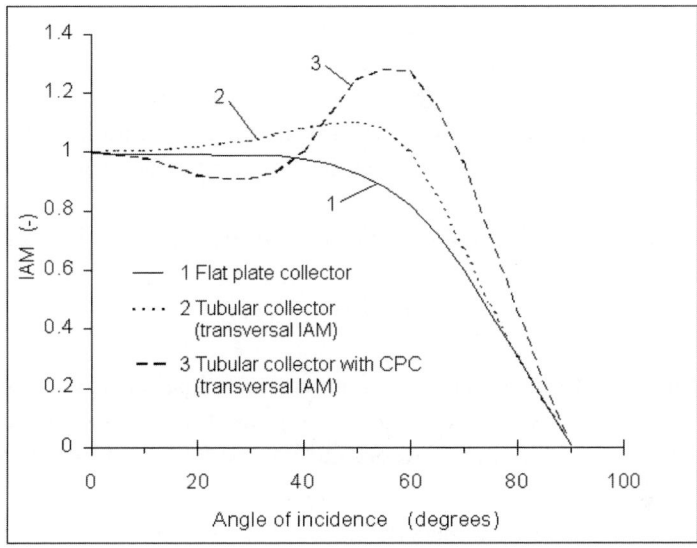

Figure 2.2-5. Incident angle modifier (IAM) for flat-plate and tubular collectors.

different IAMs are obtained for the main axes of tubular collectors (with and without mirrors).

2.2.2.3 Thermal Characteristics

Maximum efficiency ("zero-loss efficiency"), η_0, of a flat-plate collector is reached when the mean operating temperature (average of fluid inlet and fluid outlet temperatures) is equal to the ambient temperature. However flat-plate collectors are usually not designed to operate below ambient temperature.

Thermal effects and losses must be considered for operation at ambient temperatures. Incident solar radiation is not only reflected at the surfaces of the cover, but some radiation is absorbed, thus heating the cover and reducing the heat losses from the collector. The result of those losses is included in an *effective* absorptance–transmittance product $(\tau\alpha)_e$, which is of the order of 0.5 to 5% higher than the standard absorptance–transmittance product. Solar glass has a low absorptance because impurities, such as iron oxide in the raw material, are greatly reduced. Typically, the iron concentration is less than 0.1%. If so, the effect of the absorptance in the glass is less than 1%. The solar radiation is converted into heat on the absorber plate. Thereafter, heat flows to the (cooler) heat transfer medium. In a fin-and-tube absorber design (see Figure 2.2-6), heat flow is affected by the bond between the sheet metal and the tube, and the heat transfer to the fluid. All effects regarding the heat flow from the absorber plate to the heat transfer fluid are summarized in the collector efficiency factor, F'. For a simple definition, F' represents the ratio of the actual useful energy gain to the useful energy gain that would occur if the absorber surface temperature was the same as the local fluid temperature (Duffie and Beckmann, 1991). Typical absorber designs and values for F' are shown in Figure 2.2-6

The maximum efficiency of a solar collector can be expressed by equation (1).

$$\eta_0 = \dot{Q}/(GA) = F'(\tau\alpha)_e \tag{1}$$

Solar collectors operating at elevated temperatures lose energy because of heat losses. The main heat loss mechanisms are conduction, convection, and thermal radiation. The losses of any collector design may be broken down into top losses, back losses, and side and edge losses including the inlet and outlet for the heat transfer medium. In general, up to 80% of the losses are top losses, whereas the back losses are normally about 10 to 15%. Side and edge losses are normally about 10%, but may be higher because of poor thermal designs of the inlet and outlet of piping for the heat transfer medium.

The top losses for flat-plate collectors primarily result from two components. The first component is convection within the air gap between the absorber and glazing and the conduction of the air itself. The second component is the thermal radiation exchange between the absorber coating and the cover, for which the major contribution results from the thermal emittance of the absorber coating. State-of-the-art absorber coatings used in standard flat-plate and evacuated tubular collectors are made by physical vapor deposition (PVD). The thermal emittance of these coatings is less than 5% compared to that of a blackbody spectrum for 373 K. Therefore, the influence of the losses resulting from the thermal radiation exchange is only of importance at higher absorber temperatures because radiation losses increase with the fourth power of the absorber temperature.

Studies of the influence of materials properties, e.g., the emittance of the spectrally selective solar absorber coating, on the overall heat loss coefficient for the collector require detailed modeling. Regarding flat-plate collectors, this model has to treat convection and radiation between parallel plates. In all cases, an iterative process is required to balance the overall losses of the collector cover and box with the energy exchange from the absorber. An empirical equation to estimate the top losses of flat-plate collectors has been presented (Duffie and Beckmann, 1991). This equation is reasonably accurate for flat-plate collectors with one or more covers.

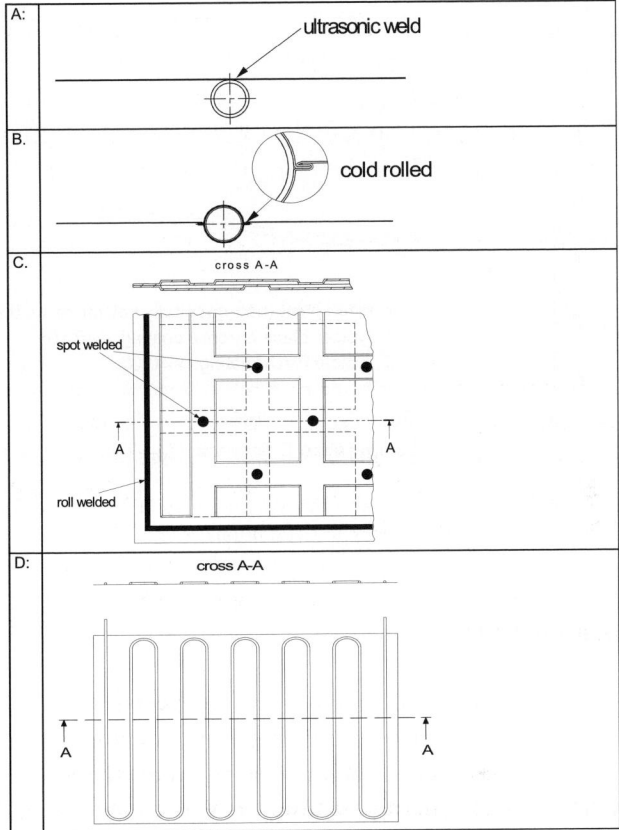

Figure 2.2-6. Absorber designs and typical F' values. (A) For bonding by ultrasonic welding, $F' = 0.92$ for a strip width of 120 mm, sheet thickness of 0.2 mm, and tube inner diameter of 10 mm, with turbulent flow of the heat transfer medium. (B) For an absorber sheet metal cold rolled around the tube, $F' = 0.92$ for a strip width of 120 mm, sheet thickness of 0.2 mm, and tube inner diameter 10 mm, with turbulent flow of the heat transfer medium. (C) For a stainless steel cushion absorber, $F' = 0.97$, using two stainless steel plates with a thickness of 0.6 mm. (D) For a copper tube soft soldered on copper sheet, $F' = 0.95$ for a tube width of 80 and 20 mm from the edge of the copper sheet to the bent copper tube; the copper tube inner diameter is 10 mm. The heat transfer medium is assumed to undergo turbulent flow.

The top losses of typical flat-plate collectors are shown in Figure 2.2-7 including different absorber coatings. The convective heat transfer coefficient used represents typical efficiency test conditions as required by international standards (ISO, EN) to include the effect of wind.

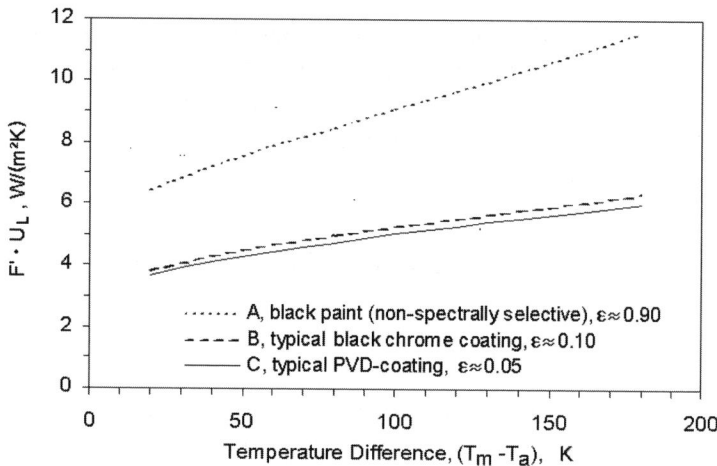

Figure 2.2-7. Top losses of three typical solar absorber coatings used in flat-plate collectors under typical efficiency test conditions.

2.2.3 COLLECTOR MODELS

Current collector models are based on early studies by Hottel and Woertz (1942). Since then, many authors have proposed different models for different applications (Duffie and Beckmann, 1991). One approach is to study the influence of material properties on the collector or system performance. For this purpose, the model must be detailed enough to consider the relevant material properties with sufficient accuracy. Equation (2) is an example for a simple steady-state collector model.

$$\eta = F'(\tau\alpha)_e K(\theta)G_K - F'U_L(T_m - T_a) \tag{2}$$

In many cases, a linear relation for the heat losses does not describe the collector losses adequately throughout the entire temperature range. Therefore, temperature dependence is introduced using the following relation:

$$F'U_L = a + b(T_m - T_a) \tag{3}$$

Collector models used to simulate thermal processes are based on parameters that can be derived by collector tests. The collector test procedures are defined by international standards (ISO 9806, EN 12975) (International Organisation of Standardization ISO, 1994; European Standard EN 12975, 2000).

In the collector model, the following terms are considered: zero-loss collector efficiency, incident angle modifier, temperature-dependent thermal losses, additional wind speed induced thermal losses, influence of infrared radiation exchange between the collector and sky, and thermal capacity. Equation (4) represents a comprehensive model for use in simulation software.

$$\frac{\dot{Q}}{A} = \eta_{0b}K_{\Theta b}(\Theta)G_b + \eta_{0d}K_{\Theta d}(\Theta)G_d - c_1(t_m - t_a) - c_2(t_m - t_a)^2$$

$$- c_3 u(t_m - t_a) + c_4\left(E_L - \sigma T_a^4\right) - c_5\frac{\mathrm{d}t_m}{\mathrm{d}t} \tag{4}$$

Standard steady-state tests, as described by international standards, do not provide methods of obtaining all of the parameters required by equation (4). In particular, the detailed information on the beam and diffuse characteristics (zero-loss efficiencies, incident angle modifier) cannot be easily obtained.

2.2.3.1 Steady-state Collector Testing

For steady-state collector testing, equation (5) shows the most commonly used equation for collector efficiency testing according to the European standard EN 12975.

$$\eta = \dot{Q}/(AG) = \eta_0 - c_1\frac{(T_m - T_a)}{G} - c_2\frac{(T_m - T_a)^2}{G} \tag{5}$$

After introducing the scaled temperature difference, $x = (T_m - T_a)/G$

$$\eta = \dot{Q}/(AG) = \eta_0 - a_1 x - a_2 G x^2 \tag{6}$$

The parameters of equation (6) are derived by curve fitting from test results (see Figure 2.2-8). The measurements are performed according to international standards ISO 9806 and EN 12975 (International Organisation of Standardization ISO, 1994; European Standard EN 12975, 2000). During clear sky conditions, the inlet temperature is kept constant for roughly four times the time constant of the collector. At least four different inlet temperatures are chosen over the operating range of the collector. During the test, an artificial wind with a constant speed of approximately 3 m/s is applied over the entire collector area.

The same biaxial test rigs are used to measure the incident angle modifier:

$$K(\Theta) = \eta_0(\Theta)/\eta_{0,s} \tag{7}$$

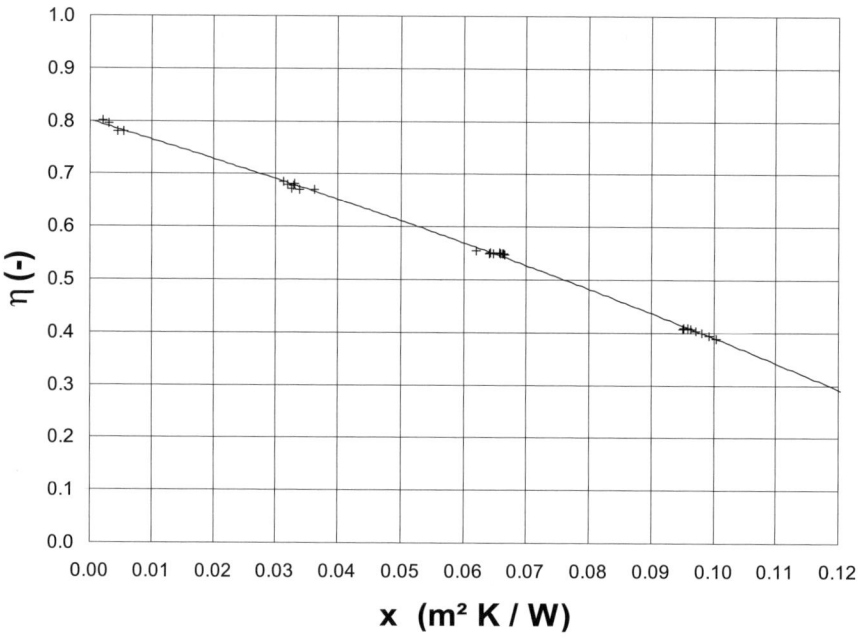

Figure 2.2-8. Second-order curve fit of typical test data measured at steady-state conditions using biaxial tracking test rigs.

Using steady-state efficiency testing, it is not possible to obtain detailed information regarding the transient behavior of the collector. The thermal performance of a system is in most cases only slightly influenced by the heat capacity of a collector. The influence of the collector heat capacity on a solar domestic hot water system for a middle European climate and an annual input oriented solar fraction of about 60% (5 m^2 absorber area), is shown in Table 2.2-1. For engineering purposes, an approximation of the heat capacity can be calculated on the basis of the measured time constant (Fricke and Borst, 1984). The determination of the time constant is described in detail in international standards, e.g., EN 12975 (European Standard EN 12975, 2000).

The collector capacity assessment is based on the following relation (Fricke and Borst, 1984).

$$t_k = \frac{(C_C/c_p\dot{m})(z - 1 + e^{-z})}{z^2} \tag{8}$$

where

$$z = (F'U_L)/(c_p\dot{m}) \tag{9}$$

Table 2.2-1. Influence of the thermal capacitance on the annual heat gain of a solar domestic hot water system (collector parameter: $\eta_0 = 0.83$, $a_1 = 3.4$ W/(m²K), $a_2 = 0.009$ W/(m²K²))

Collector type	Thermal capacity [kJ/(m² K)]	Yearly heat gain [KWh/a]	Deviation [%]
Flat-plate collector, as measured	6	2453	0
Flat-plate collector, modified capacity	2	2485	+ 1.3
Flat-plate collector, modified capacity	18	2386	− 2.7

For standard collector operation, the factor z becomes small and therefore the equation (8) can be simplified:

$$t_k \cong C_C/(2c_p\dot{m}) \tag{10}$$

2.2.3.2 *Dynamic Collector Testing*

The main disadvantage of using steady-state collector testing is that the efficiencies under beam and diffuse irradiance cannot be determined separately. In addition, the IAM for certain collector designs, for example, evacuated tubular collectors with CPC mirrors, is very difficult to evaluate. These parameters are needed as input into a computer simulation program, e.g., TRNSYS (TRNSYS, Version 15.0, 2002) and POLYSUN (POLYSUN 3, 2002). The dynamic testing approach is an elegant way to determine all necessary parameters for characterizing solar collectors. Such an approach is described as a "quasi-dynamic test method" in the EN-standard EN 12975 (European Standard EN 12975, 2000).

This detailed collector model is similar to the one described in equation (4). For any dynamic analysis, the basis of successful parameter identification is the variability of the test results concerning the parameter to be fitted. For example, if the wind speed dependency of the collector performance should be analyzed in the range from 0.5 to 10 m/s, these wind speeds must be applied during test sequences over the entire operating range of the collector.

In the EN-standard, a detailed test description of the quasi-dynamic test is described including the necessary operating conditions for the parameter identification. The mathematical tool used for parameter identification is Multiple Linear Regression. The parameter fitting procedure is based on the minimization of the error of the power output of the collector.

The dynamic approach is not yet widely used. The risk of presenting meaningless collector test results is far higher than that for the steady-state approach. Nevertheless, the risk may be kept low if steady-state periods are extracted from the quasi-dynamic test results and the steady-state parameters are compared to the results from dynamic fitting, as shown in Figure 2.2-9 (EC-Joule, 2001).

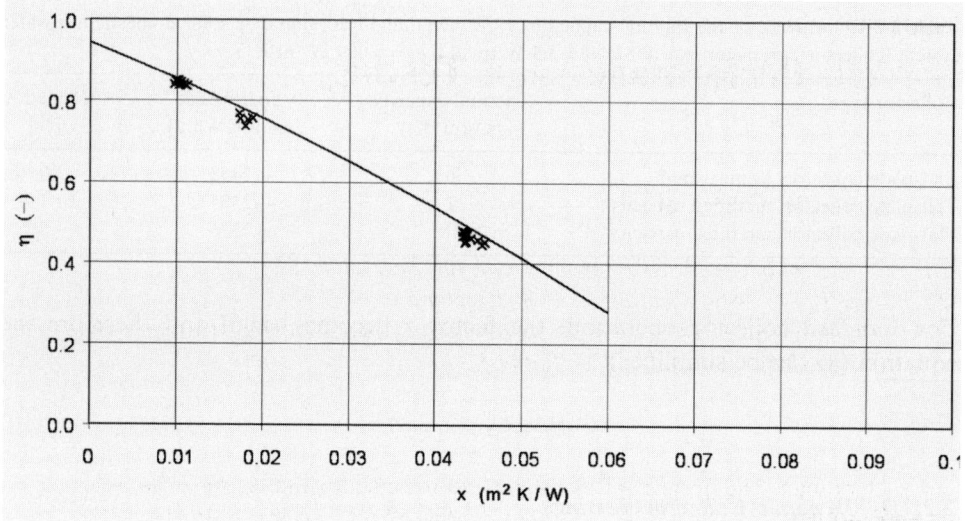

Figure 2.2-9. Comparison of collector efficiency testing using either the quasi-dynamic procedure or the steady-state approach (EC Joule, 2001). The collector tested is an uncovered metal absorber with a spectrally selective solar absorber coating. The dots are results of steady-state testing while the curved line is the result of the quasi-dynamic test.

REFERENCES

Duffie, J.A. & Beckmann, W.A. (1991) *Solar Engineering of Thermal Processes*, John Wiley & Sons Inc., New York.

EC-Joule (2001) PL 971140, *Solar Building Facades*, Task 4, Rapperswil, Switzerland.

European Standard EN 12975 (2000) European Committee for Standardization CEN, *Thermal Solar Systems and Components, Solar Collectors, Part 2: Test Methods*, Paris, France.

Fricke J. & Borst, W.L. (1984) *Energie, ein Lehrbuch der physikalischen Grundlagen, 2. Auflage*, Oldenburg Verlag, Wien.

Hottel, H.C. & Woertz, B.B. (1942) *Performance of Flat-plate Solar-heat Collectors*, Transactions of the American Society Mechanical Engineers.

International Organisation of Standardization ISO (1994) International Standard ISO 9806, *Test Method of Glazed Liquid Solar Collectors*, Geneva, Switzerland.

POLYSUN 3 (2002) *Simulation Software for Thermal Solar Systems*, Institut für Solartechnik SPF, Switzerland.

TRNSYS, Version 15.0 (2002) *A Transient System Simulation Program*, Solar Energy Laboratory, University of Wisconsin, Madison, WI, USA.

Chapter 2.3

System Performance and Testing

U. Frei[1] and H. Oversloot[2]

[1]*Institut für Solartechnik SPF, Hochschule für Technik HSR,
Oberseestr.10, CH-8640 Rapperswil, Switzerland*
[2]*TNO Building and Construction Research, Dept. Sustainable Energy and Buildings,
P.O. Box 49, 2600 AA Delft, The Netherlands*

Abstract: The purpose of this chapter is to compare and contrast the system performance of two widely used solar thermal systems using testing and simulation programs. Solar thermal systems are used in many countries for heating domestically used water. In addition to the simple thermosiphon systems, better-designed pumped systems with new operational strategies are becoming available. "Low-flow systems" achieve higher system performance than standard "high-flow systems." To compare system performance, objective criteria must be defined. The preferred criteria are the fractional solar savings that represent the solar gain compared to a nonsolar reference system. Simulation programs with different levels of sophistication are used to estimate the yearly solar gain of a particular system. Using detailed models, systems simulation is used to evaluate the performance criterion for accelerated aging tests on the basis of materials properties. CEN and ISO test methods, which are used for performance testing of the systems, are compared and applied to systems of various sizes.

Keywords: Solar thermal systems, Low-flow system, Solar fraction, Solar fractional savings, Solar system simulation, TRNSYS, Polysun, Solar thermal system test methods

ACRONYMS, ABBREVIATIONS AND NOMENCLATURE

ASHRAE	American Society of Heating, Refrigeration, and Architectural Engineers
CEN	Commission European de la Normalisation
CSTG	Collector System and Testing Group
DST	Dynamic System Testing
IAM	Incident Angle Modifier
ISO	International Standards Organization

| PC | Performance Criterion |
| SDHW | Solar Domestic Hot Water |

Q_{WW}	domestic hot water demand [kWh]
Q_{CO}	collector heat gain [kWh]
Q_{LR}	thermal losses of the nonsolar reference system [kWh]
Q_{WR}	auxiliary heating energy for the nonsolar reference system provided by a oil, gas or wood-fired boiler [kWh]
Q_{ER}	auxiliary heating energy for the nonsolar reference system provided by electricity [kWh]
Q_{HR}	auxiliary electrical energy for the pump and controller of the non-solar reference system [kWh]
Q_{LS}	thermal losses of the solar system [kWh]
Q_{WS}	auxiliary heating energy for the solar system provided by a oil, gas or wood-fired boiler [kWh]
Q_{ES}	auxiliary heating energy for the solar system, provided by electricity [kWh]
Q_{HS}	auxiliary electrical energy for the pump and controller of the solar system [kWh]
sF_i	solar fraction with respect to input [−]
sF_o	solar fraction with respect to output [−]
F_{ss}	fractional solar savings [−]

2.3.1 INTRODUCTION TO SYSTEM PERFORMANCE

Solar collectors may be used as a heat source for various applications. The most common uses of solar thermal collectors are for heating domestic water, building space in combination with domestic hot water as combined systems (Suter *et al.*, 2000), swimming pools, or in process heat for industry, cooling, and desalination applications.

In the small but growing international market for solar thermal systems, solar domestic hot water (SDHW) systems have by far the largest market share. Therefore, the following explanations relate only to SDHW applications. In some countries, solar thermal systems are most commonly used for heating domestic water. The reasons include excellent climatic conditions, expensive conventional heat sources, or simply a legal regulation.

The basic function of all systems is the same. In the collector, the incident solar radiation is transformed into heat (see Section 2.3.3). Then, using a heat transfer fluid, the heat is transported through pipes to the storage tank. In SDHW applications, the storage tank is a buffer between the varying consumption and the varying, unpredictable insolation conditions.

There are two ways to move the heated fluid from the collector to the heat exchanger of the storage tank. If the storage tank is situated higher than the collector, the thermosiphon effect can be used. Because the density of the heat transfer fluid decreases with increasing temperature, the warmer fluid flows toward the heat exchanger in the storage tank, whereas the colder and therefore denser fluid flows to the inlet of the solar collector. These simple thermosiphon systems are used in southern countries. The major advantage of these systems is their simple design, as shown in Figure 2.3-1. They need no pumps or controller. The disadvantage is the visual appearance of the storage tank situated on top of the roof. The size of the storage tank also limits the system size to single-family applications.

In all systems where the storage tank is located lower than the collectors, a pump and a controller are needed. A simple pumped system is shown in Figure 2.3-2. These systems are designed in various sizes from a 3-m^2 collector area for SDHW heating in single-family homes to several thousands of square meters for supplying a whole village with hot water.

Depending on the climate, the solar system requires an auxiliary heat source to meet the demand of the user. The auxiliary source might be a gas or oil-fired boiler, or, if only a small amount of additional energy is required, even an electric heater.

The important advantage of pumped systems is less obtrusive visual integration into the building. Modern flat-plate collectors are designed to be integrated into the roof or the facade of the building. With the appearance of the solar system becoming more and more important, architects are playing an important role in improving

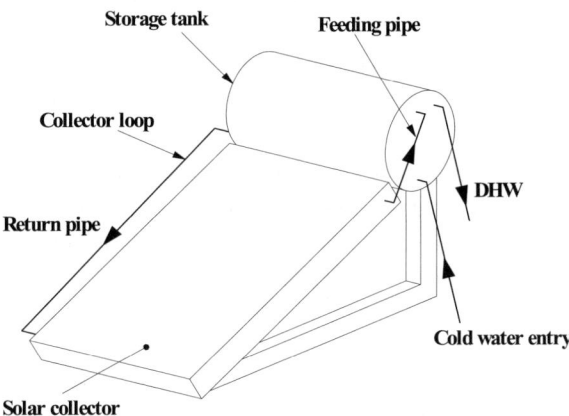

Figure 2.3-1. Thermosiphon system for SDHW heating.

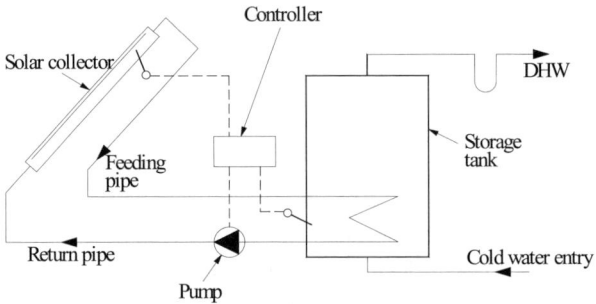

Figure 2.3-2. Pumped system for SDHW heating.

Figure 2.3-3. Integration of solar collectors into buildings.

the integration of collectors into the building. This aspect is even more important for larger systems in medium-density housing.

2.3.2 SYSTEM DESIGN

2.3.2.1 *Traditional and Advanced Designs for SDHW Systems*
Solar domestic hot water (SDHW) systems currently used in freestanding houses are mostly "factory made" systems, and are illustrate in Figures 2.3-4 and 2.3-5. This means that the system, including all necessary components, is delivered by just one supplier. This system-based approach has a number of advantages, i.e., (a) every single component, such as solar collectors, piping, fittings, heat exchanger, and the storage tank, is carefully designed and optimized to attain the highest overall system efficiency, (b) correct function, performance, durability, and reliability may be tested in the laboratory in accordance with international standards, because the

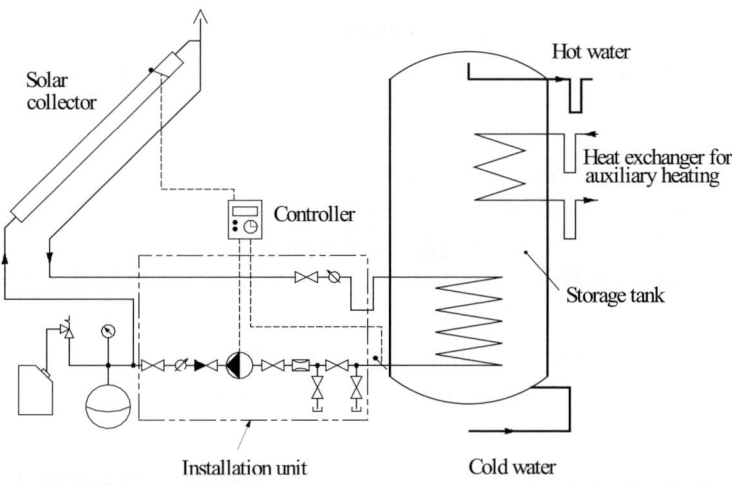

Figure 2.3-4. Traditional SDHW system.

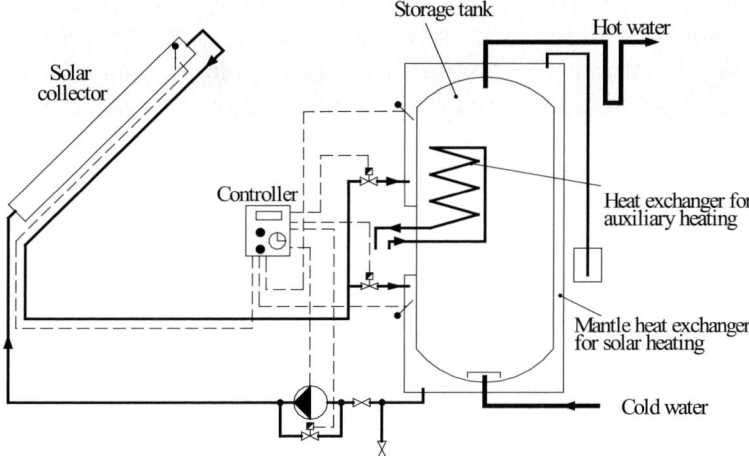

Figure 2.3-5. Advanced SDHW low-flow system.

certification of systems is important for acceptance by the consumers, and (c) the warranty, maintenance, and postsale customer service is provided by one supplier.

For maximizing the savings, further optimization requires significant changes in the solar system designs. In central Europe, the metamorphosis from traditional systems toward advanced low-flow systems took a couple of years. The results were an increase of about 25% in savings, and the price per system decreased by about the same percentage.

2.3.2.2 Performance Definitions for SDHW Systems

To compare systems and to judge the effects of optimization, meaningful definitions of performance indicators are needed. In the past, various definitions of the solar fraction have been proposed. The most common ones are the solar fraction with respect to input:

$$sF_i = Q_{CO}/(Q_{CO} + Q_{WS} + Q_{ES} + Q_{HS}) \tag{1}$$

and the solar fraction with respect to output:

$$sF_o = 1 - [(Q_{WS} + Q_{ES} + Q_{HS})/Q_{WW}] \tag{2}$$

The disadvantages of these two definitions are obvious. For the solar fraction with respect to output, the performance of the system is overestimated, because all of the storage losses are attributed to the auxiliary heat sources. In addition, the increase of the storage losses resulting from overheating in summer results in an increase of the input-oriented solar fraction (Figure 2.3-6).

For the solar fraction with respect to output, the performance of the system is underestimated because all the losses including those resulting from the conventional

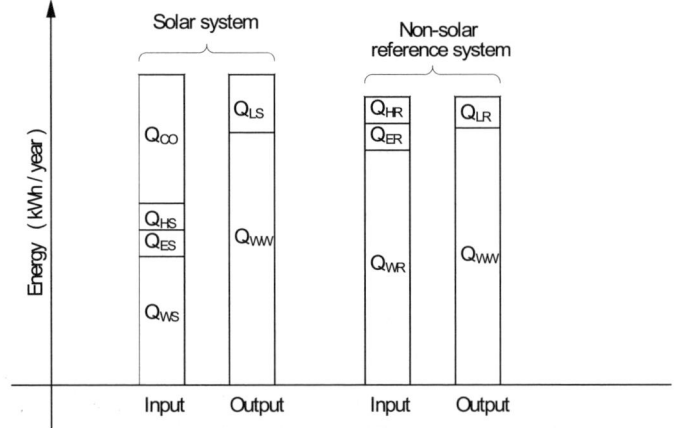

Figure 2.3-6. Solar fraction versus fractional solar savings.

heat source are attributed to the solar source only. High-storage losses resulting from poor thermal insulation of the storage tank reduce the output-oriented solar fraction.

The term, solar fractional savings (F_{ss}), is introduced to overcome the disadvantages of the solar fractions as defined above with this approach, the solar system is compared to a nonsolar reference system, and it is important to choose a relevant non-solar reference system. We calculate F_{ss} using

$$F_{ss} = [Q_{WW} + Q_{LR} - (Q_{WS} + Q_{ES} + Q_{HS})]/(Q_{WW} + Q_{LR}) \tag{3}$$

2.3.3 SYSTEM SIMULATION

Performance assessment by system simulation is a common engineering practice. It is especially of great importance to predict performance and costs to show the potential of solar technologies. Questions regarding cost/performance ratios or ecological effects can be easily investigated.

A simulation program attempts to model the thermal behavior and performance of a real physical system as closely as possible. The major advantage of simulation compared to physical system testing is that information about the influence of variations in parameters can be obtained very quickly and inexpensively. Depending on the level of detail of the model, the variations may result in the substitution of a complete component or a change in the selection of a material with different properties. In addition, the influence of weather conditions on system performance can easily be assessed. The results of the simulation studies must be validated by comparing them with experimental results. Detailed validations are used to improve models and to understand the performance of systems during in-service use.

2.3.3.1 *Simulation Studies in Accelerated Aging Testing*
Appropriate system simulation is of great importance in accelerated aging testing, including lifetime assessment. One important application is assessing microclimates using different boundary conditions. With detailed models, all relevant stress levels can be assessed. Of course, the validation of such detailed models is essential to ensure the accuracy of the results. This information is then used to define test conditions for accelerated aging tests.

The evaluation of the performance criterion (PC), which is needed to define service lifetime, is another important application. The basis for defining the end of the service lifetime can be an arbitrarily chosen value, e.g., the loss in performance can only be accepted to the chosen amount. Using detailed component models based on materials properties, simulation studies can be used to evaluate the PC function. These PC functions are used to study the degradation of materials during natural and accelerated aging. The PC function also helps us understand the relative importance of the degradation in material properties to the loss in performance of complete systems.

2.3.3.2 Simulation Programs

For solar system simulation, specialized programs have been developed. They may be divided basically into two categories, i.e., a component-based approach: TRNSYS (TRNSYS, 2002) and the system-based approach: POLYSUN (POLYSUN 3, 2002), T-SOL (T-SOL 4, 2002).

The component-based simulation programs are exemplified by TRNSYS. TRNSYS is a typical component-based simulation program that offers maximum flexibility. The different components are all separate and the user decides which ones he wants to use. Every single component is described by a model. The standard TRNSYS library contains about 100 component models and input–output routines. In addition, many user-defined components are available. For most basic components in a solar system, a variety of different model approaches are available. The models differ in terms of mathematical description and complexity.

As an overview, several solar collector models are explained. In type 1, a variety of optical modes is available to account for incidence angle modification including the ASHRAE function and user-supplied data. The solar collector does not have any capacitance. In type 52, a "matched flow collector model" is a user-supplied component. In addition to the optical modes of type 1, the "Ambrosetti" function is supported (Ambrosetti and Keller, 1985). This routine is mathematically highly complex. In contrast to type 1, the treatment of the diffuse radiation requires expertise. The capacitance is accounted for in detail. In type 132, the user-supplied single capacitance node collector model uses the Ambrosetti function for the IAM.

Both the source code and a compiled version of most of the component models are supplied with the simulation package, which also includes a graphical preprocessor program to generate the input file. Because component models may be altered or added by the user, TRNSYS offers virtually unlimited possibilities. Its use, however, requires experience and expertise. Calculation times depend on the level of accuracy and the simulation time step, but are generally much longer than that of more simple, system-based simulation programs.

The system-based simulation programs are exemplified by POLYSUN (POLYSUN 3, 2002). In central Europe, there is a well-established market for solar domestic hot water and combined systems. Although more than 100,000 systems were installed during the year 2001, only a small number of system designs were successful. In addition to establishing a larger market for solar systems, there is a need for providing easy-to-use simulation tools for engineers and planners. Such simulation programs are used not only as engineering tools, but also for marketing. Therefore these simulation tools must have very short computing times, typically less than 1 min for a one-year simulation run. This requires a system approach, in which the program code is optimized by combining various component models into a small number of predefined system configurations.

In POLYSUN, fast runtimes are also achieved with a dynamic time step. The time step may be as short as 5 s to model dynamic effects accurately, or as long as 1 h if the system remains stable during the night. Another important aspect is a simple, intuitive program interface, which enables people who use the software only once a month to perform simulations without losing much time in obtaining the desired results. A POLYSUN user interface is shown in Figure 2.3-7.

The system library of POLYSUN 3 includes the most widely used system configurations in Europe, such as SDHW systems with internal and external heat exchangers, one or two storage tanks, etc. In addition, combined solar systems may be simulated very easily because of the interface with a dynamic building simulation model HELIOS (HELIOS, 1982).

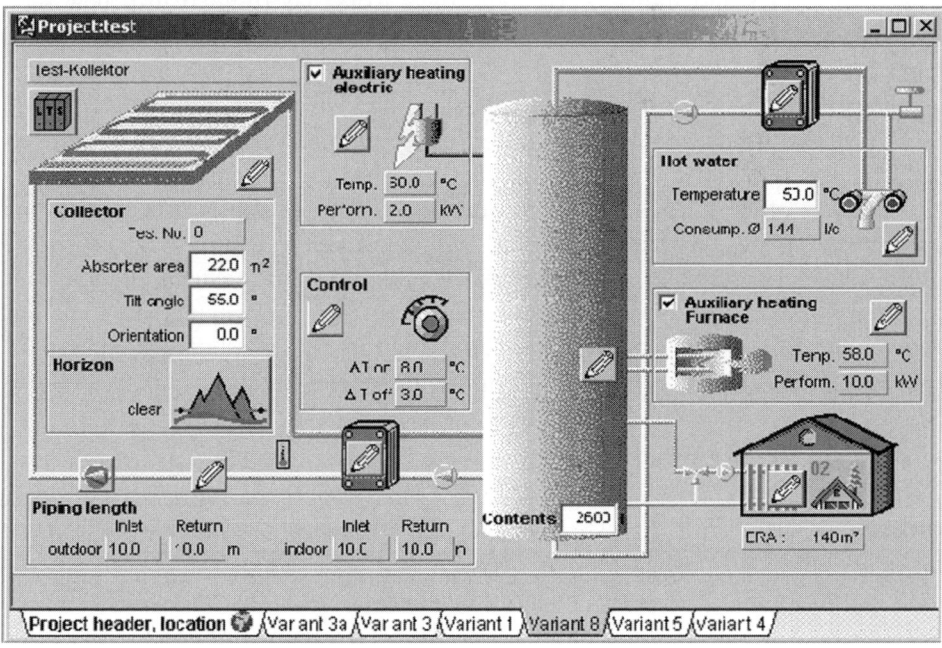

Figure 2.3-7. POLYSUN user interface.

A number of unique features, data bases and catalogs make POLYSUN a complete and easy-to-use simulation tool. These features include having access to a collector catalog including more than 150 solar collectors from major manufacturers world wide, weather data at more than 200 locations world wide, the selection of 24 building types for combined system simulations and interfaces to an external building simulation, the selection of pumps typically used in solar systems, pressure drop calculations, choosing different dimensions for the expansion vessel, making economic and ecological analysis of the system, graphical analysis of system temperatures, different languages (English, French, German), and output reports for engineers and clients in nine different languages.

2.3.4　SYSTEM PERFORMANCE TESTING

2.3.4.1　*Performance Testing using CEN Test Methods*

CEN is the responsible standardization body in the European Community. The work on standardization for testing of thermal solar energy systems and components is carried out by Technical Committee TC 312. The work of CEN/TC 312 is structured into three Working Groups, each consisting of a number of experts from research institutes and industry. In 2002, their work has resulted in a new set of standards for testing thermal solar systems in Europe. These are summarized in Table 2.3-1

The testing standards for complete solar heating systems EN-12976 (EN 12976-2: 2000) have the principle goal of establishing a common level of quality with respect to durability, reliability, and safety of the solar heating systems on the European market. One other purpose of these standards is to establish how the thermal performance is to be determined and presented on a common basis for the whole European market. These standards comprise and extend standards of the International Standard Organization, ISO. The performance test methods used for the different system types are formulated in the ISO standard 9459, part 2 (ISO 9459-2:1995), which is the collector and system testing group (CSTG) test

Table 2.3-1. CEN standards developed for testing solar heating sytems

Standard	Working field
EN-12975	Collectors, static and dynamic performance testing methods, quality of collector (fixing, thermal shock, rain penetration etc.)
EN-12976	Performance test of complete solar heating systems (factory made systems)
ENV-12977	Performance characterization of stores (storage, controller)

method and part 5 (ISO/DIS 9459-5), which is the dynamic system testing (DST) method.

Four reference locations have been chosen for which the performance is always calculated: Athens, Greece (for southern Europe), Davos, Switzerland (for mountain regions), Stockholm, Sweden (for northern Europe) and Würzburg, Germany (for central Europe and maritime climates). Data from these locations facilitate comparisons of the thermal performance of different systems for different solar insolations and microclimates.

2.3.4.2 The Dynamic System Testing Method
The general approach of the DST method is to perform a series of short outdoor tests on a SDHW system and to use a general dynamic computer model for processing the test data. This is done by using a parameter identification technique and yields a characterization of the SDHW system in terms of six to nine parameters in the model. These parameters and the same computer model are subsequently used for predicting the annual thermal performance for the energy saved by the SDHW system. The performance can be calculated for a wide range of conditions: different climates, hot water demand quantities and profiles, and desired hot water temperatures.

One of the advantages of the DST method is the "black-box" approach, which means there is no need for internal measurements or special knowledge of the system to be tested. This makes the method suitable for formalization in a standard e.g., ISO 9459-5 (ISO 9459-2:1995). More importantly, different kinds of SDHW systems can be tested with the same DST test method, solar-only, solar-preheat systems, and solar-plus-supplementary systems can be tested with this method. Other measurement methods are either limited in the types of SDHW systems that can be tested, or require extensive knowledge of the system for their use. This results in high-testing costs, because the tests have to be adapted to a specific system. Furthermore, the DST test results are independent of the location where the system is tested.

2.3.4.3 The Test Method Proposed by the Collector and System Testing Group
The approach in a test method proposed by the CSTG is to perform a short outdoor test on an SDHW system. For the CSTG test method, a correlation is used for processing the test data. This model takes into account the daily irradiation on the collector plane (input) and the daily heat delivered by the system (output) as integral (not dynamic) values. The parameter identification in this test method yields a characterization of the SDHW system in terms of three correlation parameters. As in the DST test method, these three parameters and the same correlation model are

subsequently used for predicting the annual thermal performance. This performance can be calculated for different climates and different hot water demand quantities. However, it is not flexible as far as the draw-off profile is concerned.

The CSTG test method also is a black-box approach, i.e., there is no need for internal measurements or special knowledge of the system to be tested, so the method is described in the ISO 9459-2 standard (ISO/DIS 9459-5). Various kinds of SDHW systems can be tested with the CSTG test method; however, it is limited to solar-only and solar-preheat systems. Finally, the CSTG test results are also independent of the location where the SDHW system is tested.

2.3.4.4   Comparison of the CSTG and DST Test Methods
In Table 2.3-2, the characteristics, differences, and similarities of the CSTG and the DST test methods are summarized and compared. The standard EN 12976 contains more details on which performance test method is applicable depending on the system and how to present the resulting yearly performance indicators.

2.3.4.5   Checking System Performance of Large Systems
It is not possible to carry out a set of standardized tests on large custom built systems, especially for the larger custom built solar systems. In this case, in situ monitoring is often carried out to check the functioning of a new system. In most

Table 2.3-2. Characteristics of the CSTG and DST test methods

Characteristic	ISO 9459-2 (CSTG)	ISO 9459-5 (DST)
Testing of preheat systems	Yes	Yes
Testing of systems with integrated auxiliary heating	No	Yes
Stage of development	ISO standard	Approved by TC180 and submitted to ISO/CS for printing
Development in international or European context	Yes	Yes
International Round Robin validation	Yes (1993)	Yes (Nov. 1999)
Method used in several EU countries	Yes	Yes
Accuracy	5 to 10%	About 5%
Extrapolation of performance to other climates	Good extrapolation possibilities	Good extrapolation possibilities
Extrapolation of performance to other loads	Not systematically investigated	Good extrapolation possibilities
Extrapolation of performance to other load profiles	Not possible	Good extrapolation possibilities
Test costs	Low	Low
Test period	5 to 6 weeks	2 to 3 weeks

cases, output data and solar radiation that is monitored for several weeks are sufficient to analyze the system performance and check the functioning of system against the design using modern computer tools. Predicting the performance of a solar collector system at the design stage is currently being done using tested and validated programs. At present, various programs are in use as described in section 2.3.3.2. However, most commercial programs for designing solar systems cannot be used in combination with datasets from measurements. The way to proceed here is either to use VA115 (VA115), which is a design tool for large solar systems, or to use dedicated software for the application using TRNSYS.

The VA115 program resulted from an European Altener project in cooperation among Denmark, Greece, Sweden, and the Netherlands. This program was developed with solar applications in mind and like the Polysun (POLYSUN 3, 2002) program has set standard configurations. This program also offers extendable databases with branch related tap patterns and components defined by the user. However VA115 has two special features. The first facilitates the use of tap water data and climatic data from measurements on realized systems to compare the functioning of the system against the original data at time of design. The second feature of the VA115 program is the incorporation of a genetic algorithm (Visser and Loomans, 2000) that uses a given set of requirements to find the optimal system. Optimization can be pay-back time or other output parameters. The list of requirements can be wide such as choosing from different types of collector, collector area, storage types, etc.

For nonstandard configurations, the open structure using a toolbox with a much wider selection of thermal components than solar energy only of the TRNSYS program is the way to proceed. TRNSYS was developed in the United States of America (TRNSYS, 2002). This program is best suited for applications where solar energy is related to building applications or specific industrial processes. Using TRNSYS, the user has to produce his own program by using the component library for a specifically chosen application and compiling the input to obtain a single program. The TRNSYS library is FORTRAN based. Many additional component models can be found worldwide on the web.

REFERENCES

Ambrosetti, P. & Keller, J. (1985) *Das neue Bruttowärmeertragsmodell für verglaste Sonnen-kollektoren*, ISBN-3-85677-016-X, Würenlingen.
EN 12976-2:2000, Thermal Solar Systems and Components – Factory Made Systems – Parts 2: Test Methods. European Standard, European Committee for Standardisation, rue de Stassart 36, B-1050 Brussels.

HELIOS (1982) Program Description NF-Project 4.089-0.76.04, EMPA, Dübendorf.

ISO 9459-2:1995, Solar Heating – Domestic Water Heating Systems – Part 2: Outdoor Test Methods for System Performance Characterization and Yearly Performance Prediction of Solar only Systems. International Organisation for Standarisation, Case Postale 56, CH-1211 Geneve 20, Switzerland.

ISO/DIS 9459-5, Solar Heating – Domestic Water Heating Systems – Part 5: System Performance Characterization by Means of Whole System Tests and Computer Simulation, International Organisation for Standarisation, Case Postale 56, CH-1211 Geneve 20, Switzerland.

POLYSUN 3 (2002) *Simulation Software for thermal solar systems*, Institut für Solartechnik SPF, Switzerland.

Suter, J.M., Letz, T. & Inäbnit, W. Weiss (2000) J. Solar Combisystems, IEA Solar heating and Cooling Program, Task 26, ISBN 3-905583-00-3, Switzerland 2000.

T-SOL 4 (2002) Dr. Valentin Energie Software, Berlin, Germany.

TRNSYS (2002) *Version 15.0: A Transient System Simulation Program*, Solar Energy Laboratory, University of Wisconsin, Mdison, WI, USA.

VA115, Design Tool for Large Solar Systems, VABI postbus 29, 2600 AA Delft, The Netherlands. Information on http://www.vabi.nl, email: infovabi.nl

Visser, ir. H & Loomans, dr.ir M.G.L.C. (2002) Application of the Genetic Algorithm for Optimisation of Large Solar Hot Water Systems, *Solar Energy*, **72**(5), 427–439.

Chapter 2.4

Performance Requirements and Criteria

M. Köhl

Fraunhofer-Institut für Solare Energiesysteme,
Heidenhofstr. 2, D-79110, Freiburg, Germany

Abstract: The purpose of this chapter is to summarize the concepts of system performance, performance criteria, loss in performance, and that system performance usually depends on a few important properties of the materials used. Service life prediction needs a clear definition of the end of service life. The total failure of a product is easy to detect, but a slow continuous degradation in the properties of materials, which results in a continuous decrease of the product performance, needs special criteria for the definition of service life. The evaluation of performance criteria for some materials in solar thermal systems is discussed for glazings, reflectors, and absorbers by using parametric studies on the impact of material properties on the energy performance of a system. For a specific example, the results from modeling the system performance from changes in the solar absorptance and thermal emittance are presented for flat-plate solar thermal collectors operating at ca. 45°C and 70°C.

Keywords: System performance, Performance criteria for system components, Solar absorber coatings, Service lifetime

LIST OF DEFINITIONS AND ACRONYMS

α_S	solar absorptance
$\varepsilon\,(T_K)$	thermal emittance at a temperature T_K
I_R	redirected irradiation
I_O	incident irradiation
L	maximum acceptable loss of performance caused by aging
P	performance
PC	performance criterion
R	total reflectance
R_S	specular reflectance
t	time
t_L	service life time

T	transmittance
T_K	temperature in Kelvin degrees
T_{hem}	hemispherical transmittance
T_{dir}	direct transmittance
X	any physical materials property

Materials and components have a number of different physical and technical properties that are important for their behavior in various environments, circumstances, boundary conditions, and operating conditions. Only relatively few materials properties are important to the functioning of a component and only some of the component properties are relevant for the operation of a complete system. Some materials properties can be dominant for the operation of a component or system, for example, coating adhesion, or water tightness of pipes and tubes. In these examples, only two possibilities exist; the adhesion is either good or bad or the pipes are watertight or leak. If such a failure occurs, the system performance declines drastically. These kinds of failures are catastrophic and, if they appear at the beginning of operation, are known as infant mortality failures. These failures may result from undetected fabrication flaws. The so-called "sudden death" results when a component or a system suddenly breaks down after longer operating times. Durability problems of components and systems are the subject of "Failure Mode and Effect Analysis" or "Stochastic Service Life Prediction Methodologies."

More typically, the alternative to catastrophic failures is a slowly deteriorating performance resulting from a continuous degradation in the materials properties, and corresponding losses in performance. Therefore, the performance, P, varies over time according to the changes of the properties, x_n:

$$P = P(x_n(t)). \tag{1}$$

It is necessary to know how the performance depends on the materials properties to define the maximum acceptable changes in the materials properties that result in the maximum tolerable loss in performance. The period, in which this maximum loss in performance, L, is reached, is called the service lifetime t_L of the product. The definition of L and t_L are based on economic considerations for state-of-the-art systems.

The simplest example is for a system that depends linearly on one single property. A solar reflector is a good example, where the redirected irradiance, I_R, depends on the incident beam irradiance, I_O, and the specular reflectance, R_s

$$I_R = R_s I_O \tag{2}$$

The loss of performance is directly proportional to the degradation of the specular reflectance. The service lifetime is reached when ΔR_s [$\Delta R_s = R_s(0) - R_s(t)$] reaches the limit L. Therefore, the performance criterion, PC, becomes:

$$PC = \Delta R_s < L \qquad (3)$$

where L could be set, in principle, between 0 and $R_s(0)$. In practice, the value of L may range from 0.8 R_s to 0.9 R_s [see Chapter 4.6, this volume].

A similar consideration could be applied to the transmittance, T, of a glass pane. Glass covers are used to prevent convection losses and to protect the other components from the environment. The energy gain for non-imaging solar systems is linearly proportional to the total solar transmittance of the cover and the PC based on the measured hemispherical transmittance, T_{hem}, is:

$$PC = \Delta T_{hem} < L \qquad (4)$$

In applications where directly incident radiation or a clear view through a window is required, scattered radiation does not contribute to the performance, so the direct transmittance has to be considered as the performance property:

$$PC = \Delta T_{direct} < L \qquad (5)$$

The situation is more complicated, if several materials properties influence the system performance to a comparable extent. All relevant properties have to be taken into account for an appropriate service life prediction procedure.

A typical example is a solar absorber, which is characterized by the solar absorptance, α_S, and the thermal emittance, $\varepsilon(T)$, [see Chapter 2.1, this volume]. The absorber is part of a domestic hot water system. The performance of such a system depends on α_S and $\varepsilon(T)$, which can be considered as independent variables, and a number of other parameters that are kept constant to permit comparisons to be made. The solar irradiation is taken into account by using a test reference year for a given location as input to a system simulation computer program [see Chapter 2.2, this volume]. Two alternative systems have been modeled with variable α_S and $\varepsilon(T)$, i.e. (1) a flat-plate collector system operating in Toronto, Canada for preheating water at temperatures of ca. 45°C and (2) a system operating in Zürich, Switzerland at temperatures of ca. 70°C.

The results showed good correlation, as shown in Figures 2.4-1 and 2.4-2, between the solar fraction (heat gain) and α_S and ε in the interesting ranges of $0.8 < \alpha_S < 0.98$ and $0.1 < \varepsilon < 0.3$. The changes in emittance have to be roughly half as large as the changes in absorptance to result in the same changes in the

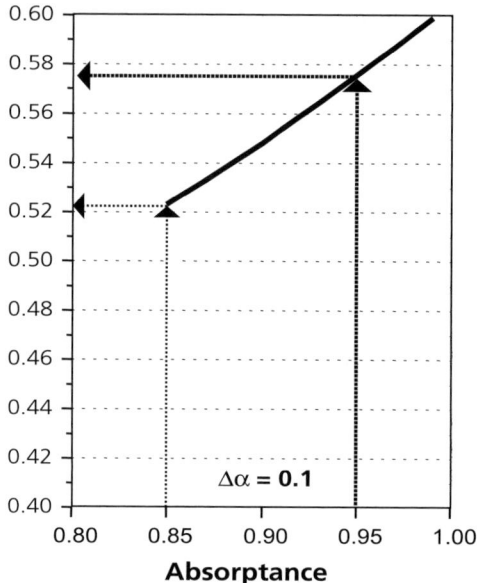

Figure 2.4-1. Solar gain of a domestic hot water system operating at the higher operating temperature in Switzerland as a function of the solar absorptance of the absorber coating for a thermal emittance of 0.1.

energy output for the system operating at higher temperatures. At the lower temperatures (Canadian system), the weighting of ε is only one-fourth of α_S. Therefore, the impact of changes in α_S and ε on the system performance P can be described as

$$PC_{\text{low}} = \Delta\alpha - 0.25 \, \Delta\varepsilon \text{ for low operating temperatures} \tag{6}$$

$$PC_{\text{high}} = \Delta\alpha - 0.5 \, \Delta\varepsilon \text{ for high operating temperatures} \tag{7}$$

in which $\Delta\alpha = \alpha(0) - \alpha(t)$ and $\Delta\varepsilon = \varepsilon(0) - \varepsilon(t)$. For equations (6) and (7), note that degradation results in a decrease in α and an increase in ε. These relations are useful for the development of solar absorber coatings, too, because these two properties are not usually completely independent because of the overlap between 2 and 3 μm of the solar irradiation and the thermal emittance. The required step-function for the reflectance of the absorber in this wavelength region is not achievable. Therefore, the optimum trade-off between high solar absorptance and low thermal emittance has to

Solar fraction

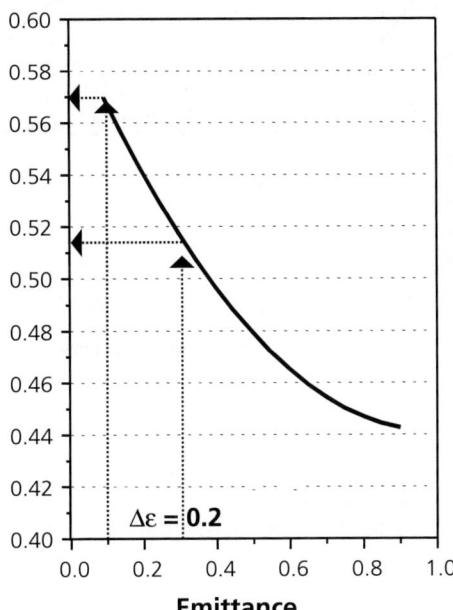

Figure 2.4-2. Solar gain of a domestic hot water system operating at the higher operating temperature in Switzerland as a function of the thermal emittance of the absorber coating for a solar absorptance of 0.94.

be identified and the application oriented performance criteria are used to optimize the coating properties.

Degradation means a monotonic decrease of the absorptance and/or a monotonic increase of the emittance as a function of aging time. Because no catastrophic failures (sudden death) are expected, a degradation limit has to be defined. It has been concluded that a maximum decrease of 5% of the solar heat again resulting from the degradation of the absorber coating could be accepted, yielding:

$$PC_{low} = \Delta\alpha - 0.25 * \Delta\varepsilon < 0.05 \quad \text{and} \quad PC_{high} = \Delta\alpha - 0.5 * \Delta\varepsilon < 0.05. \quad (8)$$

A more sophisticated approach involves integrating the performance over time, because the energy gain during the lifetime can be different for different rates of change in the performance (see Figure 2.4-3) ending at the same PC after the same t_L. The difference might be a PC in percent of the maximum

Optical Materials for Solar Thermal Sytems

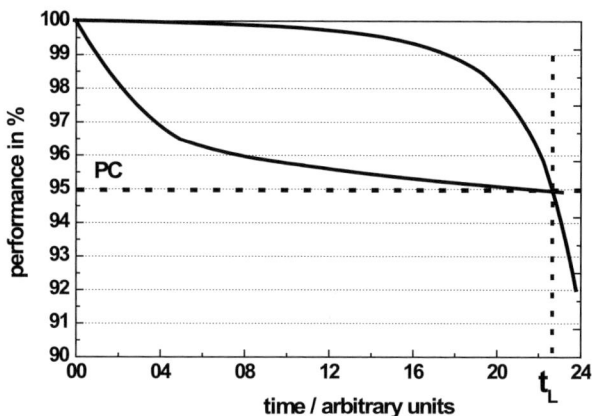

Figure 2.4-3. Different performance profiles with the same service lifetime t_L but with a significantly different average performance during this lifetime, which can be estimated from the different areas under the two curves.

performance parameter. The PC has to be changed when considering the average performance:

$$PC = P(0) * t_L - \int_0^{t_L} P(t)\,dt < L \tag{9}$$

In practice, the performance over time has to be monitored with an adequate time resolution during all tests and during operation. This drawback hinders application of the integration methods, though this procedure is clearly the most appropriate approach.

Chapter 2.5

Performance Indicators

M. Köhl[1] and G. Jorgensen[2]

[1]*Fraunhofer-Institut für Solare Energiesysteme,*
Heidenhofstr. 2, D-79110, Freiburg, Germany
[2]*National Renewable Energy Laboratory,*
Golden, CO 80401, USA

Abstract: The purpose of this chapter is to outline various ways for quantifying changes in performance. The choice of an appropriate and sensitive measure of degradation is critical to provide a rapid indication of performance loss. By relating such performance indicators to relevant environmental stress parameters, it is possible to derive correlations between outdoor and accelerated exposure test results. Several photometric performance parameters that characterize optical degradation are discussed.

Keywords: Performance loss, Response variable, Degradation indicators

LIST OF ABBREVIATIONS AND ACRONYMS

AR	antireflection
CPC	compound parabolic collector
DID	degradation indicator difference
DIF	degradation indicator factor
I	spectral irradiance ($W/m^2/nm$)
p	a photometric parameter
UV	ultraviolet
α	solar absorptance
ε	thermal emittance
ρ	reflectance
τ	transmittance
θ	collection half-angle
σ	half width of a Gaussian distribution
λ	wavelength (nm)

To assess durability, a suitable performance indicator must be found that accurately translates changes in some response variable to relevant applied environmental stresses. The response variable can be either microscopic, e.g., changes in chemical structure, or macroscopic, e.g., loss of gloss, but ideally should be easily measured and directly related to an important property of the material being tested. In quantifying degradation of optical components, a photometric parameter, p, is usually selected as the response variable. Standard methods for measuring various representations of p have been given (ASTM E903-96, 2003). For glazings, hemispherical transmittance is of interest for flat-plate collector use ($p = \tau_{2\pi}$) and specular transmittance is important for window applications ($p = \tau_s$). For absorbers, p may be some function of the solar absorptance, α, and thermal emittance, ε (Carlsson *et al.*, 2000). For reflectors, p is generally either specular reflectance, ρ_s, for imaging applications, e.g., a CPC concentrator system, or hemispherical reflectance, $\rho_{2\pi}$, for proximate relay systems, e.g., booster mirrors for flat-plate collectors. The quantities ρ_s and $\rho_{2\pi}$ are related by:

$$\rho_s(\lambda, \theta) = \rho_{2\pi}(\lambda)\left[1 - e^{-[\theta/\sigma(\lambda)]^2/2}\right] \qquad (1)$$

in which θ = the collection half-angle, λ = the wavelength of light, and $\sigma(\lambda)$ = the half width of the (assumed Gaussian) distribution of light scattered upon reflection.

A photometric performance indicator over an applicable spectral range ($\Delta\lambda = \lambda_{max} - \lambda_{min}$) can then be written as:

$$p(\Delta\lambda, t) = \frac{\int_{\lambda_{min}}^{\lambda_{max}} p(\lambda, t)I(\lambda)d\lambda}{\int_{\lambda_{min}}^{\lambda_{max}} I(\lambda)d\lambda} \qquad (2)$$

in which $I(\lambda)$ is an appropriate solar irradiance spectrum, i.e., a direct distribution for specular measurements or a global distribution for hemispherical measurements (ASTM G159-98, 2003; ISO Standard 9845-1, 1992). Broadband performance loss, weighted across the entire solar spectrum ($\Delta\lambda = 250$ nm to 2500 nm), would be much less sensitive to narrowband degradation and consequently less useful as an early indicator of changes in performance. For example, in flat-plate collector applications, antireflection (AR) coatings are used to increase the hemispherical transmittance of the glazing. Thus, a suitable measure of performance to quantify degradation would be the near normal direct-hemispherical transmittance ($\tau_{2\pi}$). For AR-coated glazings the spectral region between 600 nm to 700 nm is particularly susceptible to degradation effects; thus, selection of $\lambda_{min} = 600$ and $\lambda_{max} = 700$ nm would potentially provide a very rapid quantification of performance loss.

Performance loss in polymeric glazings is another good example. In general, polymeric materials exhibit their greatest sensitivity to stress exposure-induced loss in transmittance at the lower wavelengths in the visible part of the spectrum (Figure 2.5-1). Consequently, $\tau_{2\pi}(\lambda = 400\,\text{nm}$ to $500\,\text{nm})$ is an excellent performance indicator for such materials.

As an indication of changes in performance with weathering, two measures of degradation defined as the degradation indicator factor, DIF, and the degradation indicator difference, DID, can be used:

$$\text{DIF}(t) = \frac{\int_{\lambda_{\min}}^{\lambda_{\max}} [p(\lambda, t))/(p(\lambda, t = 0)]d\lambda}{\int_{\lambda_{\min}}^{\lambda_{\max}} d\lambda} \qquad (3)$$

$\text{DIF}(t)$ is the average ratio of some appropriate photometric quantity after weathering to the corresponding unweathered value and

$$\text{DID}(t) = \frac{\int_{\lambda_{\min}}^{\lambda_{\max}} [p(\lambda, t = 0) - p(\lambda, t)]d\lambda}{\int_{\lambda_{\min}}^{\lambda_{\max}} d\lambda} \qquad (4)$$

$\text{DID}(t)$ is the average difference between the unweathered and weathered photometric quantity. Plots of DID and DIF versus the amount of exposure then provide clear indications of degradation. For example, Figure 2.5-2 presents the DIF (hemispherical transmittance ratio) for the ultraviolet (UV) stabilized polycarbonate

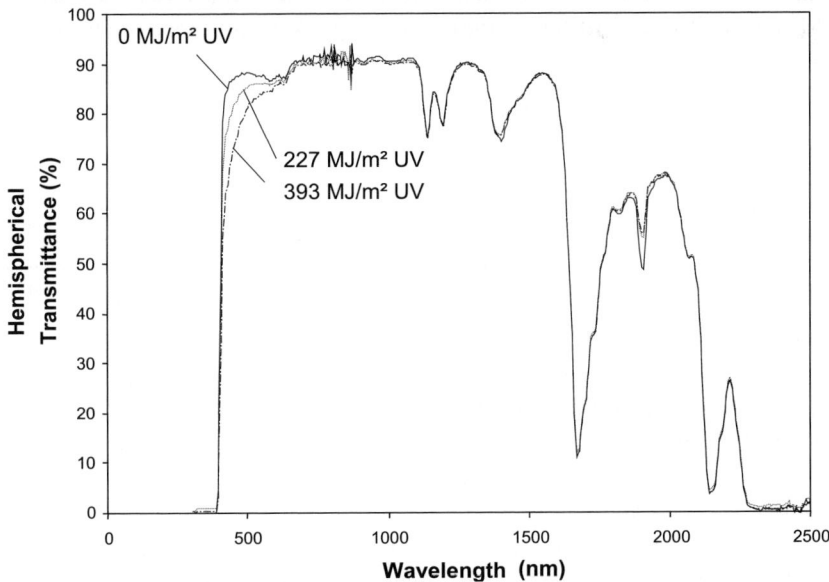

Figure 2.5-1. Change in hemispherical transmittance as a function of cumulative UV dose for UV-stabilized polycarbonate.

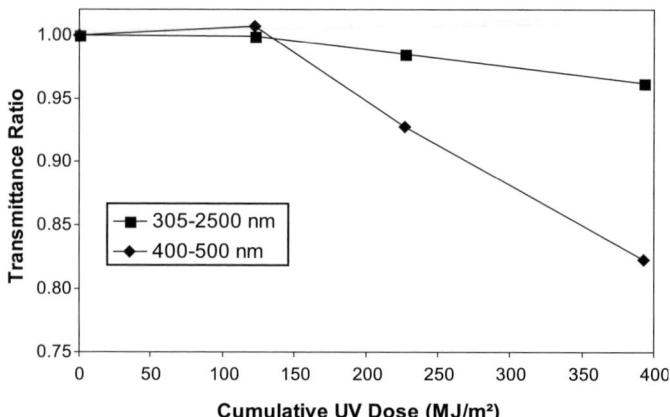

Figure 2.5-2.   Ratio of the hemispherical transmittance of a UV-stabilized poly-arbonate glazing as a function of cumulative UV dose for narrow (400 nm to 500 nm) and broad (305 nm to 2500 nm) wavelength bandwidths.

glazing material discussed in Figure 2.5-1. The DIF is plotted as a function of the cumulative UV dose exposure and the wavelength range over which the DIF is calculated. After almost 400 MJ/m^2 of UV irradiance, which is equivalent to about 2 years of outdoor exposure, the broadband ($\Delta\lambda = 305$ nm to 2500 nm) solar weighted DIF shows a slight decrease. However, the more sensitive narrow band ($\Delta\lambda = 400$ nm to 500 nm) ratio provides a much quicker and therefore more desirable indication of degradation.

REFERENCES

ASTM E903-96, Standard Test Method for Solar Absorptance, Reflectance, and Transmittance of Materials Using Integrating Spheres. *Annual Book of ASTM Standards 2003* **12.02**, American Society for Testing and Materials, West Conshohocken, PA 19428-2959 U.S.A.

ASTM G159-98, Standard Tables for References Solar Spectral Irradiance at Air Mass 1.5: Direct Normal and Hemispherical for a 37° Tilted Surface. *Annual book of ASTM Standards 2003* **14.04**, American Society for Testing and Materials, West Conshohocken, PA 19428-2959 U.S.A.

Carlsson B., Möller K., Köhl M., Frei U. and Brunold S. (2000). Qualification Test Procedure for Solar Absorber Surface Durability. *Solar Energy Materials and Solar Cells* **61**, 255–275.

ISO Standard 9845-1:1992, Solar energy – Reference solar spectral irradiance at the ground at different receiving conditions – Part 1: Direct normal and hemispherical solar irradiance for air mass 1.5. International Organization for Standardization, Geneva, Switzerland.

Part 3
Environmental Stress Conditions

Part 3 consists of four chapters. In the first chapter, an overview is given about the most important environmental stress factors that may limit the useful life of solar thermal systems. These factors are the ambient temperature, relative humidity in the air, total solar irradiation especially UV-A and UV-B, temperature of the sky, rain, wind, environmental pollutants, and microbiological contamination. In the second chapter, measurement techniques for measuring these factors are discussed in detail. In the third chapter, methods are summarized for accumulating climatic stresses data, for evaluating the data, and reporting the effects of different climates for several flat-plate collectors. Appropriate periods are usually at least one year and several different climates should be monitored at different locations and for different periods to identify the most challenging stresses. Statistical analysis tools and the interpretation of the results are discussed. In the final chapter, two models are described that can be used to optimise the construction of collectors with the most favourable microclimate. Damage of materials inside a solar collector depends on the corrosivity of the microclimate inside the collector, which is turn depends on the ambient weather and the construction of the collector.

Chapter 3.1
Environmental Stress Conditions

M. Heck and M. Köhl

Fraunhofer-Institut für Solare Energiesysteme,
Heidenhofstr. 2, D-79110, Freiburg, Germany

Abstract: The purpose of this chapter is to provide an overview of the most important environmental stress factors that may limit the useful life of solar thermal systems. Service life estimation procedures are based on the knowledge of the stress factors acting on the product. The stress factors on materials in solar thermal systems are discussed qualitatively and their quantitative assessment is discussed in the following chapters. The most important climatic stress factors are: ambient temperature, relative humidity in the air, the total solar irradiation especially UV-A and UV-B, temperature of the sky, rain, wind, environmental pollutants, and micro-biological contamination.

Keywords: Materials in solar thermal systems, Climatic stress factors, Microclimate, Macroclimate, Humidity, Thermal processes, Solar radiation, Pollutants

3.1.1 MACRO- AND MICROCLIMATES AND CLIMATIC STRESS FACTORS

The outer surfaces of the building envelope are directly exposed to stresses by weathering. The surface properties of solar thermal systems are not only degraded by weathering but also by the enhanced temperature, which can accelerate or induce degradation processes. The kind and the amount of the environmental stress conditions depend on the local climate, the interactions between the surface materials and the climate, and the function or the working conditions, and all result in a variation in stresses. These, in turn, influence the durability of the solar systems. Changes of the climate occur within different time-scales: seasonally, daily, and with periods from a few seconds to hours because of clouds and directly influence the amount of the single stress factors directly. The thermal mass or other inertial aspects of the systems may smooth the fluctuations, or may introduce additional mechanical stresses by changing the rate of thermal expansion. The orientation and inclination of a solar thermal system are also important factors.

The most important climatic stress factors are:

(1)   Ambient temperature, which influences the surface temperature;
(2)   Relative humidity in the air, which might result in very high moisture concentration at the surface and in frost/thaw cycles;
(3)   The total solar irradiation from 290 nm to 2500 nm. For most degradation processes, the relevant parts of the radiation are the highly energetic regimes of UV-A (315 nm to 385 nm) and UV-B (280 nm to 315 nm).
(4)   Temperature of the sky that influences the stress conditions on the surface by radiative heat exchange;
(5)   Rain, not only because it wets the surface but also from possible additional mechanical stresses resulting from driving rain;
(6)   Wind, which might carry abrasive particles or influence the local conditions at the surface;
(7)   Pollutants like sulphur-dioxide, nitrogen oxides, ozone etc.; and
(8)   Microbiological contamination could be important for organic materials.

The solar thermal system itself interacts with the ambient climate. Surface temperatures may change from absorbing solar radiation or emitting IR radiation. Surface temperature changes also result in changes of the relative humidity at the surface at which the changes may range from drier conditions to an increased moisture load or even dew or frost. Of course, this microclimate is strongly influenced by natural (wind) and forced convection.

Solar collectors are part of an operating system at different and varying climate conditions. The operating temperature of the absorber strongly influences the microclimate and strongly impacts the amount of stress from moisture loads. When the surfaces are protected from direct weathering by glazings, climates will have different effects. Coatings inside completely sealed units like insulating glazings or evacuated solar collectors are not exposed to moisture loads and the UV exposure may be reduced, depending on the spectral transmittance of the glazings. However, some open systems, such as ventilated windows (including transparent insulation, for example) and ventilated flat-plate solar collectors, are in contact with the ambient climate. The transient behavior of the microclimate depends on the ventilation rate, which results from the air-exchange between a glazed box and the ambient. Ventilation usually results from natural convection in the device and/or external pressure fluctuations because of wind or wind gusts. The microclimate is also influenced by materials interacting with humidity, e.g., wooden frames and insulation materials.

If the time-dependent variations of the stress factors are not causing additional mechanical stress, then collecting data for the cumulative stress over a significant time period is necessary. Appropriate periods are usually one or several years. The recorded data are usually average values for a predetermined sampling period.

The frequency of acquiring data is chosen from the typical frequency of changes of the measured property and their effect on the samples.

The assessment of macro- and microclimatic conditions, UV, T, RH, pollutants, and wind are described in Part 3, especially as a basis for evaluating the stress conditions needed for accelerated durability testing (Part 4, this volume). An overview of the most important climatic stress factors is now provided.

The intensity of UV and the other parts of the solar spectrum can change within less than one second because of cloud movements. The impact on samples can be twofold. First, the amount of UV and near-UV radiation is a degradation factor that might result in damage to chemical compounds, especially polymeric materials. Because the amount of degradation usually varies linearly with the UV intensity, the sampling period is not critical and average dose-functions can be used. However, the spectral distribution of the UV and near-UV radiation may be a major factor for the photothermal degradation of polymeric materials. The importance of radiation-induced degradation depends on the spectral sensitivity of each individual polymer, which is discussed in Chapter 4.7 and Section 6. Monitoring of the spectral irradiance is necessary in this case.

The thermal processes are slow compared with the time constants of changes in irradiation. The sample temperature can be considered as a separate degradation factor, providing that the degradation processes are either internal chemical or physical processes, or the reaction partners are a permanent part of the ambient, such as oxygen, pollutants, and water vapor. The typical time constants depend on the temperature gradients in the sample from the surface to the backside. The dominant stress resulting from an elevated absorber temperature, e.g., during the stagnation conditions when the solar thermal system is out of order, especially during mounting, vacancies of the users, or because of defects. During stagnation the absorber temperatures can reach more than 240°C in highly efficient collectors.

The effects of the humidity have a longer timescale. They strongly depend on the temperature with typical time constants ranging from minutes to hours and the weather with time constants ranging from hours to days. High humidity levels up to condensation at the surface are especially relevant for degradation processes. The actual exposure of the samples to relative humidity depends strongly on the local microclimatic conditions at the surface. Measurement techniques have to be used for determining humidity and temperature at the surface, which must not change the temperature and humidity by shading or change the radiation interchange with the ambient. Rain has to be considered for surfaces that are directly exposed to weathering, such as glazings, reflectors, or unglazed solar absorbers.

Pollutants, like SO_2, NO_X, and salt, have to be considered as another potential degradation factor. Usually humidity coupled with pollutants is needed to cause chemical corrosion. Thus, the stress levels of pollutants are strongly coupled with the relative humidity and the actual temperatures at the same time.

Wind can act as a direct degradation factor by mechanically removing particles from the surface, e.g., combined with rain or dust or sand. Wind also influences the microclimate severely by enhancing the air exchange.

Chapter 3.2

Measurements of Environmental Stress Conditions and Evaluation for Service Life Prediction

S. Brunold,[1] B. Carlsson,[2] and K. Möller[2]

[1]*Institut für Solartechnik SPF, Hochschule für Technik HSR,
Oberseestr.10, CH-8640, Rapperswil, Switzerland*
[2]*Swedish National Testing and Research Institute,
P.O. Box 857, S-501 15 Borås, Sweden*

Abstract: The purpose of this chapter is to describe some measurement techniques and methods that can be employed in the characterization of environmental stress for the durability assessment of components and materials. For service life prediction the knowledge of the environmental stress conditions is of essential importance. In recent years, a lot of research on durability assessment of absorber coatings for solar thermal collectors was made within the framework of the Solar Heating and Cooling Program (SHCP) of the International Energy Agency (IEA). In the case of solar absorbers, the environmental factors that may result in loss in optical performance are of principal interest. These were identified as high temperature, high humidity and moisture, and airborne pollutants. Measurement techniques adapted to these factors will be discussed in detail. The form of data acquisition is not restricted to time-dependent data only. Sometimes the acquisition of extreme values or dose functions is much more appropriate. Even well-defined standard specimens may be used as a sensor.

Keywords: Environmental factors, Temperature, Humidity, Moisture, Airborne pollutants, Condensation sensor, Standard specimen, Metal coupon, Frequency function, Dose function

3.2.1 INTRODUCTION

To be able to predict the expected service life of a component and its materials from the results of accelerated aging tests, the degradation factors from environmental stresses at in-service conditions need to be measured. Although the general standard IEC 60721 of the International Electrotechnical Commission (IEC 600721), which is discussed in Chapter 4.3, contains recommendations for classifying the severity of stress for various climatic, mechanical, chemical, biological, and electrical environments, it does not contain recommendations on how to measure the

101

characteristic environmental stress factors. The user must access the literature to identify suitable measurement techniques and sensors.

A large choice of environmental factors and possible damage modes are specified in Table 4.3-4 of Chapter 4.3. This table is a suitable basis for identifying stress conditions to be considered for planned outdoor tests. However, it is beyond the scope of this chapter to describe the measurement techniques for all the stress factors given in this table.

Measurement techniques that were used in the IEA (International Energy Agency) Task 10 and the IEA Working Group MSTC (Materials in Solar Thermal Collectors) absorber case study are summarized in Table 3.2-1. They will be discussed in detail in the following as an example of the factors that were considered for this study. It is extremely important to characterize the service conditions that relate to the most important degradation mechanisms identified for the materials of the component (in this case, the solar absorber) but also in terms relevant for and convertible into the test conditions for the environmental resistance tests to be used for accelerated life testing.

3.2.2 BASIC CONSIDERATIONS

The environmental factors, which may result in loss in optical performance, were of principal interest. For the study of absorber-coating materials these were identified as high temperature, high humidity and moisture, and airborne pollutants.

When installed in a single flat plate collector for domestic hot water production, an absorber coating is exposed to temperatures that may vary from $-20°C$ up to more than $200°C$. High temperature stability of the absorber coating is required but high temperature may accelerate many different degradation processes. For a selective absorber coating, high temperature increases the oxidation rate of metal that will result in a decrease in the absorptance of the coating and in an increase in the emittance.

Because the casing of a flat plate collector is usually ventilated, the absorber coating is in contact with the ambient air. Thus, humid air can flow in and out through the ventilation holes of the collector resulting in hydrolysis reactions. Sometimes, the absorber temperature is low enough that condensation of water occurs from the humid air.

Electrochemical corrosion may be accelerated by airborne pollutants even when present at very low concentrations. Because a solar collector exchanges air with the ambient, airborne pollutants will be transported from the ambient into the collector and contact the absorber plate. As a consequence, air pollutants may be important degradation factors for absorber coatings and their influence on the coating

Table 3.2-1. Techniques that were used in the IEA task 10 absorber case study for measurement of degradation factors in solar thermal collectors

Degradation mechanism	Degradation factors/ Measurement variables	Sensors
(A) High-temperature oxidation of metallic Ni particles	Temperature: Surface temperature of absorber plate	Pt sensors in holders screwed directly to the absorber plate. Heat sink compound was used to achieve a good thermal contact.
(B) Electrochemical corrosion of metallic Ni particles at high humidity levels and in the presence of sulphur dioxide	*Atmospheric corrosivity:* Measurement of corrosion mass loss rate of standard metal specimens *Air pollutants:* Measurement of sulphur dioxide concentration inside and outside the solar collector.	Metal coupons of carbon steel, zinc and copper and evaluation of corrosion mass loss according to ISO 9226 Exposed metal coupons analyzed with respect to the sulphate content of the corrosion products by EDX UV-fluoresence instrument for direct measurement of sulphur dioxide concentration in the air outside and inside the solar collector
(C) Hydratization of aluminum oxide and electrochemical corrosion of metallic Ni particles by the action of condensed water	*Humidity:* Measurement of humidity in the air-gap between the absorber plate and glazing cover of the collector *Time of condensation:* Measurement of specular reflectance of absorber surface *Surface humidity:* Measured relative air humidities converted to relative humidity on surface by use of measured surface temperatures	Capacitance humidity sensors carefully shielded from solar radiation and ambient thermal radiation. Specially designed reflectance mode condensation sensor

durability needs to be assessed. As is well known, sulphur dioxide present in air as a trace substance, greatly accelerates the electrochemical corrosion of most metallic materials at high humidity. The optical performance of an absorber coating may be adversely influenced from corrosion of the metallic substrate. Optical performance losses of selective absorber coatings pigmented with small metallic particles may also result from the oxidation/corrosion of the metal particles.

To measure microclimatic data, in situ collector test facilities were set up at different locations in Europe. Collectors from different manufacturers were operated at the same conditions as if they had been installed in a real domestic hot water (DHW) system. Additionally, collectors were operated at stagnation conditions, i.e., no thermal energy was extracted from the collector system during testing. Thus, the collectors were exposed to the maximum thermal load. The collectors, which were operated as a DHW system or at stagnation conditions, were installed on a test platform facing south with a tilt angle of 30°. This geometrical arrangement results in an average yearly solar insolation of about 1200 kWh/m² in the collector plane. The time period for the tests was up to 3 years.

The test loop shown in Figure 3.2-1 was used to simulate the conditions prevailing in a DHW system during operation. It is a typical closed loop system as usually installed in Europe. The heat transfer fluid in the loop is a mixture of ethylene glycol and water (33 vol.% ethylene glycol, 67 vol.% deionized water) with corrosion inhibitors added. Ethylene glycol is used to prevent the water from freezing. Instead of a storage tank, an air heat exchanger with a fan was used to keep the service temperature of the loop at a maximum temperature of about 50°C. The fan is automatically actuated when the global solar irradiation exceeds 200 W/m², measured in the collector plane. The fan does not operate at insolations below 200 W/m². As in a real installation, a non-return valve prevents the system from operations in the reverse mode during the night hours. The lowest temperature

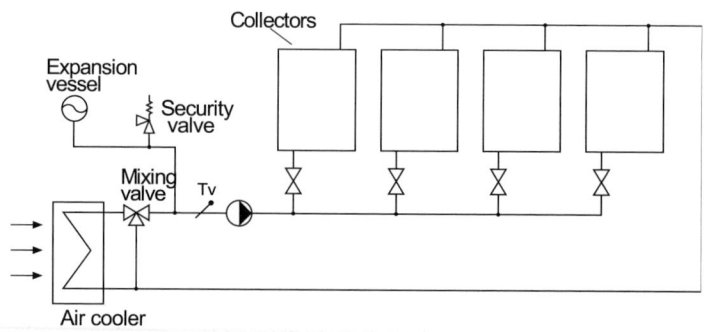

Figure 3.2-1.   Test loop to simulate the in-use conditions of a collector in a DHW system.

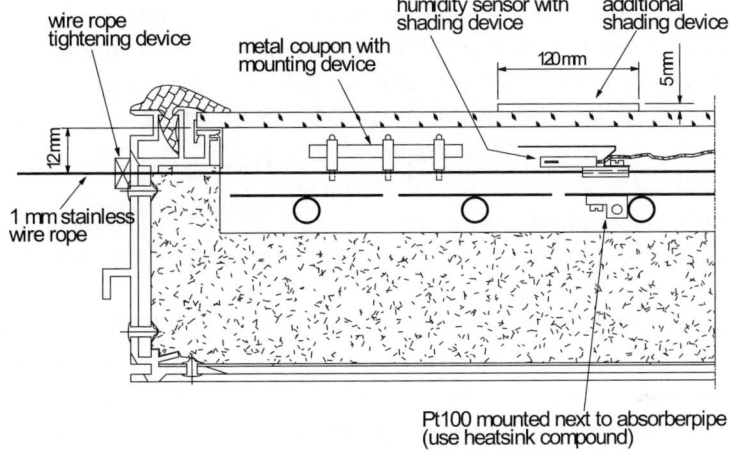

Figure 3.2-2. Cross section of a collector in test equipped with stainless steel wire ropes to hold the different sensors in the air gap.

of the collector is reached during the night because of radiative cooling into the night sky.

The most important task was to find a way for measuring the relevant stress conditions without appreciably disturbing the interaction of the object of interest, i.e. the collector, with its surrounding. For this purpose, pairs of thin stainless steel wire ropes were stretched across the collector in the air-gap between the absorber plate and the glass cover (see the cross section of the collector in Figure 3.2-2). On the one hand, the wire ropes are strong enough to be used as holder for some important sensors, which are discussed later, but on the other hand, they are thin enough not to disturb the free convection of air in the gap.

3.2.3 HIGH-TEMPERATURE STRESS FACTOR

As already discussed, high temperatures result in important degradation mechanism for solar absorbers. Thus, the temperature of the absorber plate has to be measured. For this purpose, temperature sensors are attached to the absorber plate with a special sensor holder (see Figure 3.2-3). A 4 wire Pt-100 sensor is recommended. This type of sensor is widely available in different accuracy classes and is highly reliable. The measurement technique is also thoroughly developed. Accurate measurements of high precision were routinely obtained with these sensors.

For temperature measurements, thermocouples were also considered. The main benefit of using thermocouples is the small size and, thus, a small thermal mass.

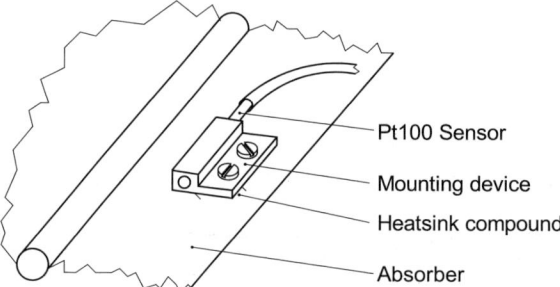

Figure 3.2-3. Measurement of the absorber temperature.

Thermocouples made of nickel and nickel–chromium alloys are widely used for the temperature range of a solar absorber, and are also usable up to more than 1000°C. They can be easily made by welding the ends of two wires of these metals together. Wires are available down to a diameter of a few hundredths of a millimetre, which results in really small and fast sensors. However, the measurement technique of these devices is error-prone because of the very small voltage that has to be measured. Because a thermo-electrical voltage is generated at each contact point with a different metal, lengthening of the sensor wires has to be done with great care. The same contact problems occur when connecting the thermocouples to the data acquisition system. Because only a temperature difference produces a thermo-electrical voltage, contact problems can be mitigated by immersing every contact at a cold junction.

3.2.4   HIGH-HUMIDITY STRESS FACTOR

To measure the humidity inside the collector, a combined humidity – temperature sensor is attached to the stainless steel wires stretched across the collector in the air-gap between the absorber plate and the glass cover (Figure 3.2-2) by a special sensor holder (Figure 3.2-4). The sensor holder includes a shading device that prevents the sensor from receiving direct radiation that would result in measuring spurious elevated temperatures. However, the shading device did not work satisfactorily and an additional metal sheet was installed above the sensor on the outer side of the glass cover. Care was taken to shade the sensor device from the sun for all seasons in the year.

   The combined temperature and capacitance humidity sensor (Rotronic I-155C) are mounted in a metallic tube with a diameter of 5 mm and a length of 50 mm. Heat exchange was minimized from the metallic case of the sensor and the low emitting

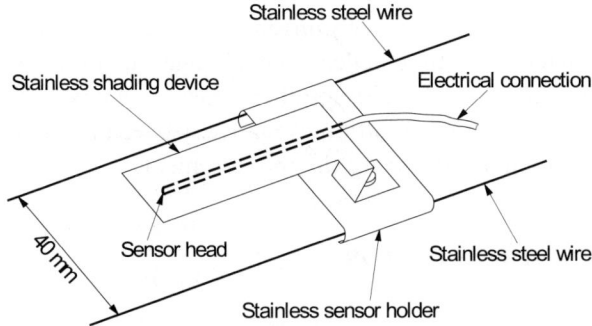

Figure 3.2-4. Measurement of humidity and temperature in the air-gap.

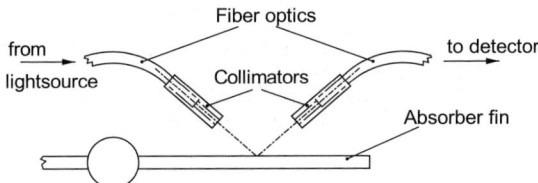

Figure 3.2-5. Reflectance mode condensation sensor (RMCS).

surface of the absorber. Therefore, the temperature measured is close to the real air temperature of the air-gap.

As an alternative measurement procedure, a dew point mirror instrument could have been used. To use this instrument, air must be removed from the collector for the measurement and then re-injected. This procedure would result in a change in the natural convection flow of air through the collector. As a result, the measured humidity would not be correct. The surface humidity of the absorber plate is calculated by converting the relative humidity of the air at the measured air temperature to a value of the relative humidity at absorber plate temperature.

To be able to measure if and to what extent condensation of water occurs at the absorber surface, a new instrument, a reflectance mode condensation sensor (RMCS) was used (see Figure 3.2-5) (Carlsson *et al.*, 1994). This instrument was developed by the Fraunhofer Institute for Solar Energy System in Freiburg (Germany). The principle is based on measuring the reflectance change when water condensation occurs on a surface. The relative specular reflectance is measured by using a small light emitting diode (LED) as a light source and a detector. The LED is pulsed to avoid measurement errors resulting from scattered light. The advantages with this method compared to more conventional measurements of corrosion or time-of-wetness meters are that reflectance changes can be detected for small quantities of condensed water. In addition, the measurement is performed directly on the surface, and the amount of radiation exchange between the absorber surface and the cover plate is not changed during the measurement.

3.2.5 POLLUTANT STRESS FACTOR

To measure the corrosivity of the atmosphere inside of the collectors, standard metal coupons were used as proposed in International Organization for Standardization no. 9226 (ISO 9226). For each measurement, a set of four standard metal specimens was used. These were (1) unalloyed carbon steel according to Swedish Standard SS 1147 (ISO 3574 CR4), (2) zinc (impurity level 0.5%), (3) copper according to Swedish Standard SS 5013-02 (ISO R1336 Cu FRTP), and (4) aluminum according to Swedish Standard SS 4007-14 (ISO 6361 AI99.5). Using special clamps (Figure 3.2-6), the standard metal specimens were attached on the adjacent stainless steel wires that were stretched across the collector in the air-gap between the absorber plate and the glass cover (Figure 3.2-2). To prevent electrochemical reactions between the different metals, a thin PTFE shrink sleeve was used to cover the wires from the clamps.

Before exposure, the specimens were carefully cleaned with a hydrocarbon solvent to remove all marks of dirt, oil, or other foreign matter capable of influencing the result from the corrosion rate determination. After drying, the coupons were weighed. After exposure, the corrosion products were removed from the exposed coupons by chemical cleaning as described in ISO DIS 8407 (ISO 8407).

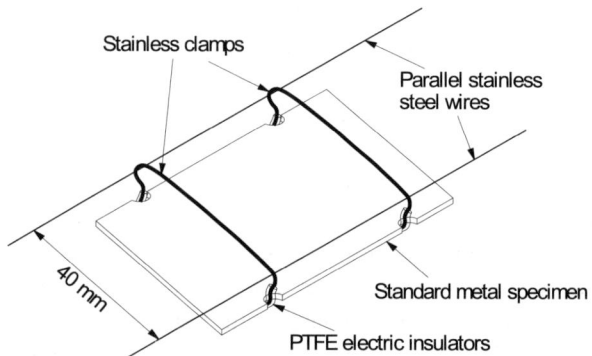

Figure 3.2-6. Mounting of the standard metal specimens used to determine the corrosivity inside the collector.

Typical metal weight losses of the standard metal specimens exposed outside of the collectors tested (but shielded from precipitation) were of the order of $50\,g/m^2$ for steel, $1\,g/m^2$ for zinc, and $5\,g/m^2$ for copper. Aluminum did not show any weight loss. The weight losses of the metal specimens inside the collectors ranged considerably among the different collectors but were clearly less than outside for all collectors.

To determine the concentration of sulphur dioxide inside the solar collectors, measurements with an UV-fluorescence instrument were made. The measurements were made to gain a better knowledge of how fast a solar collector exchanges air with the ambient during different modes of operation.

The outdoor concentration level of sulphur dioxide during one cloudy and one cloudless day of measurement was 2–3 ppb. The level inside the different collectors was the same as the outdoor concentration or even a little bit higher, apart from one exception. For the collectors with a nickel-pigmented aluminum absorber the SO_2 level was zero when the weather was cloudy. This implies that a nickel-pigmented aluminum absorber can absorb sulphur dioxide when they are relatively cold and the humidity is high.

3.2.6 VARIATIONS WITH TIME AND THE IMPORTANCE OF AMPLITUDES AND FREQUENCY OF CHANGES IN STRESS LEVELS

It is the degradation process that determines not only what degradation factors should be measured but also what data are of interest. The environmental load changes with time may be important and determine what form of time-dependent data are needed for monitoring the environmental stress factors.

Sometimes, it is sufficient to obtain only the maximum and/or minimum values of a certain environmental stress factor. The arithmetic mean value of a degradation factor is seldom useful for evaluating environmental resistance, but more often the duration of some critical time period for a particular kind of environmental stress. Corrosion occurs only when the humidity level and temperature are above critical threshold values. Accordingly, the time of wetness, which is defined as the time of the year when the relative humidity is greater than 80% and the temperature is above $0°C$, has been introduced as a simple measure for characterizing the atmospheric corrosivity of different climates. For this case, it is assumed that only the time period when materials are exposed at some critical climatic degradation conditions is important for characterizing degradation and not the history of changes in some degradation factors. However, when very complex degradation mechanisms are involved, the dose that should be measured may be hard to define properly.

In some cases, reference materials or reference components may be used, e.g., for the characterization of atmospheric corrosivity by use of standard metal specimens (see ISO 9226). For characterization of the amount of UV irradiation and effect of temperature during outdoor or artificial aging conditions, polystyrene standard specimens are used (see SAE standard J1885, Society of Automotive Engineers) (SAE J1885).

For materials with a fatigue type of degradation/failure mechanism, the number of changes (or cycles of change) in the environmental conditions is important for the service life. Concrete durability, for example, may be defined in the number of freeze–thaw cycles a material may resist. To perform and interpret the results of mechanical vibration test, information on the dynamics of the mechanical loads under service conditions are needed (see Chapter 4.3, Table 4.3-4).

For the purpose of service life prediction, data are needed about the resistance of the local environmental stress on the materials of a component. Such data may sometimes be difficult to obtain and, therefore, the environmental stress on the materials of a component has to be calculated from information on the overall environmental load (see the following chapter).

REFERENCES

Carlsson, B., Frei, U., Köhl, M. & Möller, K. (1994). Accelerated Life Testing of Solar Energy Materials – Case Study of Some Selective Solar Absorber Coatings for DHW Systems, International Energy Agency, Solar Heating and Cooling Programme Task X: Solar Materials Research and Development, Technical Report, SP-Report 1994:13, ISBN 91-7848-472-3.

IEC 600721, Classification of Environmental Conditions, International Electrotechnical Commission, P.O. Box 131, CH - 1211 Geneva 20, Switzerland.

ISO 9226, Corrosion of metals and alloys – Corrosivity of Atmospheres – Determination of Corrosion Rate of Standard Specimens for the Evaluation of Corrosivity, International Organization for Standardization, Geneva, Switzerland.

ISO 8407, Corrosion of Metals and Alloys – Removal of Corrosion Products from Corrosion Test Specimens, International Organization for Standardization, Geneva, Switzerland.

SAE J1885, Accelerated Exposure of Automotive Interior Trim Components Using a Controlled Irradiance Water Cooled Xenon-Arc Apparatus, Society of Automotive Engineers, SAE World Headquarters, 400 Commonwealth Drive, Warrendale, PA 15096-0001 USA.

Chapter 3.3
Evaluation of the Stress Conditions

M. Heck and M. Köhl

Fraunhofer-Institut für Solare Energiesysteme,
Heidenhofstr. 2, D-79110, Freiburg, Germany

Abstract: The purposes of this chapter are to summarize the methods for accumulating climatic stresses data, for evaluating the data, and reporting the effects of different climates on several flat-plate collectors. The environmental impact on materials exposed directly to weathering or indirectly to weather-dependent stress conditions as in solar systems varies over time. Therefore, it is necessary to accumulate stress data over a significant time period. Appropriate periods are usually at least one year or better still, several years to obtain the data needed for assessing the durability of a new material in comparison with others. If the data are intended to be a basis for general accelerated life tests, several different climates should be monitored at different locations and for different periods to identify the most challenging stresses. The impact of evaluation parameters like sampling time or integration intervals is considered. The effect of different climates, different materials, and different collector types for heating domestic water on the amount of stresses are shown. Statistical analysis tools and the interpretation of the results are discussed.

Keywords: Stresses, UV and solar irradiation, Temperature, Humidity and moisture, Time of wetness, Dose, Frequency distributions, Different climates, Collectors for heating domestic water

LIST OF ACRONYMS AND ABBREVIATIONS

DHW	Domestic Hot Water
TOW	Time of Wetness
UV	Ultraviolet Radiation (light)
UV-A	UV with wavelengths from 315 nm to 385 nm
UV-B	UV with wavelengths from 280 nm to 315 nm

3.3.1 INTRODUCTION

The environmental impact on materials exposed directly to weathering or indirectly
to weather-dependent stress conditions as in solar systems varies over time.
Therefore, it is necessary to accumulate stress data over a significant time period.
Appropriate periods are usually at least one year or better still, several years to
obtain the data needed for assessing the durability of a new material in comparison
with others. If the data are intended to be a basis for general accelerated life tests,
several different climates should be monitored at different locations and for different
periods to identify the most challenging stresses.

The recorded data (see Figure 3.3-1) are usually average values for a
predetermined sampling period. The frequency of acquiring data and the integration
period are chosen from the typical frequencies of changes in the measured property
and their effect on the samples. If there are no changes in the important parameters,
e.g., UV irradiation during the nights, event-driven data acquisition, which means
that no data are recorded, does not provide a continuous data set. This procedure
might be necessary, if the capacity of the data storage medium is limited. However,
the trigger level for the selected property (event) has to be selected carefully to avoid

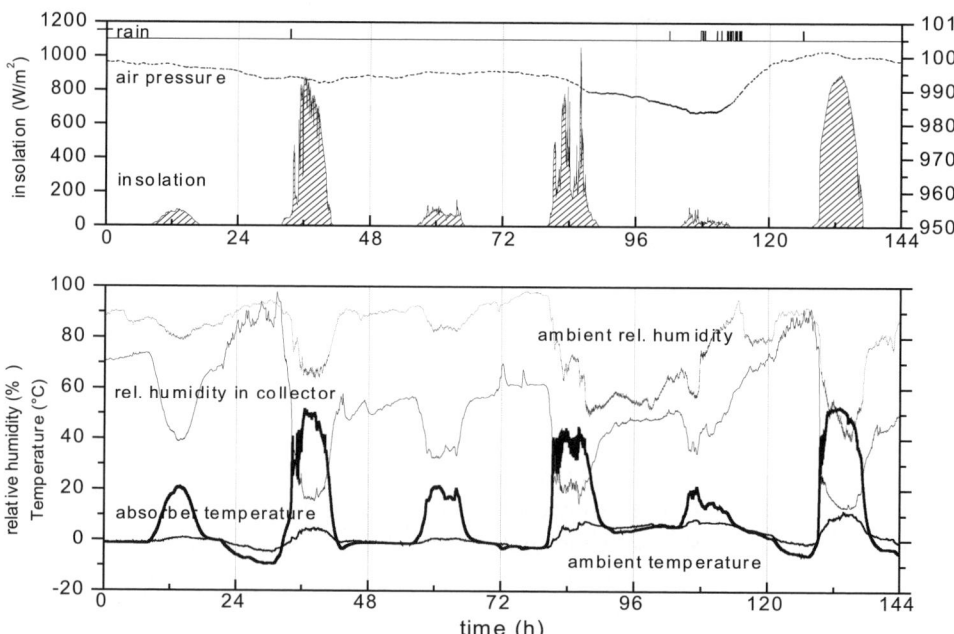

Figure 3.3-1. Example of monitored data of ambient climate and microclimate in a collector
during 6 days.

loss of important data. Evaluation of the data requires more effort for correct integration because of using random lengths and occurrences of the intervals. We now provide an overview of the evaluation of the most important climatic stress factors, i.e. UV and solar irradiation, temperature, humidity, and moisture.

3.3.2 UV AND SOLAR IRRADIATION

The intensity of UV irradiation and the other parts of the solar spectrum can change within less than one second because of moving clouds. The impact on samples can be twofold. First, the radiation is a degradation factor that might result in damage to chemical compounds, and especially organic materials. Because this is usually a linear effect, average dose functions can be used. The UV degradation of polymeric materials depends on their spectral sensitivity. Therefore, monitoring of the spectral irradiance in the UV-blue spectral ranges is especially recommended during outdoor weathering of UV-sensitive samples. A number of various UV spectroradiometers are currently available for measurements of the solar irradiation. The least expensive and most widely used devices for measuring broadband UV-A and UV-B are of limited benefit for solar materials investigations. By definition, the UV-A and UV-B spectral ranges are based on the erythema sensitivity of the human skin. Even blue light in the visible spectrum can cause damage to numerous materials, such as some polymers, paints, and varnishes. Moreover, the spectral responses are different from one type of instrument to another. This is because the increase in UV below 400 nm in the solar spectrum is very steep and the filters used for these instruments are not ideal (see Figure 3.3-2).

3.3.3 TEMPERATURE

The second major effect resulting from solar irradiation is the increased sample temperature because of absorption of the irradiation, which depends on the spectral absorptance of the surface. For sample temperatures, longer time intervals can be used for their measurement because the thermal processes are slow compared with the time constants of irradiation. The typical time constants that depend on the temperature gradients in the sample from the front surface to the back surface, are dominated by the solar absorptance, the thermal emittance, the thermal mass, the thermal conductivity, and the heat transfer coefficients of the sample. Optimizing these properties for solar applications usually results in high surface temperatures. The sample temperature can be considered as a separate degradation factor providing that degradation processes are either internal chemical or physical

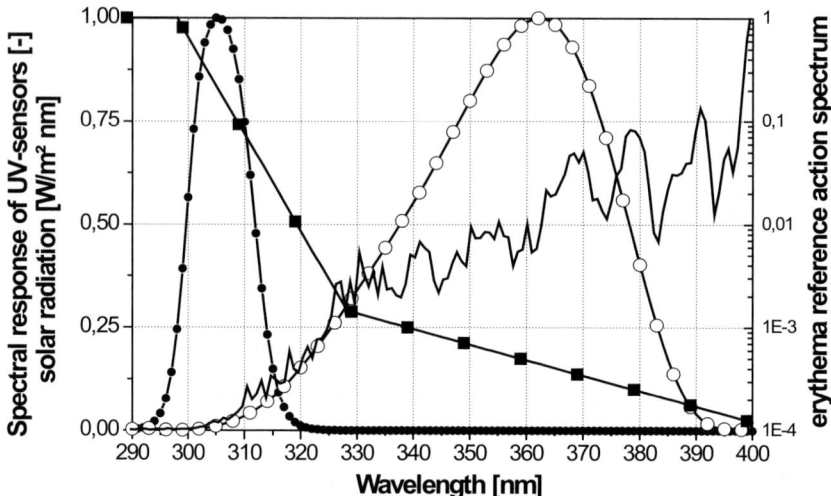

Figure 3.3-2.　Spectral response of a UV-B sensor (dots), spectral response of a UV-A sensor (open circles), CIE-IEC erythema reference action spectrum (squares), and measured solar spectrum (solid line, no points).

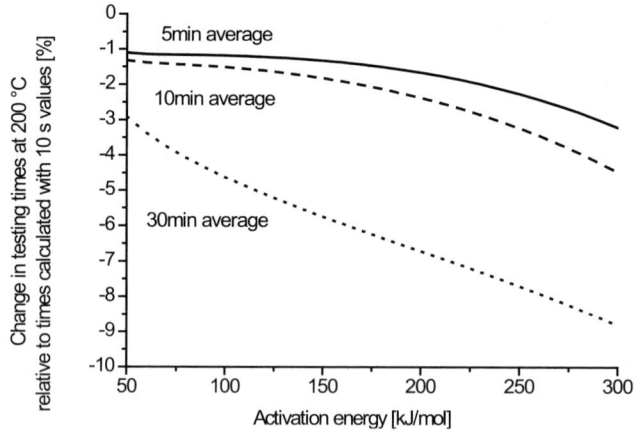

Figure 3.3-3.　Influence of the sampling intervals of the temperature on the test conditions evaluated by integration of the outdoor test conditions according to Chapter 4.5.

processes or the reactants are a permanent part of the ambient atmosphere, such as oxygen or water vapor.

The monitored data are usually evaluated by converting them into a frequency distribution by summing up the number of sampling intervals, which have a selected temperature range for each interval. This range should be larger than the

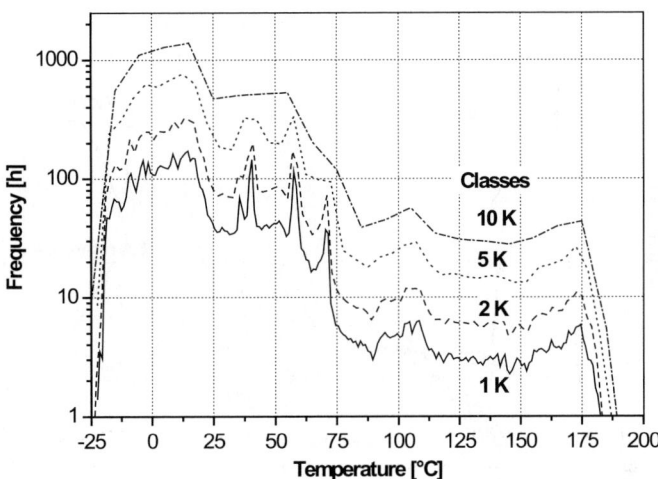

Figure 3.3-4.   Frequency per class for absorber temperatures for different integration classes.

temperature resolution of the sensors that is typically about 1 K. Attention should be paid to the correct normalization of the frequency according to the sampling intervals. In Figure 3.3-3, the test conditions are plotted as a function of the material-dependent parameter called the Arrhenius energy; the latter were calculated using the Arrhenius model described in Chapter 4.5. The exponential temperature dependence for Arrhenius-type degradation processes results in rapidly increasing reaction rates as higher temperatures are reached. Therefore, the temperature data should not be smoothed too much by averaging. The effect of using different sized sampling intervals, which are transformed into different integration classes, is shown as a function of temperature in Figure 3.3-4.

3.3.4   HUMIDITY AND MOISTURE

The rate of change in the relative humidity has a long timescale. The changes in humidity depend strongly on the temperature that have time constants ranging from minutes and hours and the weather with time constants ranging from hours to days. The monitored data are evaluated by converting them into a frequency distribution by summing the number of sampling intervals in which the relative humidity lies within a certain humidity range; the interval should be larger than the humidity resolution of the sensors. In Figure 3.3-5 the example shows the variation of the microclimate in typical collectors of domestic hot water (DHW) systems. The differences in the humidity in different collector types can already be detected after

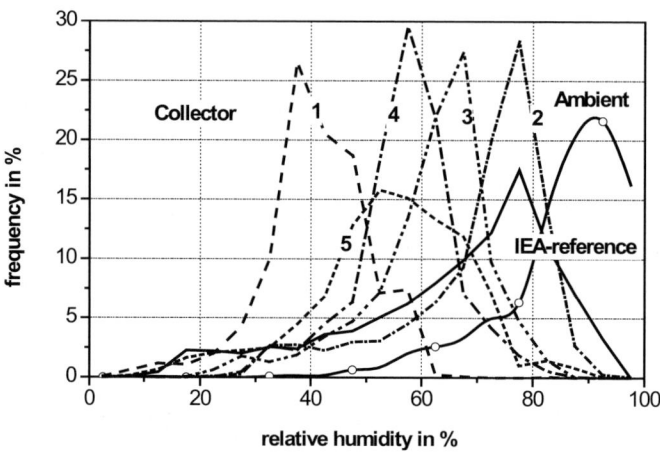

Figure 3.3-5. Frequency distribution of the relative humidity in different collectors in the same climate.

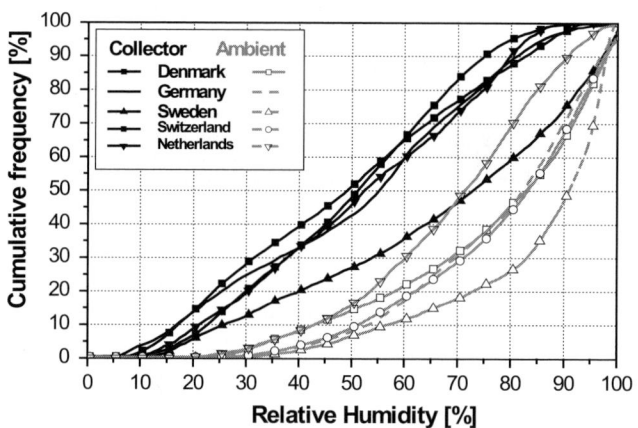

Figure 3.3-6. Cumulative normalized frequency distribution of the humidity in identical collectors in different climates.

monitoring for only three months. The most favorable climate for slowing the degradation rate is for the collector with the maximum of the frequency distribution at the lowest value for the relative humidity, e.g. collector 5 in Figure 3.3-5. Note that the microclimate in all collectors is clearly dryer than the ambient. Another way of evaluating the data is to use the cumulative normalized frequency distribution, as shown in Figure 3.3-6. The humidity levels measured in the same reference collectors are similar at the different site except for the Boras site in Sweden, where a much higher humidity was measured.

Relative humidity is commonly measured, but is often not an appropriate property for quantitative evaluation. More useful properties are the absolute humidity, x, and the partial pressure of the water vapour, p_D. The conversion from p_D to x is carried out using $p_D = \varphi \cdot p_S(T)$ for $x = 0.622 \, [\varphi \cdot p_S(T)]/[p \cdot \varphi \cdot p_S(T)]$ with the saturation vapor pressure, $p_S(T) = c_1 \exp[c_2 \cdot T/c_3 \cdot T]$ and in which the coefficients are $c_1 = 6.10780 \, \text{mbar}$, $c_2 = 17.08085$, $c_3 = 234.175°C$ and the temperature T is in °C from 0°C to 100°C. Naturally, the absolute humidity also increases with increasing temperature. By comparing the data in Figure 3.3-7 with that in Figure 3.3-8 shows that the ambient relative humidity is much higher than the relative

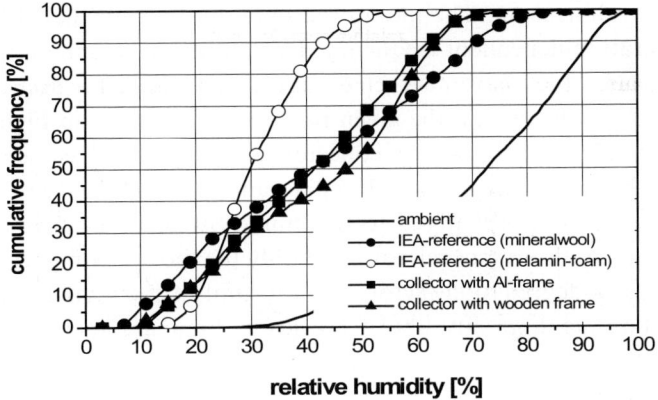

Figure 3.3-7. Cumulative normalized frequency distribution of the relative humidity in different collectors in the same climate.

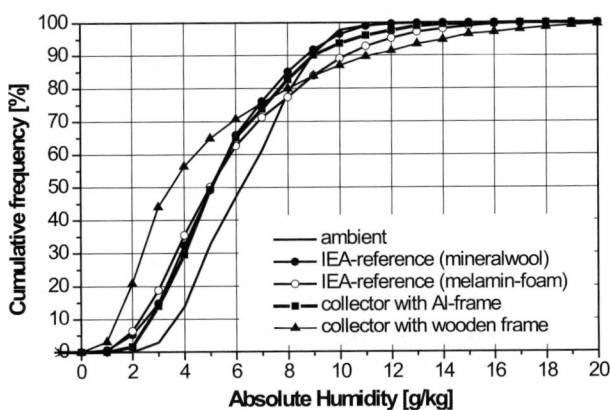

Figure 3.3-8. Cumulative normalized frequency distribution of the absolute humidity in different collectors in the same climate.

humidity in the collectors, but the absolute humidity in the collectors can exceed the maximum values for the ambient. The absolute humidity in the one collector with melamine foam as the thermal insulation material exceeded that in the ambient, but this collector had the lowest relative humidity values. This effect results from humidity storage by the insulation materials and the wooden frames combined with the higher temperatures in the collector.

Degradation factors involving moisture and pollutants are more complicated to consider because not only are their concentrations of interest, but also because the actual temperatures must be known at the time of their occurrence. This is extremely important for the design of accelerated life tests, because the kinetics of the degradation processes are dominated by the temperature (see Chapter 4.5). This opens up a third dimension. The monitored data are then evaluated by converting it into a temperature-dependent frequency distribution by summing the number of sampling intervals, which have a selected temperature range for each interval, with the humidity values also within the given range. The diagrams in Figures 3.3-9 and 3.3-10 show the differences between a collector, which was used as the reference, and another one with a dry microclimate. The diagram in Figure 3.3-11 is based on data from the reference collector exposed to the humid Swedish maritime climate.

If low humidity levels are not relevant and only the moisture impact or the time of wetness (TOW) is a degradation factor, a lower limit of the humidity level could be defined. Exposures to humidity above the lower limit are then relevant and the surface of interest is considered to be wet. This TOW has to be summed depending on the temperature, especially the surface temperature of the samples. The example in Figure 3.3-12 shows the temperature-dependent TOW (defined as periods with a

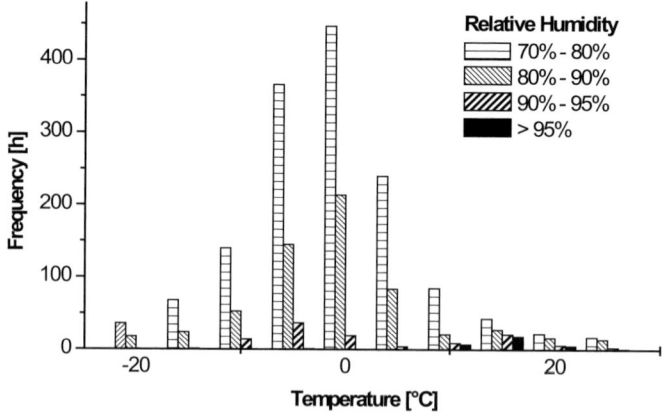

Figure 3.3-9.   Frequency distribution of the relative humidity above various levels as a function of the temperature in a collector in Freiburg (1.1.1999–1.12.1999).

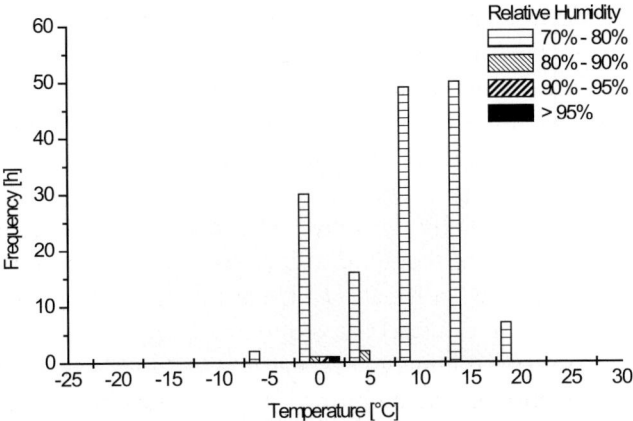

Figure 3.3-10. Frequency distribution of the relative humidity above various levels as a function of the temperature (same climate as in Figure 3.3-5 but in a different, very dry collector).

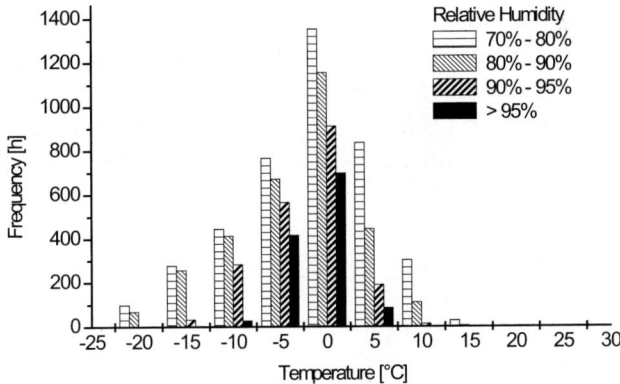

Figure 3.3-11. Frequency distribution of the relative humidity above various levels as a function of the temperature (same collector as in Figure 3.3-5 but in another climate, Boras, Sweden).

relative humidity above 99%) of a solar absorber coating in a ventilated flat-plate collector. A comparison of the degradation of a moisture-sensitive absorber coating during outdoor exposure with controlled laboratory condensation tests was used to define the TOW for solar absorber coatings in flat-plate collectors.

The moisture content on a surface depends on additional properties like the surface morphology (micro-roughness, capillarity) and the polarity, which can strongly increase the hygroscopic behavior. Condensation already occurs at relative humidity values below 100% because of these surface properties and local temperature fluctuations. Ultimately, equilibrium between the rate of condensation

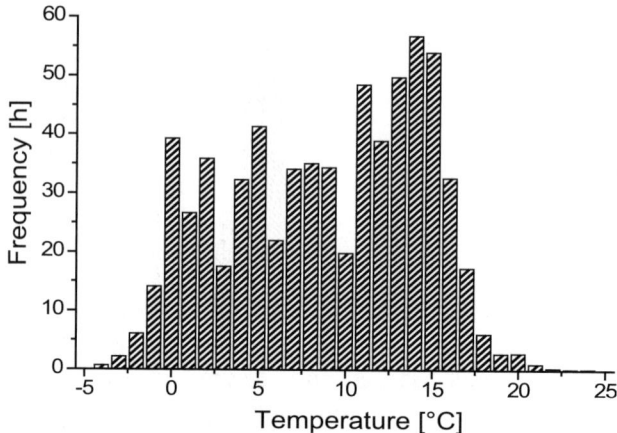

Figure 3.3-12. Temperature-dependent time-of-wetness (TOW) (r.h. > 99%) of a solar absorber coating in a ventilated flat-plate collector.

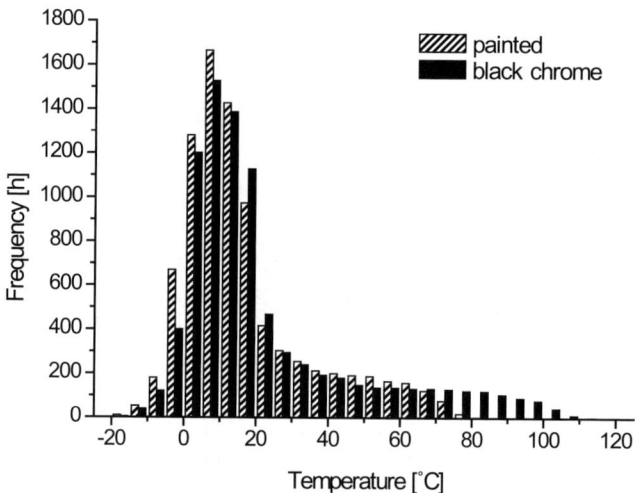

Figure 3.3-13. Temperature frequency distribution of two black surfaces oriented south 45° with different thermal emittance values (paint: 0.9; black chrome: 0.2).

and re-evaporation will be reached. Moreover, the surface temperature depends on the emittance of the surface (Figure 3.3-13) and radiative interaction of a surface with the environment, especially the cold sky. Temperature decreases of more than 5 K can easily occur during clear nights and a surface with a high thermal emittance; the decrease also depends on the orientation and the field of view of the surface. As a result, the time that dew is on the surface also varies (Figure 3.3-14). For a

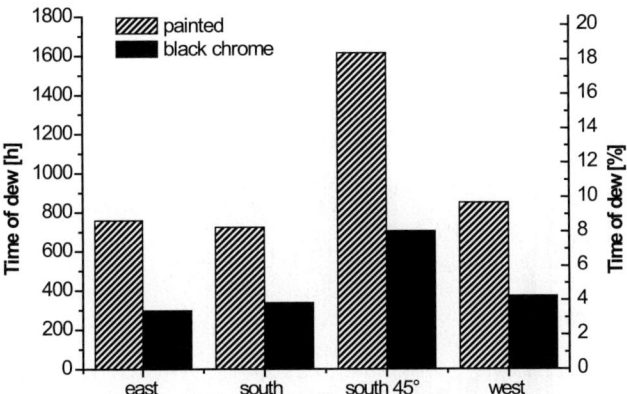

Figure 3.3-14. Time of dew frequency distribution of two black surfaces oriented south 45°, south, east and west vertically, with different thermal emittance values (paint: 0.9; black chrome: 0.2).

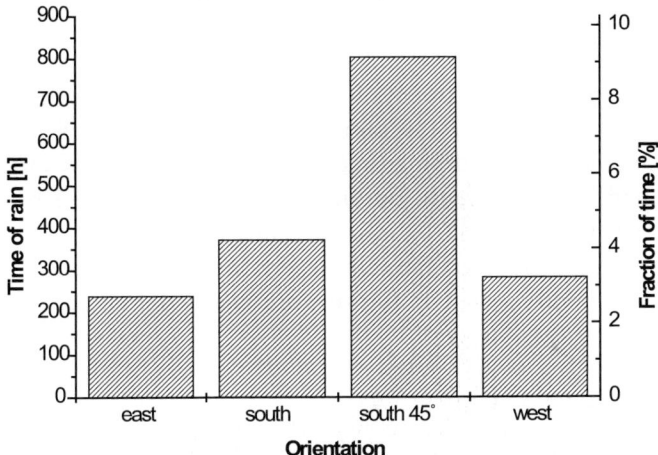

Figure 3.3-15. Time of rain frequency distribution for vertical surfaces oriented east, south and west, and south-orientated surface with a tilt angle of 45°.

relative humidity of 95%, the dew-point temperature is about 1 K and for 90%, it is about 2 K below the ambient temperature in the relevant range. The dew point can be derived from the water vapor pressure $p_{D1} = p_{D2}$, which is equivalent to $\varphi_{amb} \, p_s(T_{amb}) = p_s(T_{dew})$, with the expression given above for the saturation vapor pressure.

Materials exposed to outdoor weathering are usually wetted by rain. The time of rain (TOR) is then part of the TOW, which could easily be detected by the common rain sensors mounted in the plane of the samples. Figure 3.3-15 shows the difference

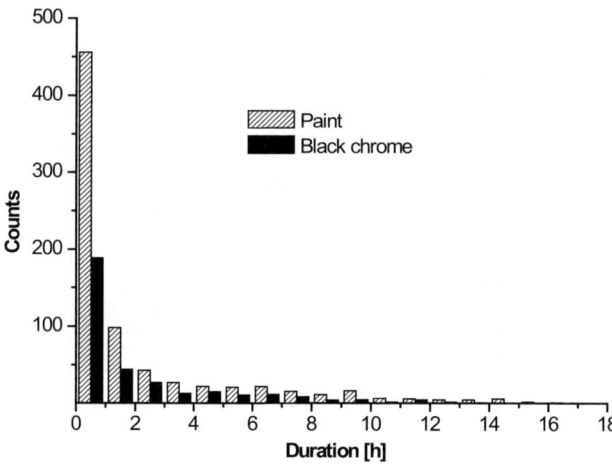

Figure 3.3-16.   Duration of dew frequency distribution of two black surfaces oriented south 45° with different thermal emittance values (paint: 0.9; black chrome: 0.2).

between the TOR for different orientations. It is more difficult to measure the time that a surface is still wet after the rain has stopped. The application of wetness sensors is limited because the drying time depends on the optical surface properties (see Figure 3.3-16). Moreover, the surface should not be heated by the sensor or shaded against solar irradiation or infrared radiation exchange with the cold sky. Obviously, these difficulties are especially severe for measuring the TOW. Using an infrared surface humidity sensor in future work is a better choice.

3.3.5 STATISTICAL ANALYSIS

The monitored data are usually not independent, e.g., the ambient humidity depends on temperature, temperature on solar irradiation, and the microclimate on the ambient climate etc. Multivariate regression analysis helps to identify such relationships, or even enables modeling and prediction of properties based on the measured time and other parameters. Figure 3.3-17 shows the regression coefficients, which are normalized by dividing the measured values by their standard deviation, for the absolute and relative humidity for different collectors. The relative humidity in the five different collectors is strongly correlated with the ambient relative humidity and anti-correlated with the absorber temperature. The absolute humidity, however, is strongly related to the ambient humidity, the absorber temperature, the ambient temperature, the solar irradiation, and also to the infrared radiation exchange with the sky. The differences in the behavior of the collectors are clearer for

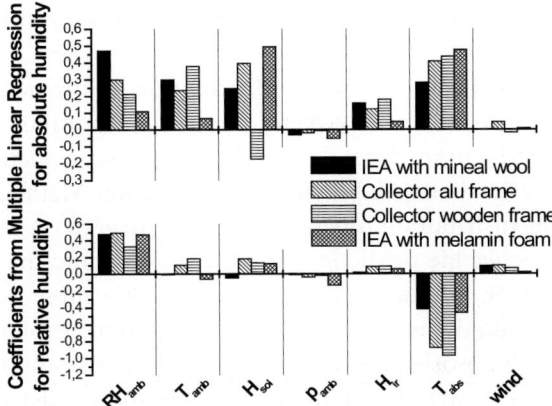

Figure 3.3-17. Multiple linear regression analysis for the relative and absolute humidity in different solar collectors depending on: ambient relative humidity (RH_{amb}), ambient temperature (T_{amb}), solar irradiation (H_{sol}), ambient air pressure (p_{amb}), infrared radiation from the sky (H_{ir}), solar absorber temperature (T_{abs}) and wind.

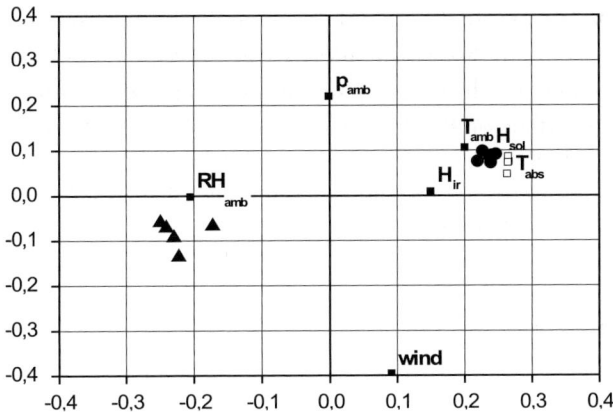

Figure 3.3-18. Correlation analysis of the relative humidity (triangles) the absolute humidity (dots), and the absorber temperatures (open squares) in the collectors with the ambient climate parameters (black squares).

the absolute humidity, because of the water storage behavior of the different materials e.g., wood, mineral wool, melamine foam used in the collectors. The correlation plot in Figure 3.3-18 shows these results in a different way. Parameters, which are clustered together, are correlated, while opposite parameters are anti-correlated, such as the relative and absolute humidity values. Wind and air pressure are anti-correlated, but have no essential impact on the other parameters.

3.3.6 MICROCLIMATE EVALUATION FOR SERVICE LIFE PREDICTION

If only the dose is important, then the distribution or frequency function of a degradation factor is of interest. By using frequency functions and knowledge of how the rate of degradation depends on the degradation factors, the extent of degradation can be estimated (see Service Life Prediction Methods, Chapter 4.5).

Sometimes it is the distribution function of a single degradation factor that is used for the purpose of service life prediction. In other cases, more than one degradation factor needs to be taken into account, which means that those degradation factors have to be measured simultaneously with an accurate time resolution. For a service life prediction of solar absorber coatings the degradation resulting from the action of temperature loads, as well as for high humidity and condensed water both air humidity and surface temperature were taken into account (see Figure 3.3-12).

Dose functions may also vary in such a way that a function of more than one degradation factor needs to be used for the purpose of service life prediction. For the photochemical degradation of a polymeric material during outdoor use, it may be that only the dose of UV radiation needs to be taken into account. The combined action of UV irradiation and thermal load should be considered and, therefore, UV irradiation and surface temperature need to be measured and evaluated simultaneously. The variable used to describe service environmental stress conditions might therefore be the product of the UV irradiation level, I, and an Arrhenius type of expression, $\exp(E_a/RT)$ (see Chapter 4.5, Table 4.5-1, (a) and (e)).

As mentioned in the previous section, reference test specimens may be used to describe environmental stress where complex degradation mechanisms are involved. In such cases, the dose function will correspond to the extent of degradation of the reference specimens. In the absorber case study, use was made of metallic test specimens of zinc as an indicator or sensor for atmospheric corrosivity.

Chapter 3.4

Correlation between Microclimate and Macroclimate

Ole Holck,[1] M. Heck,[2] and Michael Köhl[2]

[1]*Department of Civil Engineering, Building 118,*
Technical University of Denmark, DK-2800 Kgs. Lyngby, Denmark
[2]*Fraunhofer-Institut für Solare Energiesysteme,*
Heidenhofstr. 2, D-79110, Freiburg, Germany

Abstract: The purpose of this chapter is to present two models, which can be used for optimizing the microclimate inside a solar thermal collector, and the results of calculations for several types of collectors. Damage of materials inside a solar collector depends on the corrosivity of the climate inside the collector, the so-called microclimate. The microclimate depends on the ambient weather and the construction of the collector. Two models were developed, which can be used to optimize the construction of collectors with the most favorable microclimate, and are based on an analytical heat and moisture balance and on computational fluid dynamics, respectively. The models, which take into account the condensation of water vapor on the inside of collector surfaces and for the sorption of water vapor in the insulation material, have been demonstrated to predict these phenomena reliably in realistic situations. Calculated results are presented for the annual average humidity hours and the amount of condensate exceeds areal concentrations of 1 to $10\,g/m^2$ for air permeabilities up to $100\,l/h$. Results are also presented for ventilation rates at normal operating and stagnation conditions. The measured and calculated absolute humidity in a collector during a typical 24-h period are comparable.

Keywords: Microclimate, Collectors, Condensation, Model, Computational fluid dynamics

LIST OF ABBREVIATIONS

MOMIC	computer program (Modeling of Microclimate in Collectors) based on an analytical heat and moisture balance in a collector
TRY	the Danish Test Reference Year
DRY	the Design Reference Year

125

DHW   Domestic Hot Water
CFD   Computational Fluid Dynamics

3.4.1 INTRODUCTION

The collector box, which protects the inside components from the outdoor weather, is the space enclosed by the frame, the cover, and the back plate of the collector (see Figure 1-3, Chapter 1). The microclimate inside the collector box may differ from the macroclimate outside, which is determined by the ambient outdoor weather. The collector box cannot be hermetically sealed because expansion of the air results from temperature differences between the inside and outside during use. A pressure difference between the surroundings and the air gap inside the collector box will force air into or out of the collector box. The driving force for ventilation of the collector is primarily thermal buoyancy, but wind and gusts can also cause a pressure difference. In addition, humidity in the collector may be different from the humidity in the ambient air. The absolute humidity in the collector and the ambient humidity will equilibrate after high ventilation for some time. Water condenses in the collector when the temperature of the cover is lower than that of the dew point temperature of the air in the box, as frequently occurs at night. Condensation takes place on the inner surface of the glazing and if the water is not re-evaporated and removed by solar heat gain and ventilation, water can drop onto the absorber plate and may result in corrosion. High humidity, especially in combination with high absorber temperatures, may result in extensive corrosion of the metal surfaces. The collector box must be watertight to prevent rain from entering and to ensure an acceptable microclimate. Optimization of the ventilation rate, which is the number of collector volumes that are exchanged per hour by natural ventilation through the respective vents or any leakage, and the right choice of insulation and framing materials of the collector will prolong the service life for the collector and improve the financial benefit for the collector during its lifetime [see Chapter 5.2, this volume]. Modeling may be used to correlate the microclimate in the collector and the macroclimate of the surroundings. We briefly present two approaches; one is based on an analytical heat and moisture balance in a collector, and the other is a simulation using computational fluid dynamics.

3.4.2 MODEL BASED ON AN ANALYTICAL HEAT AND MOISTURE BALANCE IN A COLLECTOR

For the model, it is assumed that the air in the collector volume is well mixed. The heat is calculated per square meter of a solar collector area. Differences between the center and the corners of the collector are neglected. The temperature of the air in the collector used for the calculation is taken as the average of the cover and absorber temperatures. The thermal heat capacity of the collector is assumed to consist of that for the cover, the absorber, the working fluid, and half of the back insulation; the heat capacity of the air and half of the back insulation are neglected.

The rate of heat transfer by convection and thermal radiation is expressed as a set of linear differential equations, given as the product of heat transfer coefficients and the associated temperature differences. The error resulting from linearization is small because the heat transfer coefficient is calculated for each time interval, and the time steps are so short that the temperature changes are negligibly small. The solar transmittance–absorptance product is obtained by correcting the value at normal incidence for the actual incidence angle by using incidence angle modifier coefficients.

The ventilation of the collector is also considered in the model. The effect of wind is calculated using an empirical correlation between the wind speed and ventilation rate, which is an approximate treatment of the influence of wind. The effect of wind direction is not considered. A moisture balance provides the required information on the amount of water that condenses and evaporates on the inside of the cover. A transient model for calculating the combined heat and moisture transfer in the insulation is used, in which the insulation material is divided into three finite control volumes. An infinite resistance to heat and mass transfer is inserted at the absorber plate and at the back plate. The relation between vapor pressure and moisture content is given by the sorption curves for the insulation material. The buffer effect of moisture in the insulation materials is obtained from this calculation.

A computer program "MOMIC" (Modeling of Microclimate in Collectors) has been developed based on our model of the heat and moisture balance in a collector box (balance model) (MOMIC). Calculations using the model yield insight into the humidity and temperature conditions in the solar collector for simulated or real climatic data. An input module in "MOMIC" accepts meteorological data from data files. Three possibilities have been implemented, two for synthetic climatic data in a form used by the Danish Test Reference Year, TRY (Lund, 1985), and the Design Reference Year, DRY (Lund, 1995), and one for real climatic data. Sensitivity studies of the effect of changing collector parameters can be made by choosing a parameter, e.g., the tilt angle of the collector, collector area, width of the air gap, thickness of the air-gap, thickness of the back insulation, and maximum allowable absorber temperature or air tightness factor. Several calculations are performed with the chosen parameter varied over a suitable range.

3.4.2.1   Calculations of the Annual Average Humidity and the Areal Amount of Condensate

Two main properties have been discovered to be important for optimizing the collector to change the humidity inside the collector box. These are the capabilities of the insulation materials or other hygroscopic materials to sorb moisture and the ventilation rate of the collector box. The capability to sorb moisture includes the influence of the vapor barrier.

To identify the optimum microclimate, the accumulated presence of condensation and the period when the relative humidity attains a certain value were calculated as a function of the ventilation rate of the collector box. These results are presented in Figures 3.4-1 and 3.4-2. The assumed ventilation rate of the collector was varied in

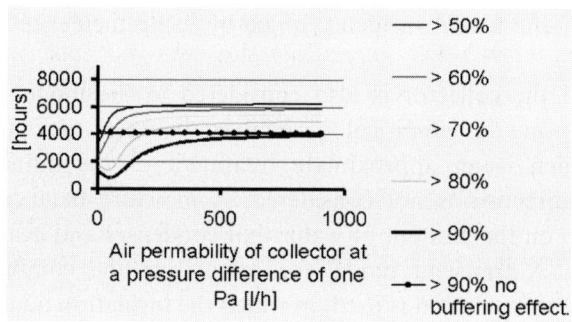

Figure 3.4-1. Calculation of the annual hours of the relative humidity above the indicated values showing the dependence on the air permeability of the collector. It is evident that an air tight collector has a less humid climate than a highly ventilated collector.

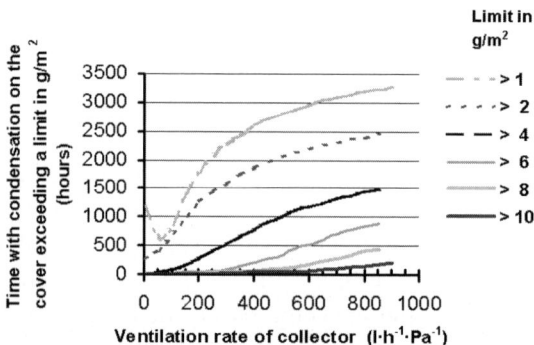

Figure 3.4-2. Calculations of the annual hours that the amount of condensation on the inner surface exceeds different areal amounts on the cover. Collectors with the lowest permeability rates show the least number of hours of visible condensation. The onset of condensation amounts greater than $6\,g/m^2$ appear at a ventilation rate of $100\,l/h$.

the range between 5 and 850 l/h/Pa. Both figures show that the annual hours of higher humidity increase for higher ventilation rates of the collector. The number of hours with humidity higher than 90% is predicted from the model with and without including the buffer effect from the insulation material. Less than 2000 h are found for a collector ventilation rate of less than 150 l/h. On average this is 5 h/day. If the hygroscopic effect from the insulation material is neglected, this number is 4000 h over the entire range of collector ventilation rates. Therefore, the influence of the insulation material or other hygroscopic materials can reduce the occurrence of high relative humidity. High humidity, especially in combination with high temperature, may result in serious corrosion of the metal surfaces. The weather data are from the Design Reference Year for Copenhagen, Denmark (Lund, 1995).

The results of the calculations for the annual hours that the amount of condensation exceeds $1 \, g/m^2$, $2 \, g/m^2$, $4 \, g/m^2$, $6 \, g/m^2$, $8 \, g/m^2$, and $10 \, g/m^2$ on the collector cover are presented in Figure 3.4-2. The simulations show that the least number of hours of high humidity and non-visible condensation occur up to 80 l/h. In all cases, the largest number of hours of condensation results at the highest rates of ventilation. Focusing on the risk of visible condensation, the lowest ventilation rate possible is preferable for Danish climates, if there is no moisture in the collector initially.

It has been shown that the humidity inside the collector depends on the wind speed and that the insulation materials can lower the relative humidity. Calculations have shown that the optimized ventilation rate is about 40 l/h for a $2 \, m^2$ collector used in domestic hot water (DHW) systems in Denmark. If a vapor barrier is placed between the insulation material and absorber, visible and heavy condensation is eliminated and protection is obtained from humidity increases resulting from outgassing from the insulation. Predictions from the calculations have to be interpreted with some care because of the approximate treatment of weather conditions and the uncertainties inherent in the use of the balance model itself. The advantage of using the balance model is the short computer time needed compared to using Computational Fluid Dynamic (CFD) models. By performing annual calculations and parametric studies, modeling transient behavior and characterizing the microclimate are easily accomplished using the balance model.

At present, collectors usually are not optimally operated with respect to humidity. Condensation on the cover of many collectors may adversely influence the long-term durability of the collector. In addition, many manufacturers are dissatisfied with the poor impression conveyed by visible condensation. There is no single solution for designing the ideal collector, but qualitative guidelines have been developed that elucidate the important effects that should be considered for optimizing the internal microclimate (Carlsson *et al.*, 1994). For example, it appears that the optimized collector should be tighter than most collectors are today for humid ambient conditions, such as those in Denmark.

3.4.2.2　Influence of the Ventilation Rate and Influence of Insertion of a Diffusion Barrier

Condensation on the inside cover surface was investigated for two days in January 1997 from the design reference year (DRY). It was a very humid period with a high risk of condensation. To quantify the areal condensation density, we defined three levels, i.e., below $1\,g/m^2$ (non-visible condensation) for which condensation is present, but the amount condensed is not visible, between 1 and $10\,g/m^2$ (visible condensation) for which condensation is visible as droplets, and greater than $10\,g/m^2$ (heavy condensation) for which water run-off occurs at about $100\,g/m.^2$ The areal density at which run off occurs depends on the slope of the collector and other factors.

The graph in Figure 3.4-3 shows the calculated areal condensation density on the inside cover surface for six collector designs. In addition, the graph shows the solar radiation profile over time. Calculations were made for designs with and without a diffusion barrier between the absorber and the insulation, and with ventilation rates of $40\,l/h$, $150\,l/h$, and $300\,l/h$. Table 3.4-1 summarizes the calculated maximum areal condensation density and the amount of time during the two days that condensation was present and visible. By comparing the collectors with and without a vapor barrier, the data obtained when using higher ventilation rates result in a greater amount of condensation over a longer period of time. By comparing the cases with and without the vapor barrier, the presence of the barrier does not significantly affect the number of hours during which condensation is present, but the areal condensation density and the amount of time with visible condensation are reduced.

The lowest areal condensation density is obtained with a collector with a vapor barrier and at a ventilation rate at $40\,l/h$. The highest areal condensation density and the longest time of wetness on the cover, which means the corrosion risk becomes a maximum, was obtained for a collector with a ventilation rate of $300\,l/h$ and without a vapor barrier.

3.4.3　FLUID DYNAMICS IN SOLAR THERMAL COLLECTORS

Computational Fluid Dynamics is the method of choice for simulating flow and transport phenomena between a collector and its environment. In this approach, the flow area is divided into small control volumes. In three-dimensional flow simulations in a collector, the flow area consists of up to 150,000 control volumes. The Navier–Stokes equations that describe the conservation of mass, momentum, and energy have to be integrated over each control volume and then formulated as a set of differential equations. For a given set of boundary conditions, a solution for these transport equations can be found iteratively, either for stationary flows with

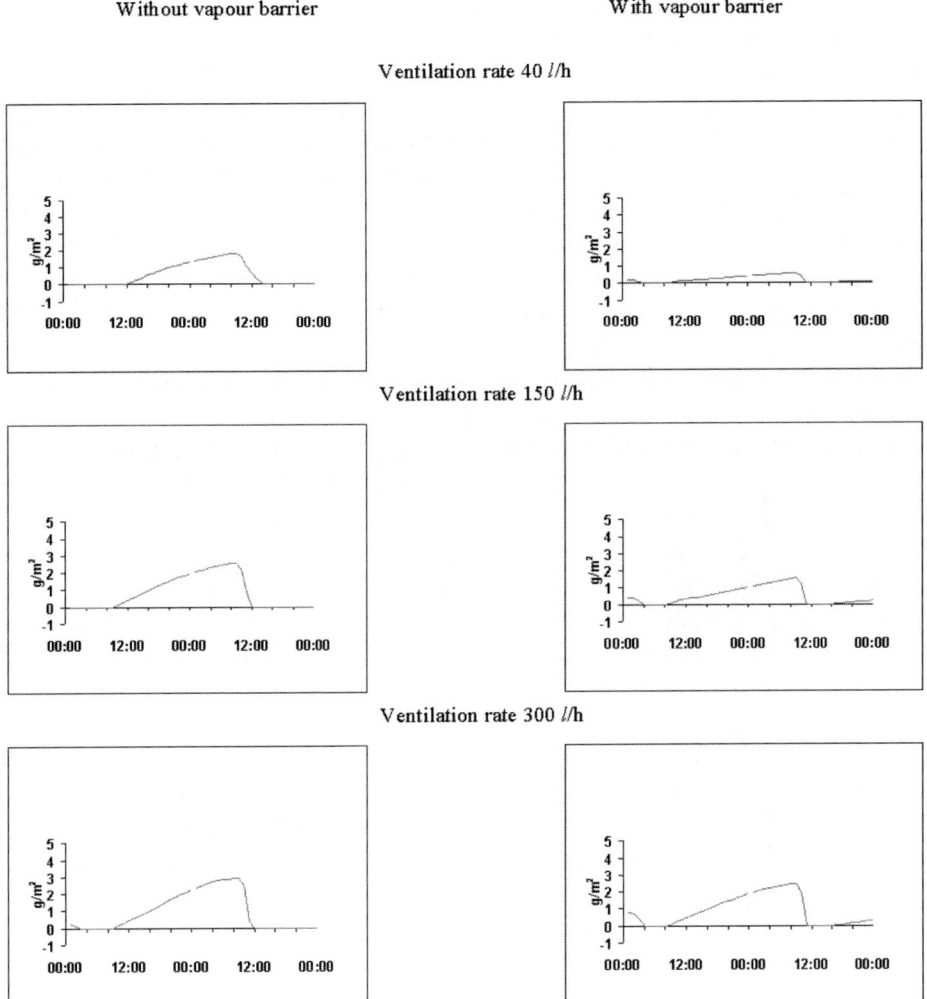

Figure 3.4-3. Calculations at three ventilation rates of the areal condensation density on the inside cover surface and the solar radiation for two consecutive days in January (DRY) 1997 (Lund, 1995). The calculations were made for collectors with and without a vapor barrier between the absorber and the insulation for a ventilation rate at 40 l/h, 150 l/h, and 300 l/h.

constant boundary conditions or transient simulations with varying boundary conditions. Additional transport equations, e.g., for water vapor pressure in air, can be set up and solved simultaneously to obtain detailed spatial and time-dependent information on the amount and rate of mass transport. To describe the behavior of the flow realistically, there is the option of implementing user-defined

Table 3.4-1. Hours of condensation present on collector surfaces and the maximum areal condensation density at three ventilation rates

Ventilation rate [l/h]	Vapor barrier	Maximum areal condensation density [g/m²]	Number of hours in the two days with condensation present on the cover	Number of hours in the two days with visible condensation present on the cover
40	yes	0.6	27	–
150	yes	1.5	27	10
40	no	1.8	26	15
300	yes	2.4	27	18
150	no	2.5	27	19
300	no	2.9	27	18

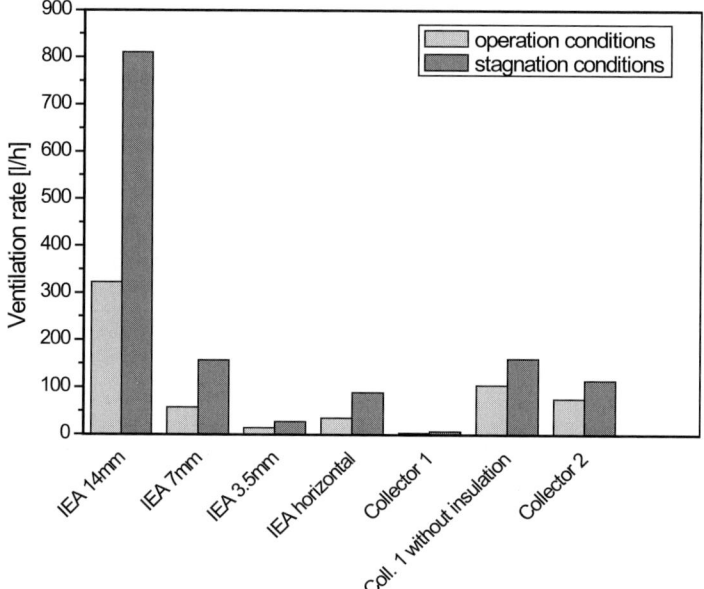

Figure 3.4-4. Calculated ventilation rates at 45°C and 120°C in l/h for different collectors (under normal operating and stagnation conditions) used for CFD simulated calculations.

models, e.g., for the condensation of water vapor on surfaces or its sorption by the insulation material.

Figure 3.4-4 and Table 3.4-2 show the computed ventilation rates of the IEA reference collector with three different sizes of ventilation hole and two client collectors. To determine the influence of the size of the ventilation holes, the original

Table 3.4-2. Calculated ventilation rates for normal and stagnation conditions for different collectors

Collector type (diameter of ventilation hole)	Normal conditions (l/h)	Stagnation conditions (l/h)
IEA (3.5 mm)	14	27
IEA (7 mm)	57	157
IEA (14 mm)	323	810
IEA horizontal	35	88
Collector 1	3	6
Coll. 1 without insulation	104	160
Collector 2	75	114

diameter of the two ventilation holes (7 mm) was changed to 3.5 mm and 14 mm, respectively. The collector of client 1 has a layer of insulation material just behind the ventilation holes that results in a very tight collector. An additional simulation calculation shows the presumed ventilation rate without the insulation material behind the holes. A comparison of the standard working conditions, i.e., absorber temperatures are limited to values below 90°C and stagnation conditions, i.e., a maximum absorber temperature of up to 180°C, shows that the buoyancy has an enormous impact on the ventilation rate.

3.4.3.1 Flow Patterns in Collectors
Each flow pattern of the modeled collectors shows a large vertical convection roll as shown in Figure 3.4-5. Air ascends along the absorber, turns around at the top of the air gap and descends again along the cover. If the ventilation holes are placed in diagonally opposite corners, an additional lateral drift results in a somewhat helical motion.

3.4.3.2 Drying of the Air Gap in the Collector
A transient simulation was developed, to determine how quickly humidity can be removed from the air gap between the collector cover and absorber by ventilation. The temperatures and ambient humidity as the transient boundary conditions were chosen for a typical sunny autumn day, combined with a high initial humidity in the collector. These conditions resulted in condensation of water vapor on the cover and the absorber immediately after beginning the simulation calculations at midnight (Figure 3.4-6). Nevertheless, the condensation started to disappear at about 8:30 am.

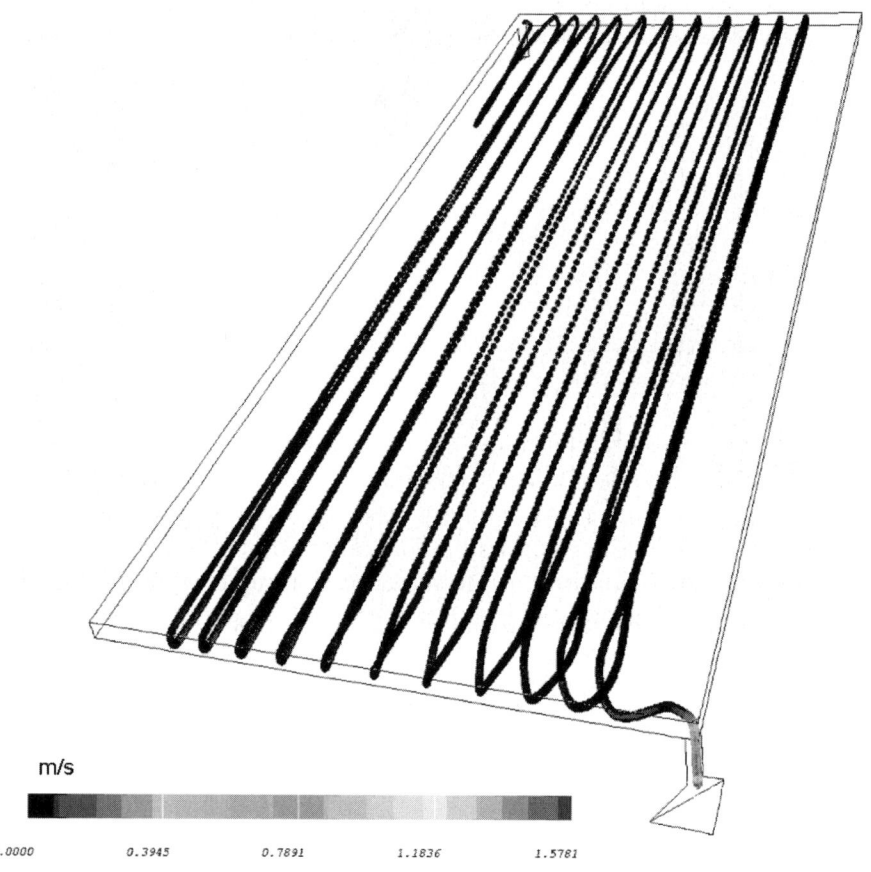

m/s

0.0000   0.3945   0.7891   1.1836   1.5781

Figure 3.4-5.   Flowlines of a particle entering the collector through a ventilation hole (lower right), circulating in the convection roll and leaving it through the other ventilation hole (upper left).

3.4.3.3   Influence of the Insulation Material on the Absolute Humidity in the Collector

Measured data for one of the tested collectors showed a correlation between the absorber temperature during the day and the absolute humidity in the air gap between the absorber and cover (see Figure 3.4-7). The collector was tight and the ambient humidity was almost constant, so the humidity in the air gap had to originate from the mineral wool used in the back insulation. The insulation material is able to sorb a large quantity of water vapor from the surrounding air depending on temperature and humidity. Measurements show that the insulation used in most of the collectors sorbs about 19 g of water per kilogram of insulation material. Thus, in equilibrium, a collector has about 53 g of water sorbed in its insulation. If only two percent of that amount is released, the absolute humidity in the air gap will increase

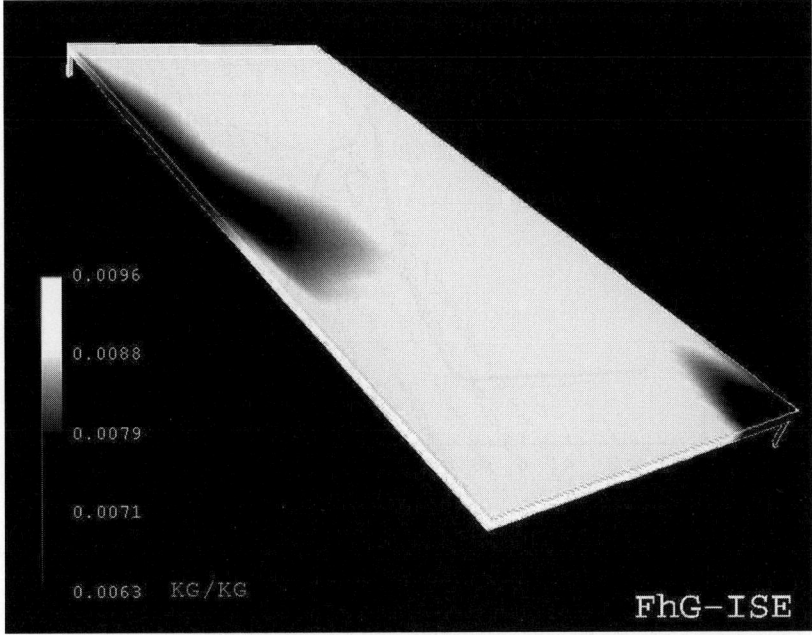

Figure 3.4-6. Absolute humidity in the air gap at 8:30 am. The absorber temperature is rising and the natural ventilation through the diagonal ventilation holes starts drying.

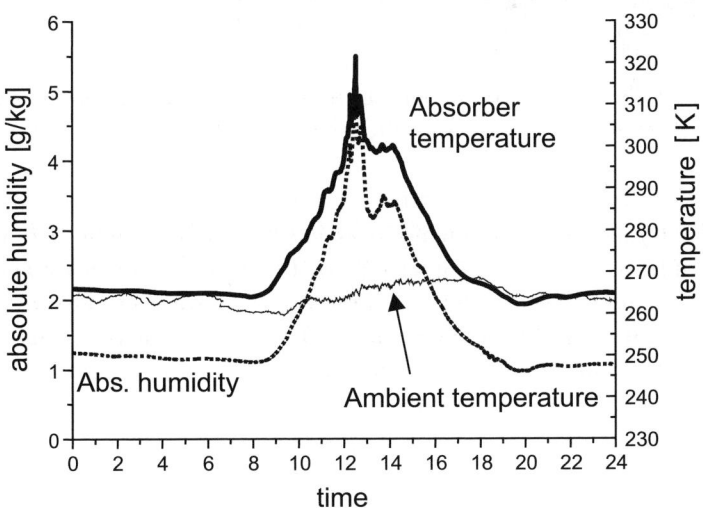

Figure 3.4-7. Measured humidity and temperature data in a tested collector during a 24-h period beginning at midnight.

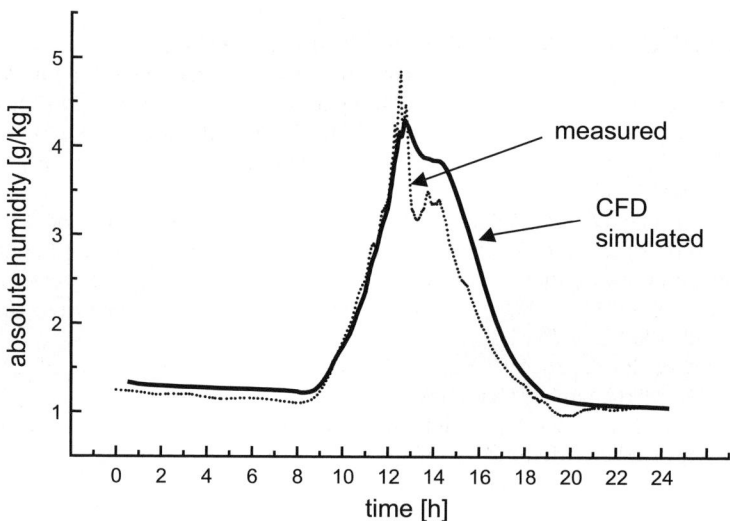

Figure 3.4-8.   Measured vs. simulated absolute humidity in a collector during a 24-h period beginning at midnight.

from $13\,g/kg$ to $28\,g/kg$ absolute humidity, and this increase has been actually observed on a summer day.

The adsorption theory of Dubinin (Sizmann *et al.*, 1988) was used to describe the effect of outgassing of sorbed water that was embedded in the simulation program in CFX-F3D. In addition, a model, which is based on Fick's law (Bird *et al.*, 1960) and that can be used to calculate the condensation of water vapor on surfaces inside a collector and its re-diffusion, was used to simulate an entire day where condensation was observed. The measured versus simulated data are shown in Figure 3.4-8. Condensation on the inside cover surface was observed.

In summary, models for the condensation of water vapor on the inside of collector surfaces and for the adsorption of water vapor in the insulation material proved their reliability for predicting these phenomena in realistic situations. By using proper boundary conditions and collector geometry, the absolute humidity and condensation can be predicted. The simulations can be used to optimize the size and position of the ventilation holes to achieve the goal of a dry microclimate inside solar collectors.

REFERENCES

Bird, R.B., Stewart, W.E. & Lightfoot, E.N. (1960) *Transport Phenomena*, John Wiley and Sons, New York, NY, USA.

Carlsson, B., Frei, U., Köhl, M. & Möller, K. (1994) Accelerated Life Testing of Solar Energy Materials – Case Study of Some Selective Solar Absorber Coatings for DHW Systems", International Energy Agency, Solar Heating and Cooling Programme Task X: Solar Materials Research and Development, Technical Report, SP- Report 1994:13.

Lund, H. (1985) The Test Reference Year (TRY), Commission of the European Communities, Report no. EUR 9765.

Lund, H. (1995) The Design Reference Year User Manual, A Report of Task 9: Solar Radiation and Pyranometer Studies, Solar Materials Research and Development, International Energy Agency Solar Heating and Cooling Programme, Report No. IEA-SHCP-9E-1, Report No. 274, Thermal Insulation Laboratory, Technical University of Denmark, DK 2800 Lyngby, Denmark.

MOMIC, software developed by Thermal Insulation Laboratory, Technical University of Denmark, DK 2800 Lyngby, Denmark.

Sizmann R. *et al.* (1988) Energiespeicherung in absorptiven Systemen, Anwendung zur Klimatisierung, Ludwig-Maximilians-Universität München, Deutschland, Nov. 1988.

Part 4

Durability Assessment and Service Lifetime Prediction

Part 4 is concerned with systematic approaches to establishing a service life prediction of components and materials, as is indicated in the first chapter, so that all essential aspects of the problem will be taken into consideration. In the general approach to service life prediction, which is the focus in this book, predictive failure modes and effect analysis (FMEA) serves as the starting point for service life prediction from accelerated life test results. The proposed methodology includes three main steps: (a) initial risk analysis of potential failure modes (b) qualification and screening testing and analysis for service life prediction, and (c) service life prediction involving mathematical modeling and life testing. The second chapter on initial risk analysis describes the necessary steps of specifying the required function (s) of the component and its minimum acceptable performance, service life, and the anticipated in-use environment, identifying important functional properties and potential failure modes. The third chapter is on qualification testing as generally employed for verification of the environmental resistance of products. The fourth chapter is concerned with explaining the accelerated aging testing adopted for screening and life testing. In the fifth chapter, a systematic procedure is described for using some of the existing and successfully applied models for analysing accelerated test data for service life prediction. The sixth chapter, in contrast, is concerned with illustrating how a phenomenological approach can be accurately used to interpret the results from a number of highly accelerated exposure tests of polymeric materials. In the seventh chapter, the purposes and methodologies are discussed for using parallel outdoor exposure and accelerated indoor testing and typical results are given for polymeric reflectors. In the last chapter, a brief overview is presented about the most widely used modern techniques for analysing optical materials. The techniques, as well as other physical measurements, are essential for establishing the rate controlling degradation or failure mechanisms that are needed for making a service lifetime prediction.

Chapter 4.1
General Methodology

B. Carlsson

Swedish National Testing and Research Institute,
P.O. Box 857, S-501 15 Borås, Sweden

Abstract: The purpose of this chapter is to give an overview of all the various activities that are needed for the assessment of the expected service life of a component and its materials in a given application. Systematic approaches to service life prediction of components and materials are needed so that all essential aspects of the problem will be taken into consideration. In almost all existing systematic methodologies for service life prediction, four basic themes appear: (a) performance analysis; (b) failure analysis; (c) laboratory aging testing; and (d) mathematical modeling for service life prediction. In one such general approach to service life prediction, which is the focus in this book, predictive failure modes and effect analysis (FMEA) serves as the starting point for service life prediction from accelerated life test results. The proposed methodology includes three main steps: (a) initial risk analysis of potential failure modes; (b) qualification and screening testing and analysis for service life prediction; and (c) service life prediction involving mathematical modeling and life testing. In the last step also procedures for validating predicted service life from long-term in-service testing are also needed.

Keywords: Service life prediction, General methodology for accelerated life testing, Failure modes and effect analysis (FMEA), Initial risk analysis, Qualification testing, Screening testing, Life testing, Performance analysis.

4.1.1   GENERAL METHODOLOGY

Many efforts have been made to develop systematic approaches to service life prediction of components, parts of components and materials so that all essential aspects of the problem will be taken into consideration (Gaines *et al.*, 1977; Frohnsdorf and Masters, 1980; Eurin *et al.*, 1985; Martin and McKnight, 1985; Sjöström, 1985; Masters and Brandt, 1987, 1989; Carlsson, 1988; Carlsson *et al.*, 1994; Martin *et al.*, 1994; ISO 15686, 2000). The requirements for such a methodology has been summarized (Sjöström, 1985): (1) it must be generic; (2) it must result in the identification of all data needed for service life prediction, e.g., in-service environmental degradation factors, possible degradation mechanisms of the material or component, quantitative performance requirements, internal maintenance methods, design features, etc.; (3) it must be based upon the use of reliable test methods or feedback data, e.g., all tests must be designed for and be of relevance to requirements dictated by item 2 above; and (4) it must provide guidance for data interpretation. Suitable methods and tools for this purpose, e.g., mathematical models, must be specified.

In almost all existing systematic methodologies for service life prediction based on accelerated life testing, four basic themes appear (Carlsson, 1988): (a) performance analysis; (b) failure analysis; (c) laboratory aging testing; and (d) mathematical modeling for service life prediction.

Performance analysis includes such elements as defining in-use performance requirements, formulating the service life requirement in terms of some functional property of the material and characterization of environmental influence on the material under service conditions which might contribute to degradation in material performance.

Failure analysis means finding interrelationships between deterioration in functional and the physical and chemical properties of materials resulting from environmental degradation factors. It may preferably comprise studies of morphological and compositional changes induced by artificial aging and theoretical interpretations in terms of dominant mechanisms of degradation of materials.

Laboratory aging testing includes evaluating the influence from different degradation factors on material durability. Tests used may simulate the effect of a single degradation factor or the simultaneous action of many degradation factors. Tests can be conducted under a constant environmental stress load or under a cyclic stress load. To ensure that the accelerated test methods will be of relevance for the prevailing in-use requirements, it is sometimes recommended to carry out long-term tests under in-service conditions in parallel with the accelerated tests. If the results from the two types of tests are not in qualitative agreement with respect to observed changes, the accelerated test is considered irrelevant.

Mathematical modeling for service life prediction means the following: (1) defining the performance requirement for the accelerated test in terms of a performance level for at least one measurable physical or chemical property of the material; (2) finding a numerical expression that relates the change in the material property selected to the environmental stress factors contributing to the degradation of the material; and (3) characterizing the environmental stress factors under service conditions to be able to extrapolate the results of the accelerated tests to service life.

In one general methodology for service life prediction, which is a focus of this book, predictive failure modes and effect analysis (FMEA) serves as the starting point for service life prediction from accelerated life test results as is illustrated in Table 4.1-1. The analysis is made at the component level. Table 4.1-1 is based on a similar scheme developed for the purpose of accelerated life testing of selective solar absorber surfaces in a joint case study of Task 10 of the IEA Solar Heating and Cooling Programme (Carlsson *et al.*, 1994).

Table 4.1-1. Failure mode analysis for planning accelerated lifetime tests for service life prediction. The key words used to describe the different steps in the initial risk analysis of potential failure modes originate from a methodology adopted for accelerated life testing of photovoltaic modules and is adapted from the work of Gaines *et al.*

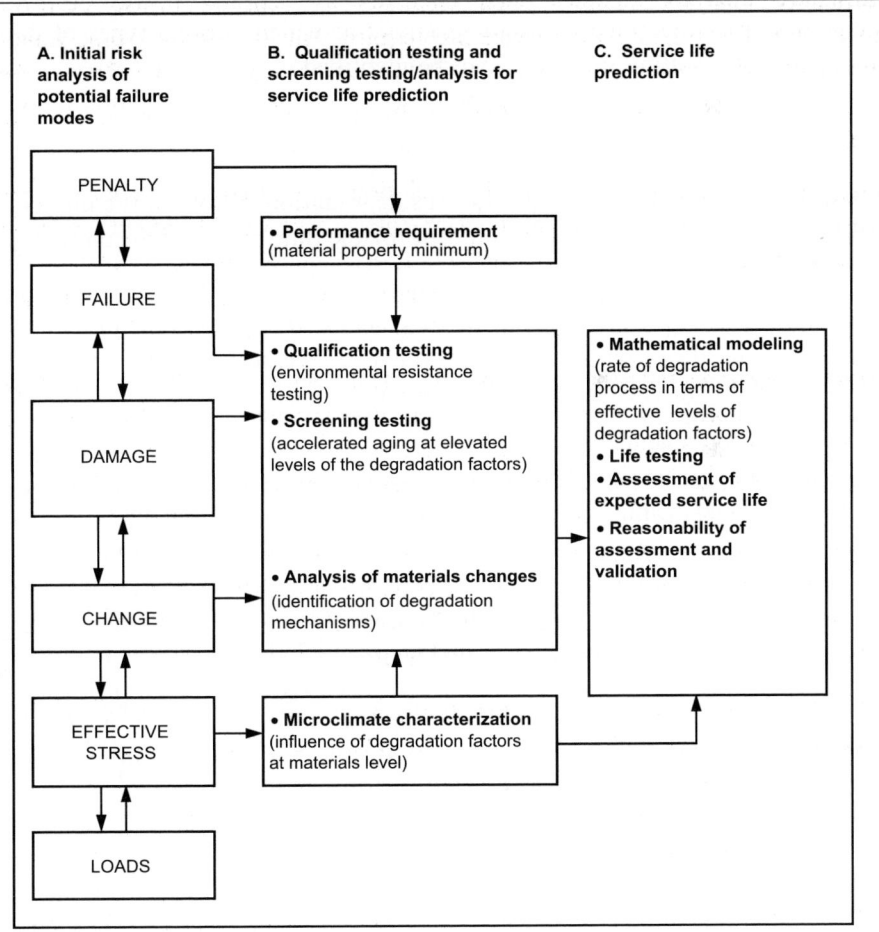

Penalty is the level at which an assessment is made of the economic effects of a component failure. Based on this assumption, it is possible to establish the minimum performance expectations that must be maintained for a given number of years.

Failure is based on the performance requirements of the components. If the requirements are not fulfilled, the particular component or part of the component is regarded as having failed. Failure in this sense accordingly means unacceptable functional performance. If the component cannot perform the design function at all, the failure may be classified as a mortal failure. Performance requirements can be formulated on the basis of optical properties, mechanical strength, aesthetic values, or other criteria related to the performance of the component and its materials.

Damage describes the stage of failure analysis at which various types of damage, each capable of resulting in failure, can be identified.

Change is related to any modification in the materials composition or structure that results in damage of the type previously identified.

Effective Stress is the level at which various degradation factors in the microclimate can be identified, which could be significant for the durability of the component and its materials. It is essential to characterize the stress levels quantitatively.

Loads, finally, is the level that describes the macro-environmental conditions (climatic, chemical, mechanical), and which is therefore a starting point for description of the microclimate or effective stress.

Each step in the scheme on the left-hand side of Table 4.1-1 may be related to the subsequent step by an appropriate deterministic or statistical relationship. The relationship should define the expected results of all the various activities involved in accelerated life testing, as indicated on the right-hand side of Table 4.1-1.

The proposed methodology includes three principal steps, which will be more fully explained in the subsequent chapters of this part of the book. Initial risk analysis of potential failure modes is discussed in Chapter 4.2. Qualification and screening testing and analysis for service life prediction are discussed in Chapter 4.3 (the characterization of microclimates is treated in Part 3 of this book). Service life prediction involving mathematical modeling and life testing is presented in Chapters 4.5 and 4.6. Long-term tests for validating the predicted service life are described in Chapter 4.7. In the cited chapters, examples are also given on how the methodology can be applied for the service life assessment of specific categories of materials such as selected solar absorber surfaces. Case studies about lifetime testing of polymeric glazing materials and reflectors are presented in Chapters 6.1 and 6.2.

REFERENCES

Carlsson, B. (Ed.) (1988) Survey of Service Life Prediction Methods for Materials in Solar Heating and Cooling, *International Energy Agency*, Solar Heating and Cooling Programme Task X: Solar Energy Materials Research and Development, Technical Report, Swedish Council for Building Research Document D16:1988, Stockholm, Sweden.

Carlsson, B., Frei, U., Köhl, M. & Möller, K. (1994) Accelerated Life Testing of Solar Energy Materials - Case Study of Some Selective Solar Absorber Coatings for DHW Systems, International Energy Agency, Solar Heating and Cooling Programme Task X: Solar Materials Research and Development, Technical Report, SP- Report 1994:13, Borås, Sweden, ISBN 91-7848-472-3.

Eurin, Ph., Marechal, J.Ch. & Cope, R. (1985) Barriers to the Prediction of Service Life of Polymer Material, In: Masters, L.M. (Eds.), *Problems in Service Life Prediction of Building and Construction Materials*, NATO ASI Series E No. 95, p. 21, Martinus Nijhoff Publishers, Dordrecht, The Netherlands.

Frohnsdorf, G. & Masters, L.W. (1980) In: Sereda & Litvan (Eds.), *Durability of Building Materials and Components, ASTM STP 691*, pp. 17–30, West Conshohocken, Pennsylvania, USA.

Gaines, G.B., Thomas, R.E., Derringer, G.C., Kistler, C.W., Brigg, D.M. & Carmichael, D.C. (1977) *Methodology for Designing Accelerated Aging Tests for Predicting Life of Photovolatic Arrays*, Battelle Columbus Laboratories, Final report ERDA/JPL-954328-77/1, Columbus, OH, USA.

ISO 15686 Building and Constructed Assets – Service Life Planning – Part 1: General Principles (2000), Part 2: Service life prediction procedures (2001), Part 3: Performance audits and reviews (2002), ISO International Standardization Organization, http://www.iso.ch, CH -1211 Geneva, Switzerland.

Martin, J.W. & McKnight, M. (1985) Prediction of the Service Life of Coatings on Steel, *Journal of Coating Technology*, **57**, 724.

Martin, J.W., Saunders S.C., Floyd, L.F. & Wineburg, J.P. (1994) Methodologies for Predicting the Service Lives of Coating Systems; NIST Building Sciences Series 172, Gaithersburg, MD, USA (available from National Technical Information Service, order number PB95-146387).

Masters, L. & Brandt, E. (1987) CIB W 80/RILEM 71-PSL, *Prediction of Service Life of Building Materials, Materials and Structures*, **20**, 155.

Masters, L.W. & Brandt, E. (1989) *Systematic Methodology for Service Life Prediction; Materials and Structures*, **22**, 385–392.

Sjöström, C. (1985) Overview of Methodologies for Prediction of Service Life, In: Masters, L. (Ed.), *Problems in Service Life Prediction*, NATO ASI Series E No. 95, p. 3, Martinus Nijhoff Publishers, Dordrecht, The Netherlands.

Chapter 4.2

Initial Risk Analysis of Potential Failure Modes

B. Carlsson

Swedish National Testing and Research Institute,
P.O. Box 857, S-501 15 Borås, Sweden

Abstract: The purpose of this chapter is to describe the important elements of an initial risk analysis of potential failure modes of a component and its materials for a given application. In this analysis all the requirements related to the performance and the durability for the intended application are defined. From a practical point of view, as well as from an economic viewpoint, an assessment of the durability or service life of a product component has to be limited in its scope and focused on the most critical failure modes. An important part of the initial step in such an assessment is therefore estimating the risk associated with each of the potential failure modes of the component. The initial risk analysis work program includes: (a) specifying the required function(s) of the component and its minimum acceptable performance, service life, and the anticipated in-use environment; (b) identifying important functional properties and methods for qualifying the performance of the component; (c) identifying potential failure modes, associated degradation mechanisms and methods for qualification of the component and its materials with respect to durability; and (d) estimating the risks associated with each different failure mode. The risk or risk number associated with each identified potential failure and associated damage mode may preferably be estimated by using the methodology of Failure Modes and Effect Analysis (FMEA). The estimated risk number is taken as the point of departure to judge whether a particular failure mode needs to be further evaluated or not. The estimated risk number is also a valuable tool to determine what kind of testing is needed for qualification of a particular component and its materials.

Keywords: Initial risk analysis, Potential failure modes, Performance requirements, Service life requirements, Functional properties, Degradation mechanisms, Performance test methods, Durability test methods, Fault-tree analysis, Failure modes and effect analysis (FMEA), Risk assessment degradation indicator

4.2.1 INTRODUCTION

Using the scheme illustrated in Table 4.2-1 the first step is to analyze the potential failure modes with the objective of obtaining (1) a list of potential failure modes of the component, and associated risks and critical component and material properties, degradation processes and stress factors; (2) a framework for the selection of test methods to verify performance and service life requirements; (3) a framework for describing previous test results for a specific component and its materials or a similar component and materials used in the component, and classifying their relevance to the actual application; and (4) a framework for compiling and integrating all data on available component and material properties and material degradation technology.

From a practical point of view, as well as from an economic viewpoint, an assessment of the durability or service life has to be limited in its scope and focused on the most critical failure modes. An important part of the initial step in such an assessment is therefore estimating the risk associated with each of the potential failure modes of the component.

The work program in the initial step of service life assessment may be structured into the following activities (Carlsson *et al.*, 1995): (a) for in-service use, specify the required function of the component and its materials, its minimum acceptable performance and service life requirement, and the anticipated in-use environments; (b) identify important functional properties defining the performance of the component and its materials, and the relevant test methods and requirements for qualifying the component to perform its designed function; (c) identify potential failure modes and degradation mechanisms, relevant durability or life tests and durability requirements for qualification of the component and its materials; (d) estimate the risks associated with each different failure mode and associated degradation mechanisms. The results of the initial risk analysis of potential failure modes and associated degradation mechanisms may be documented as indicated in Table 4.2-1.

Table 4.2-1. Examples of documention of the results from an initial risk analysis of potential failure modes

A. *Specification of end-user and product requirements*		
Function and general requirements	General requirements for long-term performance during design service time	In-use conditions and severity of environmental stress
B. *Specification of functional properties and requirements on component and its materials*		
Critical functional properties	Test methods for determining the functional properties	Requirements for functional capability and long-term performance
C. *Potential failure modes, critical factors of environmental stress and degradation*		
Failure/damage mode/ degradation process	degradation indicator	Critical factors of environmental stress/ degradation factors and severity

4.2.2 PENALTY AND FAILURE

The first activity is to specify the function of the component and service lifetime requirement in general terms from an end-user and product point of view, and from that to identify the most important functional properties of the component and its materials (see Table 4.2-2). The importance of the function of the component from an end-user and product point of view needs to be taken into consideration when formulating the performance requirements in terms of those functional properties.

For determining, the Penalty, it may be important to understand the consequences of different types of failure and to define the general performance requirements and service lifetime. Failure of a component occurs when a minimum performance level,

Table 4.2-2. Example of the results from an initial risk analysis of potential failure modes (based on the information taken from the IEA task 10 case study on selective solar absorber surfaces (Carlsson *et al.*, 1994))

A. Specification of end-user and product requirements on component

Function and general requirements	General requirements for long-term performance during design service time	In-use conditions and severity of environmental stress
Efficiently convert solar radiation into thermal energy Suppress heat losses in the form of thermal radiation	Loss in optical performance should not result in reduction of the solar system energy performance (solar fraction) by more than 5%, in a relative sense, during a design service time of 25 years	Absorber is in contact with air Air is exchanged with surroundings, so airborne pollutants may contaminate the absorber Humidity in the absorber contact air may be high if assembly is not water(rain) tight Maximum temperature permitted is 200°C

B. Specification of functional properties and requirements on component and its materials

Critical functional properties	Test method for determining functional property	Requirement for functional capability and long-term performance
		Functional capability
Solar absorptance (α)	ISO CD 12592.2	$\alpha > 0.92$
Thermal emittance (ε)	ISO CD 12592.2	$\varepsilon < 0.15$
Adhesion of the absorber to its substrate (ad)	ISO 4624	$ad > 0.5\,\mathrm{MPa}$
		Long-term performance $-\Delta\alpha + 0.25,\ \Delta\varepsilon \leq 0.05$

at which satisfactory functioning cannot be guaranteed, is no longer obtained. Thus, if the performance expectations are not fulfilled, the particular component has failed. Performance requirements can be formulated on the basis of optical properties, mechanical strength, aesthetic values, or other criteria related to the performance of the component and its materials. For failure modes characterized by a gradual degradation in performance, the consequences of failure may not be significant shortly after the minimum performance requirement is no longer met. For catastrophic types of failure modes, however, the intended functional capability of the component or some part of the component may be completely lost. Defining the performance requirements should be accompanied by an assessment of the economic effects of a component failure at that level of performance. Based on these requirements, a mean service lifetime may be defined. Alternatively, a minimum reliability expectation may be defined that must be maintained for a selected number of years (see Figure 4.5-1 in Chapter 4.5).

An example from the IEA Task 10 case study on selective solar absorbers (Carlsson *et al.*, 1994) of how these first steps in the initial risk analysis of potential failure modes may be documented is given in Table 4.2-2 (see also Figure 4.2-1).

4.2.3   FAILURE, DAMAGE, CHANGE, AND DEGRADATION INDICATORS

Potential failure modes and important degradation processes should be identified after failures have been defined in terms of the minimum acceptable performance levels. When identifying potential failure modes, it is important to distinguish between (a) failures initiated by the short-term influence of exposure to environmental stresses for which the latter represents events of high-environmental loads on the component and its materials, and (b) failures initiated by the long-term influence of exposure to environmental stresses in which the latter results in material degradation so that the performance and sometimes also the environmental resistance of the component and its materials gradually decrease. In case (a), catastrophic failures occur, whereas in case (b), materials degradation may result not only in gradual types of failures but also in catastrophic types of failures.

In general, many kinds of failure modes exist for a particular component and even the different parts of the component and the different damage mechanisms, which may lead to the same kind of failure, can sometimes be quite numerous. An overview of common failure mechanisms experienced by materials has been given (Dasgupta and Pecht, 1991). Discussions of specific mechanisms are also available for corrosion (Tullmin and Roberge, 1995), electromigration (Black, 1969; DiGiacomo, 1982; Nitta *et al.*, 1993; Young and Christou, 1994; Krumbein, 1995), diffusion (Li and Dasgupta, 1994), and photodegradation (Al-Sheikhly and Christou, 1994).

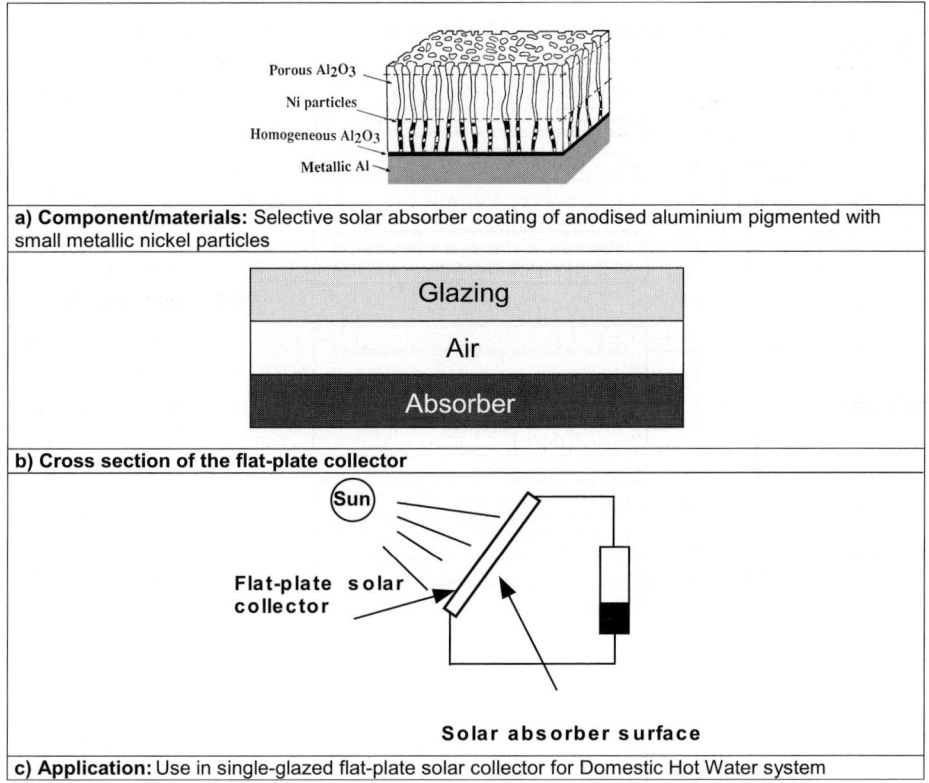

Figure 4.2.1. Principal components of the selective solar absorber system studied in IEA Task 10 for use as single glazed flat-plate solar collectors to be installed in domestic hot water systems (Carlsson *et al.*, 1994).

Fault-tree analysis is a tool that provides a logical structure relating failure to various damage modes and the underlying chemical or physical changes. The use of fault-tree analysis to describe the potential degradation pattern of an organic coating system for metals is illustrated in Table 4.2-3.

The objective of fault-tree analysis is to identify potential failure and damage modes, associated degradation mechanisms or mechanisms that result in material degradation and damage, and the associated critical factors of environmental stress or degradation factors.

For the purpose of service life prediction, it is important also to select suitable degradation indicators for the different potential failure modes so that failure may be assessed properly. If possible, the degradation process responsible for each critical failure may also be followed (see the example in Table 4.2-3).

Table 4.2-3. Example of using a fault-tree analysis to represent potential relationships of failures, damage, and change to identify suitable degradation indicators and critical environmental stress/degradation factors for organic coatings on metals (Carlsson *et al.*, 1995)

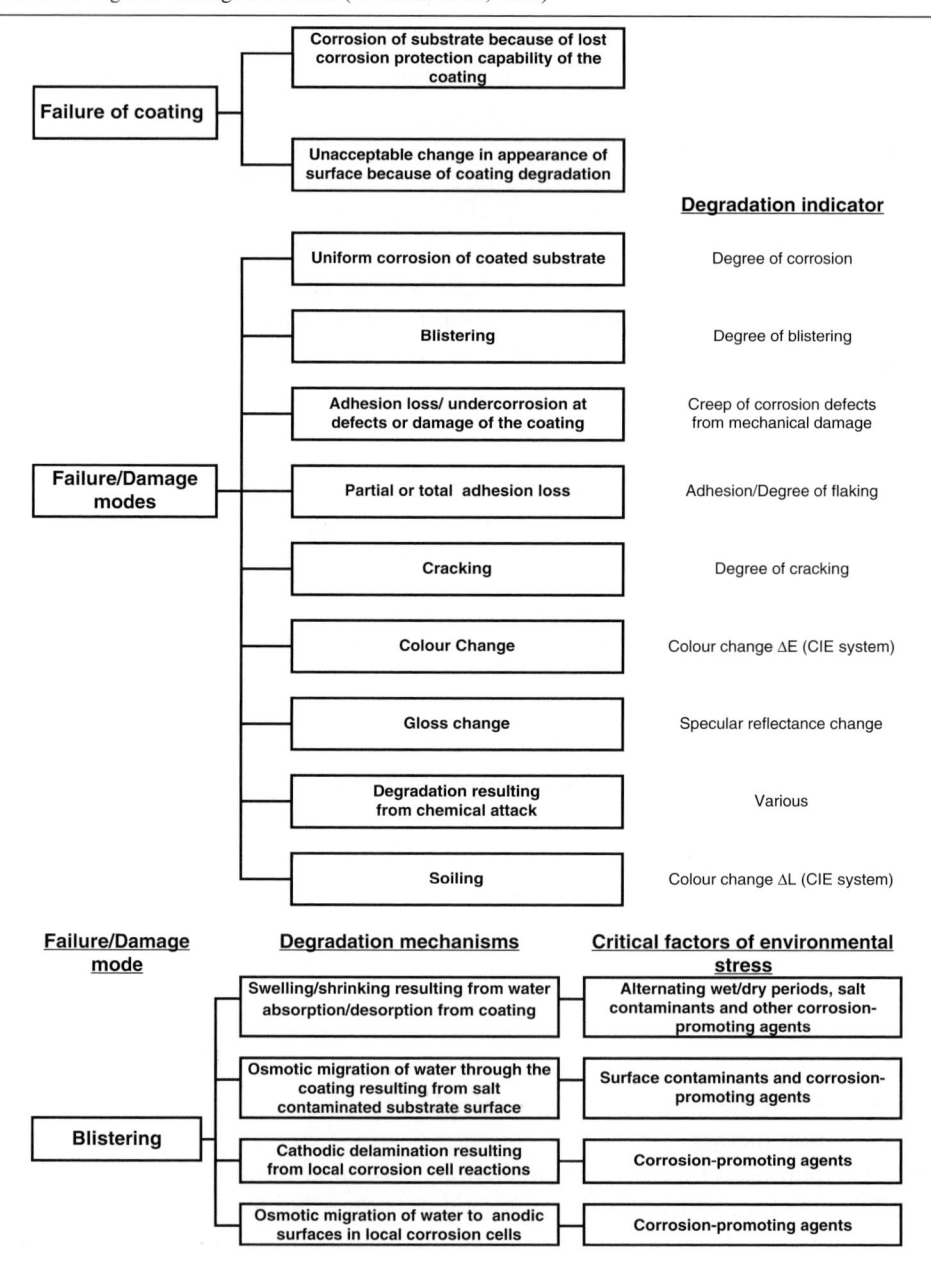

4.2.4 LOAD, EFFECTIVE STRESS, AND DEGRADATION FACTORS

To identify potential failure modes, the type of environmental stress factors and their severity under service conditions must be known. For the purpose of service life prediction, in-use conditions representing a worst case may be selected. Alternatively, in-use conditions may be determined by measuring the environmental stresses under varying service conditions and selecting data from the most representative case as a basis for service life prediction. In the initial phase of service life prediction, however, the most important issue is to identify the most critical in-use conditions and environmental stress factors that may contribute to material degradation and result in failure. Potential degradation mechanisms, failure and damage modes are identified on the basis of this knowledge (see Table 4.2-3).

A literature research is recommended after potential failure modes, degradation processes and critical factors of environmental stress have been initially identified. The objective of this search is to find reports on durability and available service life data on similar components and materials used in a similar application as those being investigated. Alternatively, durability or service life for the specific component and materials of the component in other applications and in-use environments as the investigated object are also sought (see Table 4.2-4).

4.2.5 RISK ANALYSIS ACCORDING TO FAILURE MODES AND EFFECT ANALYSIS

The risk or risk number associated with each identified potential failure and associated damage mode can be estimated by use of the methodology of Failure Modes and Effect Analysis (FMEA) in a simplified way; see IEC Standard (1985) and Failure Mode and Effect Analysis (FMEA) (1993) for reviews of the FMEA methodology. The estimated risk number is taken as the point of departure to judge whether a particular failure mode needs to be further evaluated or not. The estimated risk number may also be used to determine what kind of testing is needed for qualification of a particular component and its materials.

The risk number associated with a particular failure mode is estimated by using the following factors: Severity (S), Probability of occurrence (P_O), and Probability of escaping detection (P_D). The risk number RPN is the product of all these factors,

Table 4.2-4. Examples of docuemention of the available service life data of relevance to the investigated case

Component/Materials	Available service life data	Remark

Table 4.2-5. Rating scales for FMEA analysis with respect to servity, probability of escaping detection, and probability of occurence in the assessment of the risk numer for a particular failure/damage mode. An example is given in the lower table for organic coatings on metals (see Table 4.2-3)

Severity	RPN	Probability of detection	RPN	Probability of occurrence	RPN
No effect on product	1	Failure that always is noted. Probability for detection > 99.99%	1	Unlikely that failure will occur	1
Minor effect on product but no effect on product function	2-3	Normal probability of detection 99.7%	2-4	Very low probability for failure to occur	2-3
Risk of failure in product function	4-6	Certain probability of detection >95%	5-7	Low probability for failure	4-5
Certain failure in product functioning	7-9	Low probability of detection >90%	8-9	Moderate probability for failure to occur	6-7
Failure that may affect personal safety	10	Failures will not be found - cannot be tested	10	High probability for failure to occur	8-9
				Very high probability for failure to occur	10

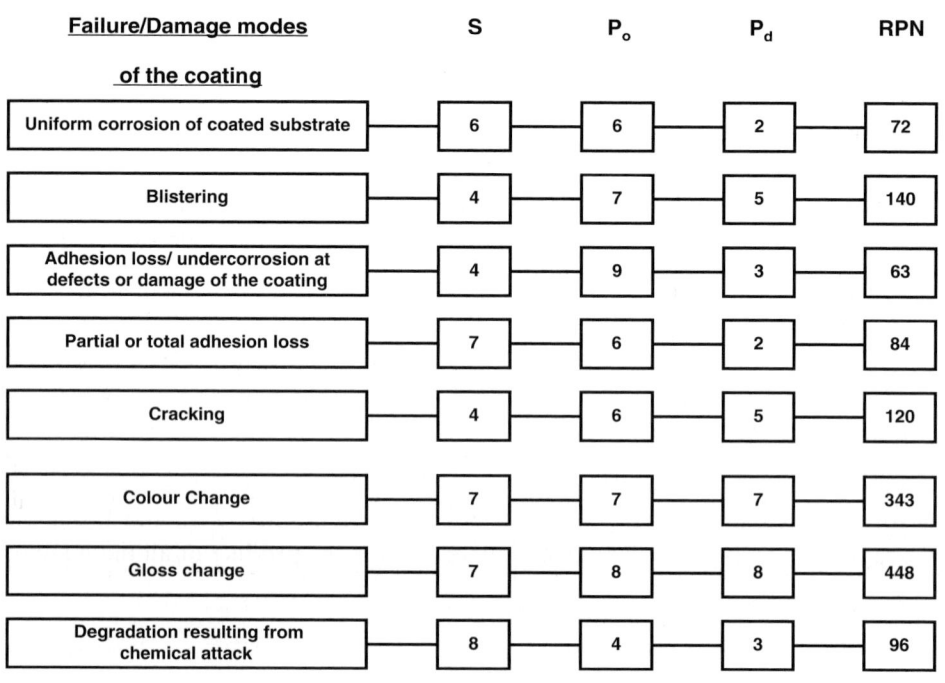

Failure/Damage modes of the coating	S	P_o	P_d	RPN
Uniform corrosion of coated substrate	6	6	2	72
Blistering	4	7	5	140
Adhesion loss/ undercorrosion at defects or damage of the coating	4	9	3	63
Partial or total adhesion loss	7	6	2	84
Cracking	4	6	5	120
Colour Change	7	7	7	343
Gloss change	7	8	8	448
Degradation resulting from chemical attack	8	4	3	96

Table 4.2-6. Risk assessment of potential failure modes

Failure mode/degradation process	Severity (rating number)	Probability of occurrence (rating number)	Probability of discovery (rating number)	Rating number for risk

Table 4.2-7. Example of the results of an initial risk analysis with information from the IEA task 10 case study on selective solar absorber surfaces (see also Tables 4.2-2 and 4.2-5)

C. *Service reliability/service life, cost of failures and maintenance*

Failure/damage mode/degradation process	Degradation indicator	Critical factors of environmental stress/degradation factors and severity	Estimated risk of the failure/damage mode from FMEA (see Figure 4.2-2)			
Unacceptable loss in optical performance	PC[1] Adhesion		S	P_O	P_D	Risk RPN
(A) High-temperature oxidation of metallic nickel	Reflectance spectrum Vis-IR	High temperature	7	2	8	112
(B) Electrochemical corrosion of metallic nickel	Reflectance spectrum Vis-IR	High humidity, sulphur dioxide (atmospheric corrosivity)	7	5	5^2	175
(C) Hydratization of aluminum oxide	Reflectance spectrum IR	Condensed water, temperature	7	7	4^2	196

[1]PC = performance criteria = $-\Delta\alpha + 0.25\Delta\varepsilon$.
[2]P_D value resulted from possible failure of the glazing.

i.e., $RPN = S \cdot P_O \cdot P_D$. The first factor, **Severity**, is a measure of the consequences of a particular failure for safety and economic reasons, when the component and its materials are treated as part of a product or system. For rating Severity, a scale with ten degrees may be used as defined in Table 4.2-5. The second factor, **Probability of occurrence**, is a measure of how probable it is that failure will occur according to the particular mode during the design service life of the component and its materials. A ten-degree scale may also be used for rating, as defined in Table 4.2-5. The third factor, **Probability of escaping detection,** accounts for the probability that a damage or failure mechanism will escape detection that could have prevented failure. The ten-degree scale may also be used here, as defined in Table 4.2-5.

The risk assessment may be documented as shown in Table 4.2-6. In the risk assessment of potential failure modes, the relevance of durability and life data found in the literature for the specific component and its materials must be considered. The risk assessment is most advantageously performed by a group of experts.

The reasonableness of setting the design service life of the component or parts of the component at the same level as that of the product may also be questioned during the risk assessment. During maintenance or repair work, the component or parts of the component may be replaced, which may considerably lower the requirement on the service life of the component or some of its replaceable parts. The risk assessment made at this initial stage of service life prediction is only qualitative in nature. The main purpose of the risk assessment is to limit the scope of the service life evaluation to focus on the most important failure modes. The rating numbers may be used principally as an aid to reduce the number of critical failure modes in the subsequent evaluation of the service life of the component. An example of the result of an initial risk analysis based on information from the IEA Task 10 absorber surface case study referred to in Table 4.2-2 is shown in Table 4.2-7.

REFERENCES

Al-Sheikhly, M. & Christou, A. (1994 December) How Radiation Affects Polymeric Materials, *IEEE Trans. Reliability*, **43**, 551–556.
Black, J.R. (1969 April) Electromigration – A Brief Survey and Some Recent Results, *IEEE Trans. Electron Devices*, **ED-16**, 338–347.
Carlsson, B., Berglund Åhman, A. & Jutengren, K. (1995) Assessment of Service Life by Accelerated Testing – Methodology for Qualification of Rust Protective Paint Systems; SP Swedish National Testing and Research Institute, SP-Report 1995:65, ISBN 91-7848-592-4 (Swedish), SE-50115 Borås, Sweden.
Carlsson, B., Frei, U., Köhl, M. & Möller, K. (1994) Accelerated Life Testing of Solar Energy Materials - Case Study of Some Selective Solar Absorber Coatings for DHW systems, International Energy Agency, Solar Heating and Cooling Programme Task X: Solar Materials Research and Development, Technical Report, SP- Report 1994:13, SE-50115 Borås, Sweden.
Dasgupta, A. & Pecht, M. (1991) Material Failure-mechanisms and Damage Models, *IEEE Trans. Reliability*, **40**, 531–536.
DiGiacomo (1982) Metal Migration (Ag, Cu, Pb) in Encapsulated Modules and Time-to-fail Model as a Function of the Environment and Package Properties, *IEEE Proc. Int'l Reliability Physics Symp.*, 27–33.
Failure Mode and Effect Analysis (1993) Instruction manual from Volvo Car Corporation, In: Britsman, C., Lönnqvist, Å. & Ottosson, S.O., Failure Mode and Effect Analysis, Ord&Form AB, ISBN 91-7548-317-3, Stockholm, Sweden.
IEC Standard, Publ. No.812 (1985) Analysis Techniques for System Reliability – Procedure for Failure Mode and Effect Analysis (FMEA), CH-1211 Geneva 20, Switzerland.
Krumbein, S.J. (1995 December) Electrolytic Models for Metallic Electromigration Failure Mechanisms, *IEEE Trans. Reliability*, **44**, 539–549.
Li & Dasgupta, A. (1994, March) Failure-mechanism Models for Material Aging due to Interdiffusion, *EEE Trans. Reliability*, **43**, 2–10.

Nitta, T. *et al.* (1993 April) Evaluating the Large Electromigration Resistance of Copper Interconnects Employing a Newly Developed Accelerated Life-test Method, *J. Electrochem. Soc.*, **140**, 1131–1137.

Tullmin, M. & Roberge, P.R. (1995 June) Corrosion of Metallic Materials, *IEEE Trans. Reliability*, **44**, 271–278.

Young, D. & Christou, A. (1994, June) Failure-mechanism Models for Electromigration, *IEEE Trans. Reliability*, **43**, 186–192.

Chapter 4.3
Qualification Testing

B. Carlsson

Swedish National Testing and Research Institute,
P.O. Box 857, S-501 15 Borås, Sweden

Abstract: The purpose of this chapter is to describe the methodology generally adopted for qualification testing of products and components with respect to environmental resistance. For assessing the durability of a product, component, or material, it is important to distinguish between (a) the resistance to the short-term influence of exposure to environmental stresses representing cases of high-environmental loads on the component and its materials, and (b) the resistance to the long-term influence of exposure to environmental stresses that may result in the gradual degradation of the materials. For characterization and specification of the severities of stresses in the first case, the standard IEC 60721 may be used as the starting point. In this standard the severity for environmental factors in different applications is given as the maximum value to which the product may be exposed. IEC has also published a standard on basic environmental testing procedures, IEC 60068-2, which only treats test methods. The environmental resistance of a product may be verified by using a number of available standard environmental resistance tests related to the action of mechanical, climatic, chemical, and biological environmental stress factors. Some of the tests can be applied directly for qualification of products. However, in the case of long-term influence to environmental stresses and gradual degradation of the materials, appropriate accelerated life tests have to be tailor-made for qualification testing and correspond to the expected in-service environmental conditions, but at enhanced stress level.

Keywords: Qualification testing, Environmental engineering, Severity of stress, Environmental resistance testing, Mechanical stress factors, Climatic stress factors, Chemical stress factors, Biological stress factors

LIST OF ACRONYMS

SEES	Swedish Environmental Engineering Society
IEC	International Electrotechnical Commission, http://www.iec.ch.
MIL STD	Department of Defense Standards, Philadelphia, PA 19111-5094, United States.
IEEE	Institute of Electrical and Electronics Engineers Inc., http://standards.ieee.org.

ISO   International Standardization Organization,
                       http://www.iso.ch.
SAE   Society of Automotive Engineers Standards,
                       http://www.sae.org.
SP   Swedish National Testing and Research Institute,
                       http://www.sp.se.
VDA   Verband der Automobilindustrie E.V., http://www.vda.de.

4.3.1 INTRODUCTION

The combination of imposed stresses and analysis of the failure or damage modes forms the basis for identifying suitable durability tests as summarized in Table 4.3-1 for the organic coating system on metal analysed as described in Table 4.2-3. Sometimes relevant durability tests for the application considered can be found and used directly for qualification testing as described in Table 4.3-1. However, durability tests for accelerated life testing usually have to be tailor-made, as we describe in more detail in this chapter.

4.3.2 QUALIFICATION TESTING

For assessing the durability of a product, component or material, it is important to distinguish between case (a) the resistance to the short-term influence of exposure to

Table 4.3-1. Identification and ranking of suitable durability tests for organic coatings on metals based on risk assessment of potential failure and damage modes from Table 4.2-3 (given in brackets is the risk number estimated for different failure/damage modes from Table 4.2-3)

Durability test methods		Failure/damage modes (RPN)
Weatherability test involving exposure to solar irradiation, moisture, temperature, and humidity changes, acid rain precipitation	⇒	• Cracking (120) • Colour change (343) • Gloss Change (448)
Wet/dry cycling test involving intermittent condensation on coating surface	⇒	• Blistering (140) • Total or partial adhesion loss (84)
Chemical compatibility test or series of test	⇒	• Degradation caused by chemical attack (96)
Corrosion test involving cyclic variation in humidity and exposure to corrosion-promoting agents	⇒	• Uniform corrosion of coated substrate (72) • Blistering (140) • Adhesion loss/under corrosion at damage of the coating (63)
Combined corrosion and stone chip test		• Adhesion loss/under corrosion at damage of the coating (63)

environmental stresses representing cases of high-environmental loads on the component and its materials and case (b) the resistance to the long-term influence of exposure to environmental stresses that may result in the gradual degradation of the materials. In (b), the performance and sometimes also the environmental resistance of the product, component, and materials gradually decrease. In case (a), catastrophic failures occur, whereas in case (b) gradual failures usually occur, but catastrophic failure may also occur.

4.3.2.1 Short-term Environmental Influence Characterized by High-environmental Loads

For qualification testing for case (a), gradual degradation in the material resulting from aging does not affect the results of the test. The general practice used in environmental resistance testing of products is recommended (see, for example, the SEES Environmental Engineering Handbook (Handbook in Environmental Engineering, 1998)).

For environmental factors, characterization and specification of the severities of stresses, the standard, IEC 600721 (IEC 60721-1), published by the International Electrochemical Commission for classification of environmental conditions is used as the starting point (see Tables 4.3-2 and 4.3-3).

The philosophy for using this environmental classification is that it is necessary to have a general reference frame when specifying the environmental loads on products. The IEC 60721 is commonly used for the environmental classification. IEC has also published a standard on basic environmental testing procedures, IEC 60068-2 (IEC 60068-2), which only treats test methods.

The objective of using IEC 60721 is to assess the requirements for the environmental resistance of products and for the establishment of the environmental life-cycle profile for a product. However, in most cases, it is necessary to perform a careful tailoring process to correctly specify the severity of the environmental stresses. For further guidance, a methodology for the tailoring

---

Table 4.3-2. classification of environmental conditions according to IEC 60721

IEC 60721-1 Classification of environmental parameters and their severities
Introduction to the standard
IEC 60721-2 Environmental conditions appearing in nature
Temperature and humidity, precipitation and wind, air pressure, solar radiation and
temperature, dust, sand, salt mist/wind, earthquake vibrations and shocks, fauna and flora
IEC 60721-3 Classification of groups of environmental parameters and their severities
Storage, transportation, stationary use at weather protected locations, stationary use at weather
exposed locations, ground vehicle installations, ship environment, portable, and nonstationary use

Table 4.3-3. Classification of environmental stress according to the three-character code of IEC 60721-3

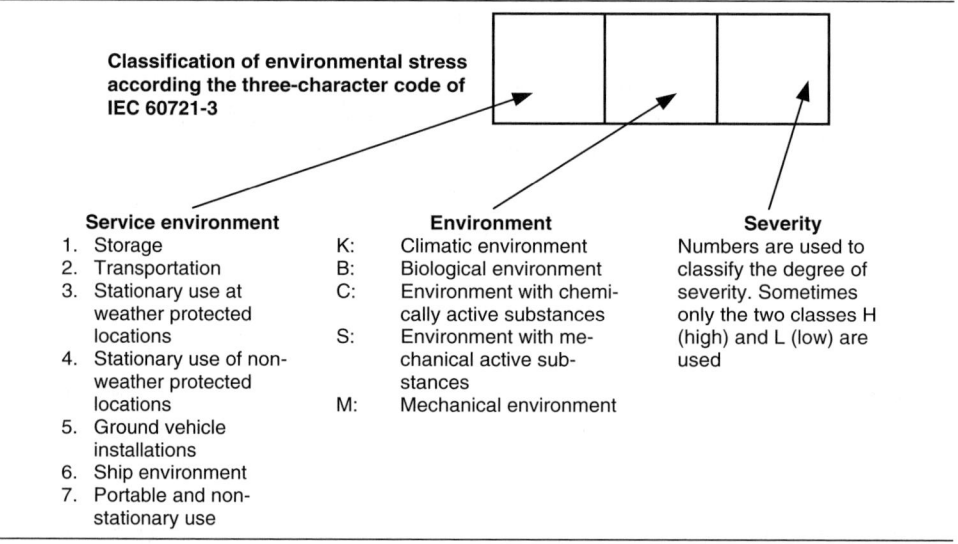

Classification of environmental stress according the three-character code of IEC 60721-3	

Service environment	Environment	Severity
1. Storage	K: Climatic environment	Numbers are used to
2. Transportation	B: Biological environment	classify the degree of
3. Stationary use at weather protected locations	C: Environment with chemically active substances	severity. Sometimes only the two classes H (high) and L (low) are
4. Stationary use of non-weather protected locations	S: Environment with mechanical active substances	used
5. Ground vehicle installations	M: Mechanical environment	
6. Ship environment		
7. Portable and non-stationary use		

process and the relevant test methods are described in American MIL-STD 810 (MIL STD 810E, 1989).

In IEC 60721, the severity for each environmental factor is the maximum value to which the product may be exposed. The severity generally represents conditions that will be exceeded with 1% probability. The classification does not provide information about the duration or statistical distribution of the stress factor. Frequently, the severity for an accelerated test cannot be selected from the values given in the standard. Therefore, the standard can only be used as a guide for determining product resistance to the short-term environmental influence of extreme events.

The environmental resistance of a product may be verified by using a number of available standard environmental resistance tests as listed in Table 4.3-4. The content of Table 4.3-4 has been extracted mainly from the information given in the SEES Environmental Engineering Handbook (Handbook in Environmental Engineering, 1998). Most tests given in the table are from the standard IEC 60068-2 or from the American MIL-STD 810 (MIL STD 810E, 1989). Some of the tests given, however, cannot be applied directly for testing at the component or material scale, but the table illustrates the large number of tests that are sometimes needed for assessing the environmental resistance of a product. The IEC technical committee, TC 104, "Environmental Conditions, Classification and Methods of Test" is developing the processes for transforming the condition classes of IEC 60721-3

Table 4.3-4. Some standard tests for assessing environmental resistance of products against environmental stress factors

Environmental factor	Specification	Examples of standard test methods (Standardization bodies)[1]	Examples of possible damage modes/mechanisms
Mechanical			
• Mechanical sine vibration	Direction for verification, mounting of test object, frequency, amplitude and duration should be specified.	(a) IEC 60068-2-6, Test Fc: Vibration (sinusoidal) (b) MIL-STD-810E, Method 514.4 "Vibration"	Fracture Permanent deformation Fatigue Wear
• Mechanical random vibration	Acceleration density as a function of frequency, vibration level where function is required, duration of test.	(a) MIL STD-810 E, Method 514.4 "Vibration" (b) ISO 2041, Vibration and shock	Crack initiation, crack growth Material loss (wear) Plastic flow Creep
• Mechanical random sounds	Sound level, frequency contents and duration of test. Type of sound field, measuring and analysis methods should also be specified.	(a) MIL STD 810 E Method 515.4 "Acoustic noise" (b) IEC 68 Test Fg: Vibration acoustically induced	Injuries and discomfort to people from high air pressure levels Vibration damages in mechanical structures resulting from sound waves
• Mechanical shock waves	Maximum pressure of the shock pulse, its duration and number of pulses	(a) IEC 60068-2-27, Test Ea: Shock (b) MIL STD 810 E, Method 516.4 "Shock"	Mechanical damage resulting from the pressure of the shock wave on the surface of product Vibration damage in mechanical structures resulting from the shock wave
• Earthquake	Acceleration level, frequency contents, duration, number of earth-quakes the product must withstand	(a) IEEE STD 344, Recommended practice for seismic qualification of nuclear power generating system	Mechanical (effect of high amplitude of single load, fatigue resulting from repeated load cycles)

(Continued)

Table 4.3-4. Continued

Environmental factor	Specification	Examples of standard test methods (Standardization bodies)[1]	Examples of possible damage modes/ mechanisms
• Shock in vehicles	Response spectrum of shock. If the test is carried out with a standard pulse its peak acceleration and duration should be stated	(b) IEEE STD 501, Seismic testing of relays (c) IEEE STD 693, Recommended practice for seismic design of substations (a) IEC 60068-2-27, Test E a: Shock (b) MIL STD 810 E, Method 516.4 "Shock"	Fatigue or wear Instantaneous fracture Plastic deformation
• Linear acceleration	Acceleration values in different directions must be stated as well as duration.	(a) IEC 60068-2-7 Test Ga: Acceleration, Steady state (b) MIL-STD-8100 Method 515.3 "Acceleration"	Fracture Plastic deformation (plastic flow)
• Angular acceleration	Acceleration magnitude and duration, its direction relative to the test object.	(a) IEC 60068-2-7, Test Ga: Acceleration, Steady state (b) MIL-STD-810 D, Method 515.3 "Acceleration"	Fracture Plastic deformation (plastic flow)
• Free fall	Fall height, orientation of test object, if the object shall be packaged or not during the test	(a) IEC 60068-2-32, Test Ed: Free fall (b) MIL-STD 331 A, Tests 103.1,111.1, and 117 (c) DEF STAN07-55, Test A9 "Free fall"	Mechanical (fracture, permanent deformation)
• Drop and topple connected to bench work	See free fall	(a) IEC 60068-2-31 Test Ec: Drop and topple, primarily for equipment-type specimen	Mechanical (fracture, permanent deformation)

Test	Standard	Parameters	Effects
• Shake	(b) MIL-STD-810 E, Method 516.4 Shock: Procedure IV "Transient drop", Procedure VI "Bench handling" (a) MIL-STD 810 E Method 516.4 "Shock" (b) IEC 60068-2-29, Test E b: Bump	Mounting of the test object in the shaker, mode of operation of test object during test	Fracture Permanent deformation Fatigue
• Bounce	(a) MIL-STD 810 E, Method 514.4 "Vibration; Category 3 Loose cargo transport" (b) IEC 60068-2-55, Test Ee: Bounce	Mounting of the test object during test, if the object shall be packaged or not during the test, movement of the platform of the bounce machine.	Fracture Permanent deformation Fatigue
• Tumbling	(a) IEC 60068-2-32 Test Ed: Free fall (b) MIL-STD-331 Test 102 "Tumble"	Number of falls, see free fall	Mechanical (fracture, permanent deformation)
Climatic • Precipitation	(a) MIL-STD-810 E, Method 506.3 "Rain" (b) IEC 60068-2-18, Test R : Water	Rain: Rain intensity, drop size, duration, angle of impact Snow: The load which a snow cover may constitute for an object Hail: Specified only exceptionally Dew and frost: See "Moisture" and "Ice formation" (Composition of water should also be specified e.g., salts, pH)	Mechanical (permanent deformation) Erosion Chemical (see "Moisture") Malfunctions of electric components

(Continued)

Table 4.3-4. Continued

Environmental factor	Specification	Examples of standard test methods (Standardization bodies)[1]	Examples of possible damage modes/ mechanisms
• Moisture	Humidity and temperature, duration, alternatively number of cycles for cyclic tests	(a) MIL-STD-810 E, Method 507.3 "Humidity" (b) IEC 60068-2-3, Test Ca: Damp heat, Steady state (c) IEC 60068-2-30 Test Db: Damp heat, cyclic	Metals: Corrosion Polymeric materials: Fatigue because of swelling and shrinking resulting from water sorption/ desorption, chemical degradation caused by hydrolysis, chemical degradation resulting from extraction of stabilizers from material surface Malfunctions in electric function
• Water	Dripping water: Fall height, duration Spraying: Water flow, pressure, nozzle diameter, duration Immersion in water: Water depth, duration	(a) IEC 60068-2-18, Test R: Water (b) MIL-STD 810 E, Method 506.3 "Rain"	Most damage mechanisms mentioned under "Moisture" and "Precipitation" apply also to "Water"
• Ice formation	Degree of icing (thickness of ice coating), procedure to accomplish icing, e.g., rain/cooling, water spray/ cooling, duration	Build up of ice layers (a) MIL-STD 810 E, Method 521 "Icing/Freezing rain" (b) ISO 2653, Ice formation, Test C Frost appearance (c) IEC 60068-2-39, Z/AMD: Combined sequential cold, low air pressure and damp heat test	Mechanical resulting from volume expansion during freezing Mechanical because of increased load from an ice coating

• Temperature	Temperature histogram or effective mean temperature and duration, maximum and minimum temperatures	(a) IEC 60068-2, Part 2, Test B: Heat (b) MIL-STD 810 E, Method 501.3 "High temperature"	Mechanical because of thermal expansion and thermo mechanical stresses Physical resulting from phase change reactions or irreversible changes in material properties with temperature (physical ageing) Chemical (all chemical reactions are influenced by temperature)
• Temperature change	Temperatures and time characteristics of the temperature cycle, number of cycles in test	(a) IEC 60068-2-14, Test N: Change of temperature (b) MIL-STD 810 E, Method 503.3, "Temperature shock"	Fracture/cracking resulting from thermo-mechanical stress
• Solar irradiation	Spectrum of light source and intensity of light at the surface of the test object, surface temperature, time characteristics of the test cycle, and test duration.	(a) IEC 60068-2-5, Test Sa: Simulated solar radiation at ground level (b) IEC 60068-2-9, "Guidance for solar radiation testing" (c) MIL-STD 810 E, Method 505.3 "Solar radiation"	Chemical and physical changes resulting from an increased surface temperature, see "Temperature" Photochemical degradation caused mainly by UV radiation, which may result in deterioration of mechanical properties, optical properties (e.g. color, gloss) of materials.
• UV irradiation and outdoor climate	Spectrum of light source and intensity levels in test cycle, black standard temperature, humidity frequency of water spraying (composition of water), duration of test expressed as light dose.	(a) ISO 4892, Plastics – Methods of exposure to laboratory light sources (b) SAE J 1960, Accelerated exposure of automotive exterior materials using controlled-irradiance, water-cooled xenon-arc lamp	As above but also damage resulting from the combined action of UV radiation, temperature, moisture, and in SP method 2710 also of acid rain

(Continued)

Table 4.3-4. Continued

Environmental factor	Specification	Examples of standard test methods (Standardization bodies)[1]	Examples of possible damage modes/mechanisms
		(c) SP method 2710, Accelerated weathering test with simulation of acid rain	
• Low atmospheric pressure	Lowest pressure, limits for pressure change and pressure cycling, time characteristics of pressure cycle, duration of test	(a) IEC 60068-2-13, Test M : Low air pressure (b) MIL-STD 810 E, Method 500.3 "Low pressure"	Mechanical because of pressure gradients Malfunctions in electric components like contacts, resistors and transformers Wear resulting from evaporation of lubricants
• Sand and dust	Reference to standard method and suitable severity of the test in terms of test dust (hardness, size distribution, concentration), particle velocity, temperature and humidity	(a) IEC 50B (Sur.) 286 (Test L) "Dust and sand" (b) MIL-STD 810 E, Method 540.3 "Dust and sand"	Wear Malfunctions in electric components resulting from sand penetration
• Wind	Maximum wind speed, limits for wind cycling	(a) US Army Material Command AMPC 706-116, Engineering Design Handbook, Environmental Series, Part Five, Natural Environmental Factors, 1975	Mechanical damage resulting from high-mechanical load

Chemical

- Salt mist and saline atmosphere

  Reference to standard method, duration of test or alternatively required corrosivity dose in terms of metallic mass loss of reference metals to be used in the standard test

  (Tests (a)–(c) are mainly for quality control)

  (a) ISO 9227, Salt spray tests
  (b) IEC 60068-2-11, Test Ka, Salt mist
  (c) IEC 60068-2-52, Test Kb, Salt mist cyclic
  (d) ISO 11474, Accelerated outdoor test by intermittent spraying of a salt solution
  (e) VDA 611-415 Prüfung des Korrosions-schutzes von Kraftfahrzeuglackierungen bei zyklisch wechselnder Beanspruchung
  (f) ISO/DIS 16701 Accelerated corrosion test involving exposure under controlled conditions of humidity cycling and intermittent spraying of a salt solution

  Corrosion of metals and alloys Degradation of organic coatings which may give rise to various kinds of damage, see Figure 2:2 Malfunction in electric components

- Sulphuric atmosphere

  Reference to standard method, duration of test

  (a) IEC 60068-2-42, Test c: Sulphur dioxide test for contacts and connections
  (b) IEC 60068-2-43, Test d: Hydrogen sulphide test for contacts and connections

  See below

(Continued)

Table 4.3-4. Continued

Environmental factor	Specification	Examples of standard test methods (Standardization bodies)[1]	Examples of possible damage modes/ mechanisms
• Mixtures of pollutants in industrial and marine atmospheres	Reference to standard method, duration of test or alternatively required corrosivity dose in terms of metallic mass loss of reference metals to be used in the standard test	(a) IEC 60068-2-60, Part 2, Test Ke, Flowing mixed gas test (b) SP method 2499 Accelerated corrosion testing involving alternate exposure for corrosion-promoting gases, neutral salt spray, and drying (c) ISO 10062, Corrosion tests in artificial atmosphere at very low concentrations of polluting gas(es)	Corrosion of metallic materials including metals protected by organic and inorganic coatings Acidic hydrolysis of polymeric materials
• Contaminating liquids	All possible contaminating liquids the product will come in contact with during its life cycle should be specified and severity limits given	(a) ISO 175 Plastics – Methods of test for the determination of the effects of immersion in liquid chemicals (b) ISO 1871 Rubber, vulcanized – Determination of the effect of liquids	Chemical (possible degradation mechanisms are dictated by the specific contaminating liquid)
• Explosive atmosphere	Requirements for intrinsic safety stated in standards	(a) IEC Publication 79, Electric apparatus for explosive gas atmospheres	

• Fire	Prediction of the resistance to fire is dictated by its use and controlled by a number of regulations both regarding design and use	(a) IEC Publication 695, Fire hazard testing	Damage from a fire may originate from the high temperature, water for extinguishing the fire and corrosive gases (hydrogen chloride, sulphur dioxide) produced by the fire
Biological			
• Micro-biological organisms	Kind of micro-biological organisms that are of relevance for the product durability are listed, e.g. bacteria, single-cell fungi, single-cell algae, protozoa	(a) IEC 60068-2-10, Test J: Mould growth	Microbiological corrosion of metallic materials Discoloring of organic coatings due to algae and fungi growth Rot attack of wood due to mould fungus
• Macro-biological organisms	Kind of macro-biological organisms that are of relevance for the product durability are listed, as e.g. rodents, birds, insects, dogs	(a) US Army Material Commend, AMCP 706-116, Engineering design handbook	Attack on wood by e.g. termites, boring beetles, ants Material losses due to gnawing, e.g. of insulation, caused by rats and mice Corrosion induced by dog urine or bird droppings

[1]IEC International Electrotechnical Commission, http://www.iec.ch; MIL STD Department of Defense Standards, Philadelphia, PA 19111-5094, United States; ISO International Standardization Organization, http://www.iso.ch; VDA Verband der Automobilindustrie E.V., http://www.vda.de; SP Swedish National and Research Institute, http://www.sp.se

Table 4.3-5. General procedure for qualification or acceptance testing based on accelerated testing

Qualification or acceptance lifetime testing
(1) Identify critical failure mode and degradation processes to be assessed by using accelerated testing.
(2) Select appropriate accelerated tests to study the importance of each failure mode.
(3) Select suitable functional property to be used as degradation indicator and from the performance requirement, evaluate the lowest tolerable level of this to define failure.
(4) Choose most appropriate set of test conditions for the accelerated test.
(5) From the service life requirement and information on the in-use conditions, estimate the acceptable failure time for the accelerated durability tests.
(6) Perform the tests and determine actual failure time of test specimens. Analyze test specimens for the expected degradation resulting from the assumed failure mode.
(7) From the results obtained, conclude whether the tested component or material shall be considered qualified or not with respect to its durability.

into test conditions for suitable methods of IEC 60068-2. Some of the tests in Table 4.3-3 are useful for assessing the long-term environmental influence on materials as described in Section 4.3.2.2.

4.3.2.2 Long-term Environmental Influence Resulting in Gradual Degradation of the Materials

For qualification testing with respect to long-term environmental influence, the general procedure outlined in Table 4.3-5 may be followed. In this case, appropriate accelerated life tests have to be tailor-made for qualification testing and correspond to the expected in-service environmental conditions, but at enhanced stress levels. First, this requires that the test must be able to reproduce the same kind of degradation mechanism as is observed under in-service conditions. Secondly, the acceleration factor, i.e., the ratio between the time to reach a certain extent of degradation in-service and the time to reach the same extent of degradation in the accelerated test, must be independent of the extent of degradation at which this comparison is made.

The principal difficulty after an appropriate accelerated test has been identified is to determine the ratio between the accelerated testing time and the in-service time for a component or parts of the component in a particular application. The design service life of the component or parts of the component has to be converted into an acceptable failure time for the component or parts of it in the accelerated test. Because this ratio is not easily obtained, it is rare that an accelerated test can be found during qualification testing. Lifetime tests, for service life prediction therefore usually need to be developed.

REFERENCES

Handbook in Environmental Engineering (1998) Swedish Environmental Engineering Society (Book can be ordered from SP Swedish National Testing and Research Institute, P.O.Box, 857, SE-51015 Borås, Sweden).

IEC 60721-1 Classification of environmental parameters and their severity, -2 Environmental conditions appearing in nature, -3 Classification of groups of environmental parameters and their severities; International Electrotechnical Commission, http://www.iec.ch.

IEC 60068-2 Basic Environmental Testing Procedures, International Electrotechnical Commission, http://www.iec.ch.

MIL STD 810E (1989) Environmental Test Methods and Engineering Guidelines, Department of Defense Standards, Philadelphia, PA 19111Ù5094, United States.

Standardization bodies referenced to in Table 4.3-4:

Chapter 4.4

Accelerated Indoor Testing

S. Brunold,[1] M. Köhl,[2] K. Möller,[3] and B. Carlsson[3]

[1]*Institut für Solartechnik SPF, Hochschule Rapperswil HSR,*
Oberseestrasse 10, CH-8640 Rapperswil, Switzerland
[2]*Fraunhofer-Institut für Solare Energiesysteme,*
Heidenhofstr. 2, D-79110, Freiburg, Germany
[3]*Swedish National Testing and Research Institute,*
P.O. Box 857, S-501 15 Borås, Sweden

Abstract: The purpose of this chapter is to elucidate the role of accelerated indoor testing in identifying marginal products with screening testing, accelerating the degradation rates of materials and components for durability testing, and making a service lifetime prediction for a product or solar system. Aging processes can be accelerated either by increasing the concentration of reactive components or by increasing the degradative stress intensities, such as ultraviolet light, temperature, relative humidity, and cycle frequencies. To illustrate the application of some of the aging processes, the degradation of absorber coatings in solar thermal collectors was chosen. Elevated temperatures, higher humidities, and increased concentrations of reactive pollutants were chosen as the accelerating parameters. The experimental design used for studying degradation of absorber coatings is presented. A brief description is given of the accelerated testing equipment used. The effect of temperature fluctuations on the results was analyzed using the Arrhenius relation. Measurements that are useful for monitoring the changes in materials and their performance at various stages of degradation are briefly summarized. The change in the performance criterion of a selective solar absorber with time is given for 255 and 285°C, different relative humidities, and a pollutant concentration of 1 ppm, which corresponds to 20 to 500 times the concentrations found outdoors.

Keywords: Arrhenius law, High-temperature test, Condensation test, Elevated pollutant concentrations, Screening testing, Accelerated life testing, Service lifetime prediction, Solar thermal collector, Solar absorber degradation, Accelerated life testing equipment

4.4.1 INTRODUCTION

One of the needs for a service life prediction is to conduct accelerated tests under carefully controlled conditions. The first requirement is to apply the same type

175

of stresses that will be encountered during in-service use, but at increased intensities to accelerate the degradation processes. The second requirement is to determine exactly the integrated loads during testing that are needed for modeling and extrapolation of the stresses in the service life prediction. To accelerate the aging of a material in the laboratory, two methods are usually used: In the first, the ambient concentration of the reactive component is increased. This can be done if its natural concentration during in-use conditions is small, as is the case for airborne pollutants, such as sulfur dioxide, which is actually present in air at concentrations ranging from only a few parts per billion (ppb) to a highest limit of 70 ppb (Graedel and Schwarz, 1977). In the second method, acceleration can be achieved by simply raising the temperature, if the degradation is caused by oxidation and/or hydratization and hydrolysis. Oxygen and/or water vapor already occur in the atmosphere in relatively high concentrations, so no significant increase in their concentrations is possible. Increasing the temperature accelerates numerous activated chemical processes, and normally results in an increased rate of degradation of materials. This is by far the most commonly applied method in accelerated life testing (ALT). In the second method, acceleration can also be achieved by elevating the other stresses such as UV, relative humidity, and cycle frequency.

In the case of service lifetime testing of solar absorber coatings, use is made of both methods for accelerating degradation. Increasing the temperature above in-service conditions is applied to test the resistance of solar absorbers to high temperature, high humidity, and condensation. Increasing the concentration of the reactive component is used to test the resistance of solar absorbers to airborne pollutants.

All of the following discussion is related to the methods of accelerated life testing of solar absorber coatings as they were developed and applied within Task 10 of the International Energy Agency (IEA) and refined in the IEA Working Group Materials in Solar Thermal Collectors (MSTC) (Carlsson *et al.*, 1994; Carlsson *et al.*, 2000). In particular, it was assumed that the temperature dependence of the degradation processes can be described by Arrhenius behavior. Nevertheless, most of the considerations are of a general character and may be useful for planning accelerated life testing of other materials.

Considerable demands are made on the laboratory equipment used for ALT. The test specimens have to be exposed to a well-defined and accurately controlled environment. In particular, all dominating environmental degradation factors have to be controlled, measured, and logged with high reliability and high accuracy. In addition, the testing equipment must by able to function perfectly for longer times than the duration of the accelerated testing time.

4.4.2 ACCELERATION BY INCREASING THE TEMPERATURE

For testing solar absorber coatings, degradation was accelerated by increasing the temperature for (1) examining the thermal stability of the absorber at high temperatures, and (2) testing the absorber stability in a high-humidity environment or repeated condensation cycles.

4.4.2.1 *High-Temperature Testing*

In the test procedure of the IEA Working Group MSTC (Brunold *et al.*, 2000; Carlsson *et al.*, 2000), the stability of the absorber materials was examined by performing multiple constant-load tests, i.e., tests at different constant temperatures for defined periods of time. From a mathematical point of view, it would be desirable for the timing diagram of the test to resemble a square wave, i.e., heating the sample instantaneously to the desired testing temperature, holding the temperature accurately at this temperature without any fluctuations caused by the controller, and cooling it to the ambient temperature instantly after the testing time at the elevated temperature is completed. However, this idealized test procedure is not realistic and, as we will see in some cases, not even desirable.

High-temperature tests are preferably performed in circulating-air ovens. In this type of a chamber, all installations placed inside as well as all interior walls are coupled to the same turbulent stream of hot air. The same is true for the test samples and temperature sensors placed inside, so the temperature of the samples is essentially at the temperature measured by any temperature sensor inside the oven.

In radiative ovens, the heat exchange depends principally on the infrared (IR) optical properties of the samples and sensors on the one hand and the interior of the furnace on the other. The measured temperature will only be the same as the true sample temperature if the optical properties of the samples and the sensors are the same and if the geometry for the radiant heat exchange is similar. Because the optical properties of the sample will probably change with time because of degradation, precise control of the sample temperature will only be possible by measuring the sample temperature directly. Because different samples will have different IR optical properties, it is not possible to test more than one type of coating simultaneously.

It is strongly recommended that circulating-air ovens be used, even at high temperatures, when testing materials with special optical properties have to be investigated. For solar absorber testing, the oven must be able to attain a temperature of up to 300°C and maintain the set temperature within ±1 K. Ovens with a high maximum temperature, and therefore a high-heating power, tend to generate fluctuations with a high amplitude. Based on to the Arrhenius relation, even a symmetrical variation around the set temperature results in a higher effective mean testing temperature.

To achieve a well-defined testing time period, the time needed for heating and cooling should be approximately zero. However, this might result in an undesirable degradation of the sample because of thermal shocks. As a compromise, it has been proven reasonable to place the samples into the oven after the test temperature has been reached. For cooling, the rate of temperature decrease should be at least 10 K/ min for the first 100 K of cooling. This can be achieved by taking the samples out of the hot furnace at the end of the test and placing them onto an insulating material. We note that other materials with high-thermal conductivity and/or high-thermal mass would act as a heat sink with the risk of a thermal shock. Another possibility is to open the oven door automatically and cool the inside using a fan.

Some examples of different high-temperature tests at 300°C are represented in Figure 4.4-1. The duration of all tests was chosen to be 75 h. The samples were placed into the hot oven at time zero. The results of the four tests differ. In the *exact test procedure*, the test followed the prescribed test procedure, i.e., the temperature was maintained at exactly 300°C for 75 h and cooled at a temperature decrease of 10 K/min from 300°C to ambient room temperature. For the +0.5 K *temperature shift*, the test was the same as for the *exact test procedure*, except the testing

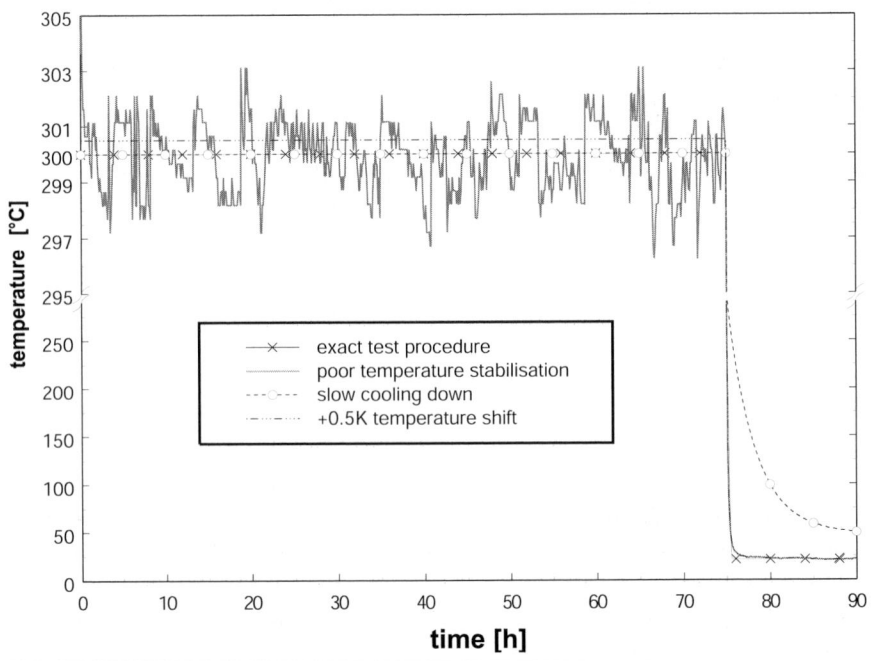

Figure 4.4-1.   Temperature profile for four different 75 h tests at 300°C.

temperature was 300.5°C instead of 300°C. For the *slow cooling down* the test was the same as for the *exact test procedure*, except the cooling was only 1 K/min (too slow) from 300 to 200°C because of a failure in the "forced-cooling system." Finally, in the *poor temperature stabilization*, a poorly operating temperature controller resulted in fluctuations of ±3.5 K about a mean value of 300°C but cooling was as prescribed in the procedure.

For an ideal test, i.e., a step from ambient room temperature to 300°C followed by 75 h of constant 300°C and then a step back to ambient room temperature, the effective mean temperature T_{eff} would be exactly 300°C. For nonideal actual conditions, the effective mean temperature of the tests performed will differ. Because the temperature sensitivity of the degradation mechanism varies with the activation energy E_A of the process, the actual effective mean temperature of the tests performed will be a function of E_A.

With the assumption that the temperature dependence of the degradation processes can be described by the Arrhenius law relationship, the effective mean temperature corresponding to an ideal test of exactly 75 h at 300°C can be calculated for the tests performed by an integration:

$$\exp\left(-\frac{E_A}{R \cdot T_{eff}}\right) = \int_0^{t_{end}} \exp\left(-\frac{E_A}{R \cdot T(t)}\right) \cdot \frac{1}{t_{eff}} \cdot T(t) \cdot dt$$

where E_A is the Arrhenius activation energy (kJ mol^{-1}), R is the ideal gas constant ($R = 8.314$ J K^{-1} mol^{-1}), T_{eff} is the effective mean temperature for the ideal test (K), t_{eff} is the time duration at T_{eff} at the ideal test, and $T(t)$ is the time-dependent function for the temperature during the real test (K).

The results of the four tests are summarized in Figure 4.4-2. As can be seen, the effects on the effective mean temperature resulting from the nonideal test procedures are not too large, i.e., the deviation is clearly less than 1 K. However, if several conditions differ simultaneously from the ideal, the errors can be additive. Thus, great care needs to be taken when carrying out the tests.

To predict the lifetime of a solar absorber, a large number of tests at different temperatures is necessary, and could result in unaffordably high costs in a long-term test program. Thus, the standard test procedure actually used is only useful as a qualification test of the candidate material (Brunold *et al.*, 2000; Carlsson *et al.*, 2000). We define qualification to mean that the material tested will have at least a lifetime of 25 years. No statement can be made to indicate whether the lifetime will be 25 years, 100 years, or even more.

The high-temperature tests were begun at a temperature of 250°C. Depending on the extent of degradation resulting from this load, new samples had to be exposed either at a temperature of 300°C or 220°C. After exposure to the second temperature,

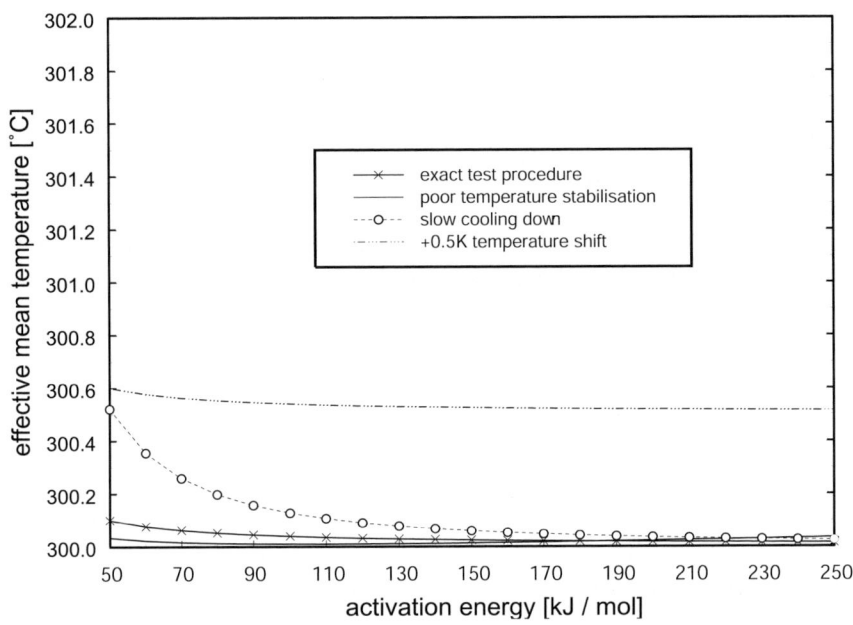

Figure 4.4-2. Effective mean temperature for the four different tests in Figure 4.4-1.

a conclusion can be made about the thermal stability and qualification of the tested absorber.

As explained above, the effective mean temperature of a particular test may differ from the prescribed level by several tenths of a Kelvin degree as a result of nonideal test conditions. Because the experimental setup will be the same for both temperatures, the resulting errors will be of the same order and of the same sign. This means, for example, that slow cooling as well as poor temperature stabilization will increase the effective mean temperature by a few tenths of a Kelvin degree, regardless of whether the testing temperature is 220, 250, or 300°C. Inaccurate calibration of a Pt-100 temperature sensor will result in a temperature shift that is almost the same as for a set-point temperature of 220, 250, or 300°C. As a consequence, this "systematic" deviation of the effective mean temperature will result in a change in the qualification limit. The borderline case of a minimum lifetime of 25 years will be shifted, depending on the activation energy of the degradation process, as can be seen in Figure 4.4-3.

4.4.2.2 Condensation Testing

Condensation tests are performed in a climate cabinet that must be able to generate $95 \pm 3\%$ relative humidity at temperatures ranging from 35 to 65°C with an accuracy

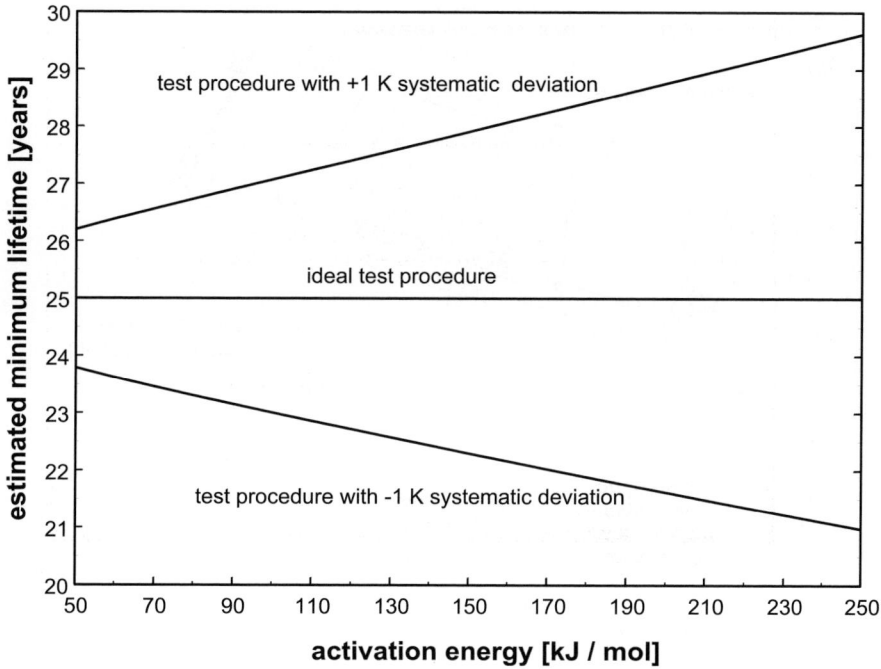

Figure 4.4-3. Change in the minimum lifetime resulting from test errors when using the high-temperature test procedure (Brunold *et al.*, 2000; Carlsson *et al.*, 2000).

of ±1 K. The absorber samples have to be placed inside the climate cabinet mounted on a sample holder and positioned at an angle of 45° relative to the horizontal. Because the cabinet temperature must be kept 5 K above the specified testing temperature, which is the sample temperature, permanent condensation of water will occur on the surface of the samples.

Figure 4.4-4 shows a suitable sample holder and Figure 4.4-5 shows a photograph taken during a condensation test. The liquid coolant is cooled by a thermostatic bath and flows through channels in the cooling block, which is made of aluminum. To prevent bimetallic corrosion, the cooling block is coated with a thin PTFE layer. The samples are attached to the sample holder with a heat sink compound. The sample temperature is controlled using a Pt-100 thin-film sensor that is bonded to the surface of the sample holder. The temperature has to be controlled to an accuracy of ±0.5 K. Considerations concerning possible testing errors are analogous to the high-temperature test procedure. However, different temperatures and error margins will influence the qualification limit of 25 years to a slightly different magnitude, as can be seen in Figure 4.4-6.

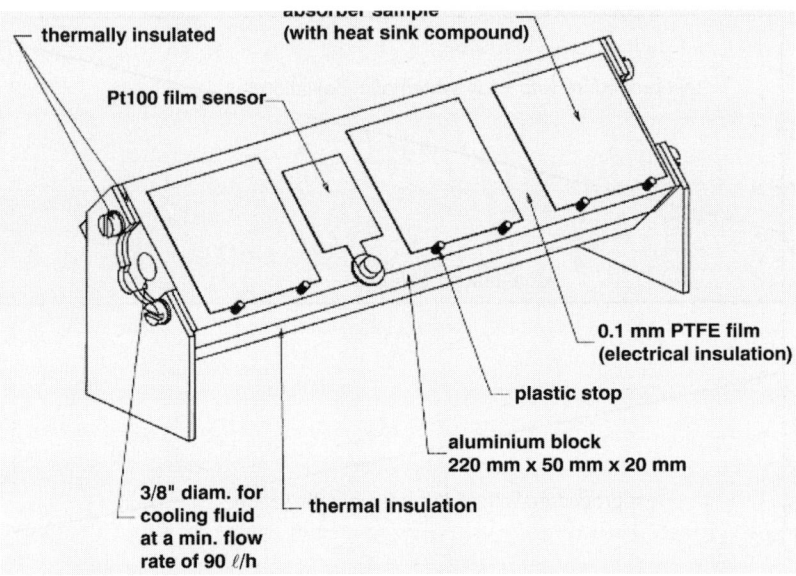

Figure 4.4-4. Suitable sample holder for condensation tests.

Figure 4.4-5. View of absorber samples during a condensation test.

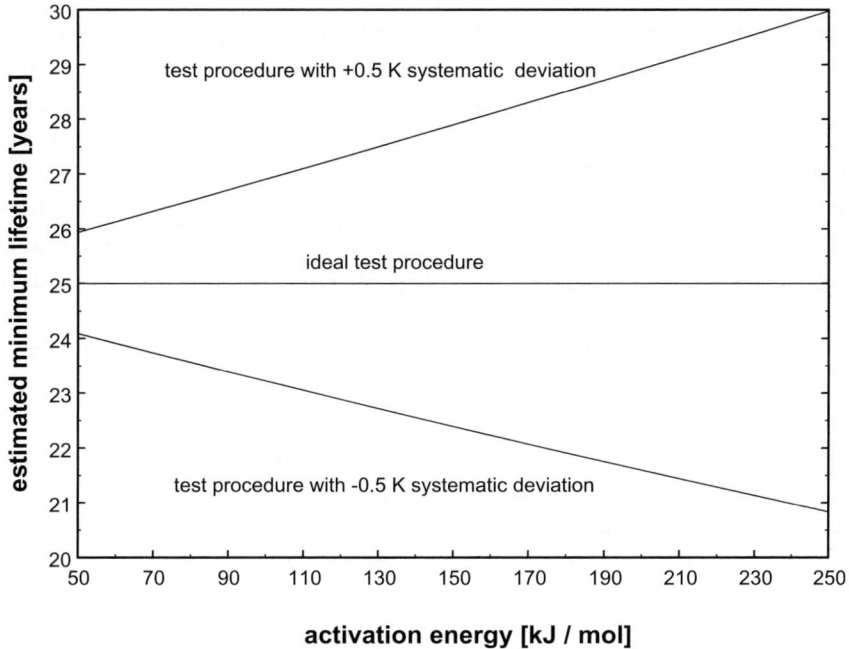

Figure 4.4-6. Change in the minimum lifetime resulting from test errors when using the condensation test procedure (Brunold *et al.*, 2000; Carlsson *et al.*, 2000).

4.4.3 ACCELERATION BY INCREASING THE CONCENTRATION

It is well known that atmospheric corrosion is strongly correlated to high-humidity levels and the time-of-wetness (Ailor, 1982). Furthermore, it is also well known that air pollutants such as sulfur dioxide and nitric oxides strongly accelerate corrosion rates (Ailor, 1982). Moreover, as discussed in Chapter 3.2, sulfur dioxide has been found inside solar collectors and at concentrations corresponding to the amounts measured outdoors. For that reason a quite extensive investigation was undertaken regarding sulfur dioxide induced corrosion of solar absorber coatings in IEA Task 10 and was continued in the work of the IEA Working group MSTC. A series of long-term laboratory exposure tests were conducted to study the combined effect of sulfur dioxide, high air humidity, and temperature on the degradation of solar absorber coatings.

One important outcome of the investigations was that it is quite difficult to obtain reproducible results concerning both the degradation of absorber coatings and metal weight loss of metal coupons, which were exposed simultaneously. For that reason deposition rates of sulfur dioxide onto the surface of the solar absorbers were measured as a function of sulfur dioxide concentration in the surrounding air and also as a function of relative humidity (RH), at well-controlled exposure conditions at Chalmers University of Technology (Eriksson *et al.*, 1991). One of the most important results of the investigation is that the increase in the deposition rate is quite low below 95% RH. At about 95% RH, the slope of the curve becomes very steep (see Figure 4.4-7). To reach reasonably high-acceleration factors, the tests have to be conducted at high-humidity levels, i.e., in the region in Figure 4.4-7 where there is a dramatic change in the slope of the deposition curve.

Because of the complex nature of the degradation resulting from the presence of sulfur dioxide, it is consequently difficult to formulate a model to describe the kinetics of it. Even if the kinetics of the chemical reactions could be quantitatively described in terms of temperature dependence, humidity dependence, and sulfur dioxide concentration dependence nothing would have really been gained. One of the most difficult modeling problems is related to the sorption of sulfur dioxide on the surface of the absorber coating, as discussed above. Furthermore, the complex correlation between the deposition rate of sulfur dioxide, the relative humidity, and the temperature makes controlling the accelerated aging test conditions inside the climatic chamber very difficult. One way, however, to circumvent these problems is to adopt a comparative testing procedure. Therefore, metal coupons were used to determine the atmospheric corrosivity in the accelerated sulfur dioxide tests and inside the solar collectors

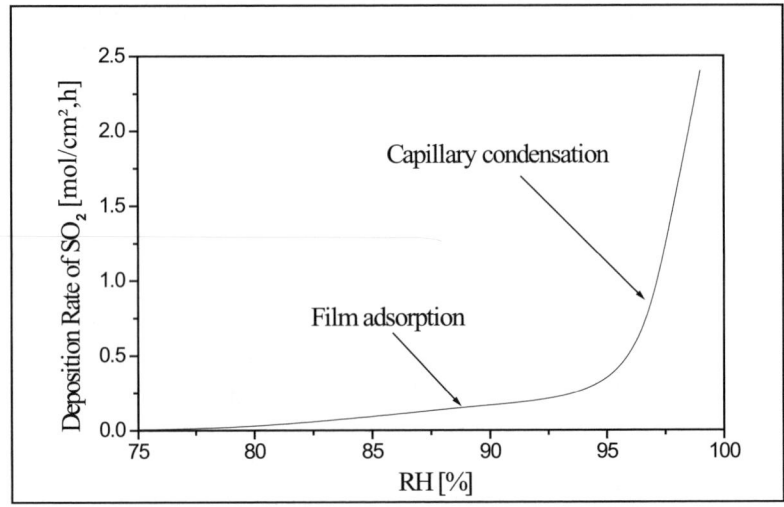

Figure 4.4-7.   Deposition rate of sulfur dioxide at a concentration of 1 ppm onto a nickel pigmented aluminum oxide solar absorber coating as a function of relative humidity.

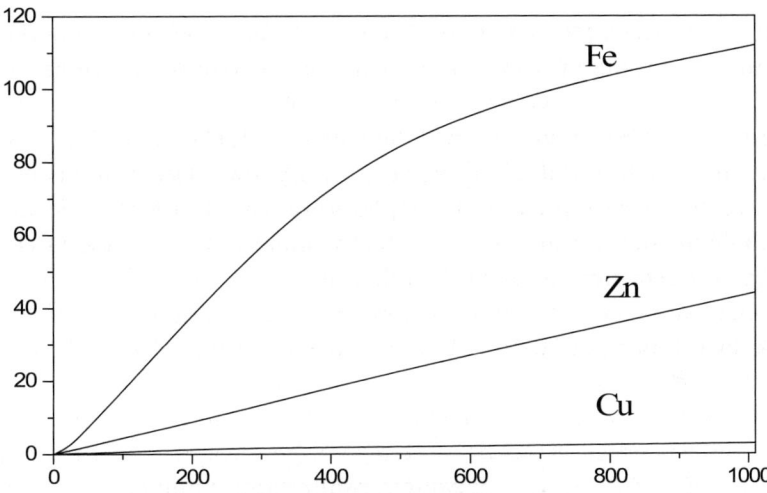

Figure 4.4-8. Metal weight loss of standard metal coupons exposed to 1 ppm sulfur dioxide, 95% RH, and 20°C versus exposure time.

tested. The change in the solar absorptance, the thermal emittance, and the PC-function of the coatings, thus, are considered being a simple function of the corrosion rate of the different metal coupons used. The metals used were aluminum, copper, iron (carbon steel), and zinc.

The total metal mass losses of the coupons after 25 years of exposure inside solar collectors have been predicted from extrapolation of data from a large number of exposures inside different solar collectors extended over several years. The metal that best met the requirements of linearity over time and appropriate sensitivity to the corrosivity both inside collectors and in accelerated tests is zinc (see Figure 4.4-8). Aluminum (not included in the figure) especially, but also copper, react too slowly. Steel on the other hand corrodes too rapidly and results in the formation of a thick oxide layer that retards the corrosion rate. Another important property is the similarity in corrosion behavior of the metals commonly used as pigments in absorber coatings, such as nickel and chromium. This requirement is also fulfilled by zinc.

Two different types of collectors have also been identified in the investigations conducted over the years. Type I collectors are tight and have a very controlled and low-ventilation rate, whereas Type II collectors have a more or less uncontrolled and high-ventilation rate. The predicted (25 years) total metal weight loss of zinc coupons in Types I and II collectors is 2.7 and 7.8 g/m^2, respectively.

In an accelerated sulfur dioxide atmospheric corrosion test, both test specimens as well as zinc coupons are exposed side by side. The optical properties of the absorber and the metal weight loss of the coupons are measured at appropriate time intervals.

When the metal weight loss reaches 2.7 or 7.8 g/m², depending on the type of collector to be used, the PC-function must not exceed 0.05 (or any other acceptable value).

The test procedure employed was similar to that described in ISO 10062 "Corrosion tests in an artificial atmosphere at very low concentrations of polluting gases." In each test performed, test samples were placed in a climatic cabinet and exposed to circulating air at a specified temperature (20°C), humidity (RH 95%), and a fixed concentration of sulfur dioxide (1 ppm).

Test equipment must be made of inert materials, e.g., Teflon or glass, to avoid or minimize adsorption of sulfur dioxide on surfaces other than that of the test samples

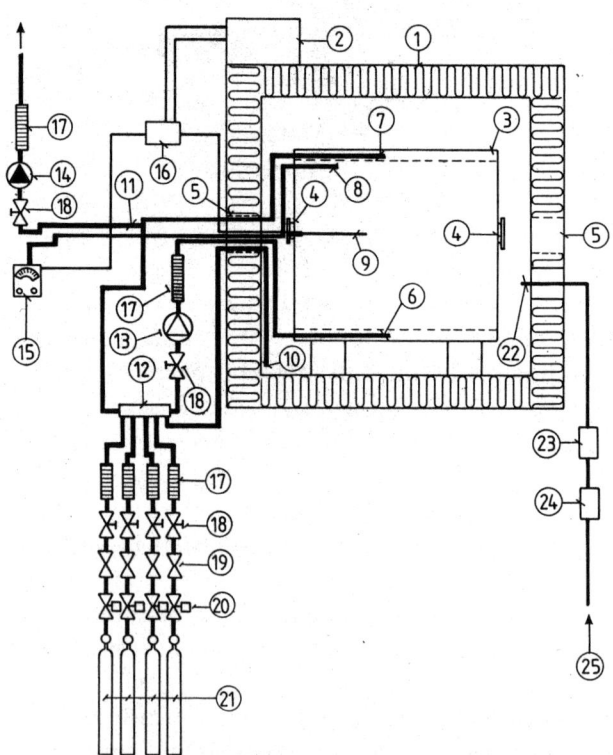

Figure 4.4-9. Climatic cabinet used for accelerated exposure test in high-humidity air containing sulfur dioxide. 1. Climate chamber, 2. External control, 3. Inner chamber (Teflon walls), 4. Lead-through, inner chamber, 5. Lead-through, climate chamber, 6. Inlet for circulation, 7. Outlet for circulation, 8. Analysis point (movable), 9. Temperature sensor, 10. Intake of conditioned air, 11. Outlet from system, 12. Mixing tube, 13. Circulation pump, 14. Evacuation pump, 15. Gas analyzer, 16. Connection to logger, 17. Flow meters, 18. Regulation valves, 19. Shut off valve, 20. Magnetic valve, 21. Gas bottle, 22. Intake of clean air to climatic chamber, 23. Particle filter, 24. Active coal filter, 25. Air inlet.

and the metal coupons. An inner chamber with Teflon covered walls was, therefore, installed in the climatic cabinet (see Figure 4.4-9). The airflow and sulfur dioxide injection system must be designed to ensure uniform test conditions in the working space. The test atmosphere in the working space is obtained by continuously introducing the necessary quantity of sulfur dioxide into a damp airflow to obtain the required concentration. Sulfur dioxide and conditioned air are mixed outside the cabinet. After injection of sulfur dioxide, the airflow is then mixed with a flow of recirculated test atmosphere and the resulting gas flow admitted into the inner chamber or working space of the cabinet. To ensure uniform test conditions in the working space, the test atmosphere is supplied to the working space from the bottom and the outlet is placed at the top. Perforated plates are placed in front of the openings to assure uniform airflow through the working space. The temperature uniformity in the working space should be better than $\pm 1\,K$ and the uniformity of the relative humidity better than $\pm 3\%$. The damp airflow should be within the tolerance for the specified temperature and relative humidity and be free of water droplets or aerosols. The air used is introduced to the outer chamber of the cabinet, after it has been filtered and purified by activated charcoal and a particulate filter.

During the sulfur dioxide exposure tests, the absorber coating samples are vertically placed in a special Teflon stand or holder at a distance of 3 cm from each other. The working space should not be filled with test specimens in such a way that they shield one another, disturb the uniformity of airflow across the chamber or cause a change of sulfur dioxide concentration in the circulating air of more than 10% during passage through the working space. The concentration of sulfur dioxide in the airflow is adjusted to the correct level at the beginning and during testing, based on measurements of it at sampling points placed close to the inlet and outlet of the airflow through the working space.

When measuring the extent of degradation after a certain test time, the test specimens are removed from the working space and placed in desiccators over silica gel. If, however, the test conditions are such that condensation would occur on the specimen surfaces upon removal from the working space, the humidity of the cabinet is first reduced before the test specimens are removed.

The climatic cabinet and procedures used for testing are described in more detail in Carlsson *et al.* (1994).

4.4.4 TESTING WITH SOLAR OR UV-RADIATION

The simulation of temperature and moisture stresses is an easy task compared to the simulation of solar irradiation or its UV-component in accelerated laboratory testing. No light source exists with the same spectral distribution as sunlight, even when stationary conditions like for standard solar spectra. e.g., ASTM E891, are assumed. A number of different lamps or combinations of lamps are used in accelerated weathering equipment or in commercial testing chambers (Weathering Testing Guidebook). All spectra are more or less different from the natural sunlight resulting in different amounts of degradation compared to natural weathering or even different degradation modes.

Polymeric materials are often sensitive to the wavelength of the incident light, which is called spectral sensitivity. The energy of the light increases strongly with decreasing wavelength ($E \sim 1$/wavelength) and the energy above a certain threshold might be enough to destroy chemical bonds in specific polymeric compounds (see Table 4.7-1, Chapter 4.7). Therefore, the spectral distribution of the light is crucial for testing and attention has to be paid to emission lines in the short wavelength range below 450 nm (see Figure 4.4-11). The radiation should always be measured at the sample location with an appropriate spectroradiometer and an integrating sphere because the reflectors of the lamps and the reflecting walls of the cabinet influence the spectral distribution of the light as well as aging of the lamps.

Comprehensive durability studies should include the investigation of the spectral sensitivity by using edge filters (Andrady *et al.*, 1992) (see Figure 4.4-10) to cut off the most energetic part of the solar spectrum (see Figure 4.4-12). The filters can be placed directly on the samples, but then a quartz glass should be placed on the reference for providing the same thermal conditions, or at some distance to allow air

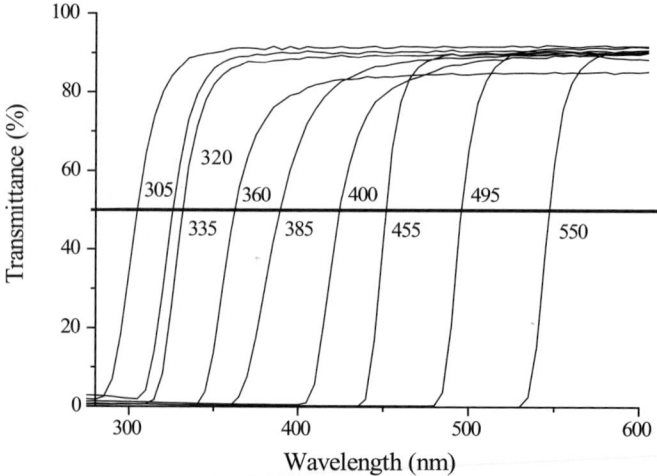

Figure 4.4-10. Spectral transmittance of edge filters suitable for investigations of spectral sensitivities of materials. The numbers indicate the transition wavelength in nanometers given by the manufacturers.

Table 4.4-3. Experimental design for screening tests for unglazed solar facades

	Temperature	Wetness	Corrosion	UV-radiation	Temperature with humidity and/or pollutants
High value (+)	150°C	Condensation	SO_2+NO_X+Salt	One sun	60°C
Low value (−)	80°C	Dry	SO_2+NO_X	Dark	40°C
Number					
1	−				
2	+			−	
3	−	−		+	
4	+			+	
5		+		−	−
6		+		−	+
7		+		+	+
8			+	−	+
9			−	−	+
10			+	−	−

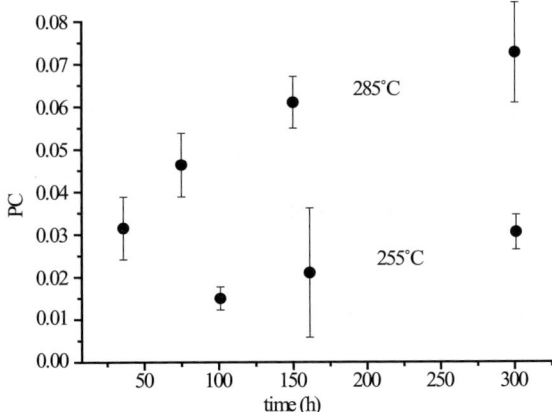

Figure 4.4-13. Change in the performance criterion (PC) over time for a sputtered solar absorber coating at 255 and 285°C. The symbols are average values from three different samples and the range bars show the extremes of the PC at each time.

data should be good enough for further processing for assessing the service life. The acceleration during condensation testing was slower, but the effect of the increasing the test temperatures can be clearly seen in Figure 4.4-14. The changes in PC were more drastic and started obviously after an introduction period, in which the same degradation process was active, but without affecting the optical properties relevant for PC. Therefore, the kinetics of this process could be modeled with one single parameter.

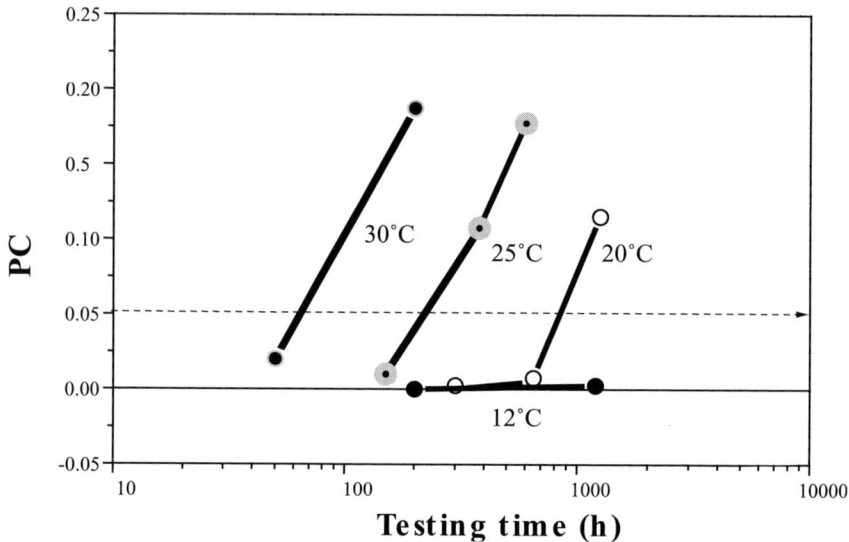

Figure 4.4-14. Example from accelerated testing of a nickel-pigmented anodized aluminum absorber surface during condensation tests at a sample temperature of 12, 20, 25, and 30°C, respectively.

When multiple environmental stresses contribute to performance degradation, an appropriate damage function that includes all relevant factors must be formulated. An approach to modeling the effect of individual stresses is presented in Chapter 4.5. A discussion of how this methodology can be extended, and examples of how to apply multifactor damage functions to specific materials such as glazings and polymers is provided in Chapter 4.6.

REFERENCES

Ailor, W.H. (Ed.) (1982) *Atmospheric Corrosion*, John Wiley and Sons, New York.

Andrady, A.L., Searle, N.D. & Crewdson, L.F. (1992) Wavelength Sensitivity of Stabilized and Unstabilized Polycarbonate to Solar-simulated Radiation, *Polymer Degradation and Stability*, **35**, 235–247.

Brunold, S., Frei, U., Carlsson, B., Möller, K. & Köhl, M. (2000) Accelerated Life Testing of Solar Absorber Coatings – Testing Procedure and Results, *Solar Energy*, **68**(4), 313–323.

Carlsson, B., Frei, U., Köhl, M. & Möller, K. (1994) Accelerated Life Testing of Solar Energy Materials – Case Study of some Selective Solar Absorber Coating Materials for DHW Systems; A Report of IEA Task X; Solar Materials Research and Development; February 1994; SP-Report 94:13; ISBN91-7848-472-3; (Report available from Swedish National Testing and Research Institute, P.O. Box 857, S-50115 Boras, Sweden).

Carlsson, B., Möller, K., Köhl, M., Frei, U. & Brunold, S. (2000) Qualification Test Procedure for Solar Absorber Surface Durability, *Solar Energy Materials and Solar Cells*, **61**, 255–275.

Eriksson, P., Johansson, L.G. & Svensson, J.E. (1991) Department of Inorganic Chemistry, Chalmers University of Technology, Gothenburg, Report for IEA SHC task 10 B (in Swedish).

Graedel, T.E. & Schwarz N (1977) Air Quality Reference Data for Corrosion Assessment, *Mat. Perf.*, **16**(8), 17.

Weathering Testing Guidebook, Atlas Material Testing Technology LLC, infoa@atlas-mts. com.

Chapter 4.5

Service Life Prediction from Results of Accelerated Aging

B. Carlsson

Swedish National Testing and Research Institute,
P.O. Box 857, S-501 15 Borås, Sweden

Abstract: The purpose of this chapter is to describe in a systematic way some of the existing and successfully applied models for analyzing accelerated test data for service life prediction. Deterministic mathematical models based on the time-transformation function approach are reviewed and shown applied in cases where (1) a single degradation factor or combination of degradation factors are involved in materials degradation, (2) the levels of stress are constant or varying with time, and (3) more than one degradation reaction contribute to the overall degradation in performance. To the procedures for applying some nonmechanistic phenomenological models for service life prediction is also reviewed and the principles of using probabilistic models in accelerated life testing is also outlined. The planning process for developing accelerated testing programs for service life prediction is described and methods are outlined for making a reasonability assessment and validation of predicted service life. The examples given in the chapter show the great applicability of the general methodology for accelerated life testing. Their usefulness and validity are also confirmed by comparing predicted results with actual measured data for samples exposed under real in-service conditions. Consequently, highly abbreviated testing times at elevated stress conditions can be substituted for long-time exposures at lower stress levels in durability assessment. Successful application of a service lifetime prediction methodology will facilitate much shorter development cycle times for new products and allow improvements to be identified and readily incorporated into new products prior to market introduction.

Keywords: Service life prediction, Accelerated testing, Deterministic mathematical models, Time-transformation functions, Arrhenius equation, Eyring–Flood equation, Effective mean values of stress, Probabilistic mathematical models, Failure rate equations, Weibull life distribution, Selective solar absorbers

LIST OF ACRONYMS, SYMBOLS AND SUBSCRIPTS USED

Symbol	Definition
A	material dependent constant in equation (19)
A, B, C	degradation mechanisms of the nickel pigmented anodized aluminum coating (see, for example, Table 4.5-3)
A_s	acceleration factor
a_{Sn}	acceleration factor with the respect to the stress factor S_n
B	parameter in the Eyring–Flood equation (see equation (5))
b	material dependent constant (see equation (g), Table 4.5-1)
C	material dependent constant in equation (21)
c	concentration of chemical substance
c	characteristic life parameter in the Weibull life distribution, (see equation (26))
$C(S)$	constant factor in overall stress function $g(S)$ (see equation (7))
CDF	cumulative probability density function
d	constant in equation (a), Table 4.5-1
E_a or E	Arrhenius activation energy (see equation (a) in Table 4.5-1)
f	frequency function or probability density function
f	function for performance-versus-time relationship (see equation (32))
F(t)	cumulative probability density function
FDIS	final draft international standard
G	material dependent constant in the Eyring equation (see equation (b), Table 4.5-1)
$g(\underline{S})$	stress function
$g(\underline{S}_D)/g(\underline{S}_s)$	time-transformation function
H	humidity
h(t)	failure rate or hazard rate function
HDPE	high density polyethylene
$h_n(S_n)$	factor of the stress function $g(\underline{S})$ with respect to influence from stress factor S_n (see equation (2))
I	light intensity
ISO	International Standardization Organization
L(t)	damage function resulting from photochemical degradation (see equation (15))
LD(t)	cumulative damage function resulting from photochemical degradation (see equation (15))
n	number
n	material dependent constant in equation (i), Table 4.5-1

n, m	reaction order constants (see equation (j), Table 4.5-1)
m	shape factor parameter in the Weibull life distribution, see equation (26)
P	performance property
p	material dependent constant related to photochemical degradation (see equation (e) in Table 4.5-1)
PC	performance criterion function
PDF	probability density function
q	material dependent constant in equation (h), Table 4.5-1
R	ideal gas law constant
RH	relative humidity
$\underline{S}$	vector specifying a set of environmental stress factors
t	time
T	absolute temperature (K)
UV	Ultraviolet
UV-B	Ultraviolet light between 280 and 315 nm
V	electrical potential
w	material dependent constant in the Eyring equation (see equation (b), Table 4.5-1)
$\underline{X}$	vector specifying a set of inherent property parameters of a material
y	specific time period

Greek

α	solar absorptance
α	proportion of components susceptible to manufacturing defects (see equation (24))
ε	thermal emittance
ϕ	quantum yield
λ	wavelength
τ	failure time

Subscript	*Definition*
c	constant in time
c	critical, for example critical humidity level RH_c
Co	corrosion
D	accelerated test
d	manufacturing defect (see equation (24))
EFF	effective mean value
f	failure
H	humidity

M	metal, see equation (k), Table 4.5-1
N	normal competing failure modes
p	period
s	in-service
T	temperature
tow	time-of-wetness (see equation (d), Table 4.5-1)

4.5.1 INTRODUCTION

Service life prediction may be based on accelerated life testing data and calculated by extrapolating the test results using mathematical models. Accordingly, accelerated aging procedures must use one or more of the degradation factors encountered during in-service, but the stresses are higher than those applied during in-use conditions. The advantage is, of course, that considerable time is saved by applying accelerated life testing compared with testing at in-service conditions.

Despite the fact that a large number of laboratory test methods are available for determining the durability properties of materials, only a few are helpful for reliably predicting service life. Among the problems encountered, the following are especially relevant: (a) recommendations are seldom given for relating the results of short-term tests to in-service performance; (b) degradation mechanisms of materials are complex and not well understood, so that it is difficult to design meaningful short-term tests; (c) degradation factors affecting service life, especially those involving natural weathering, are difficult to quantify, e.g., many current tests therefore do not take into account all of the important factors, and the factors that are taken into account are seldom quantitatively related to actual exposure conditions; and (d) mathematical models, for relating results of laboratory tests to actual performance, are difficult to develop because of the complexity of the mechanisms resulting in material degradation.

Two different kinds of approach to mathematical modeling exist, i.e., the deterministic and the probabilistic or statistical approach. The most pronounced problem of both is to find numerical expressions to describe accelerated life data in terms of environmental stress factors so that all factors in the environment contributing to degradation are taken into account. Another significant problem may be that the mechanism of degradation or the rate of degradation may be different at the elevated stress levels chosen for the accelerated testing program compared with those applicable during in-service conditions. In such cases, the numerical expressions formulated to model accelerated life data may result in an erroneous extrapolation to in-use conditions. In the probabilistic approach, both variations in material properties and environmental influence are taken into consideration. This implies that a probabilistic approach to accelerated life testing may be preferable.

Accelerated life data are most often only observations of the changes in appearance or are measured changes in some physical or chemical property after a certain testing time. As a result, most of the literature related to accelerated aging of materials reports results of comparative tests and is generally incomplete as far as service life prediction is concerned

However, in some literature, which will be considered in this chapter, the accelerated aging data are reported in more detail and numerical expressions are provided for performance-versus-time relations, of which two principal categories can be distinguished. In the first category, the main emphasis is placed on numerically fitting the experimental data from the accelerated tests to an algebraic function of environmental stress factors whose values for the anticipated in-service conditions are known. This category is not, generally based on a physical and chemical model for describing the degradation process and is, therefore, called nonmechanistic. The disadvantage of the first category of performance-versus-time expressions is that extrapolation of test results to in-service conditions will be uncertain because the numerical expressions derived may not be representative for low-stress levels. The second category, which is based on a physical and chemical model, is often limited because of the complex influences of all the degradation factors. In either case, predictions of service life are much less certain as the number of mechanisms of degradation increases. In this chapter the principal focus is on the using mechanistic models for service life prediction, whereas using the nonmechanistic models or phenomenological models for service life prediction is the principal subject in Chapter 4.6.

4.5.2 MATHEMATICAL MODELING

Mathematical treatment to evaluate the service life of a component and its materials from the results of accelerated tests involves (a) defining the performance requirement for the component or materials in terms of a level for at least one measurable functional property or performance variable, i.e., the degradation indicator, (b) finding a numerical expression that relates the change in the degradation indicator to the environmental stress factors, which contribute to the degradation in the performance of the component and its materials, i.e., the degradation factors, and (c) characterizing the degradation factors under service conditions to be able to extrapolate the results of the accelerated tests to a service life for the component and its materials. Two different kinds of approaches to mathematical modeling of lifetime data can be found in the literature, which are the deterministic approach and the probabilistic or statistical approach. The difference between the two, however, seems to be merely a question of how to formulate the service life requirement for the pertinent component or its parts.

In the deterministic approach, the service life requirement can be formulated in such a way that the variation in service life is not taken into account, so the mean or median service life will have a certain value. However, the service life requirement can also be formulated in such a way that a certain population fraction should survive beyond a given age. In this case, a probabilistic approach has to be used, which also takes into account the distribution in service life. The two approaches are compared in Figure 4.5-1.

In the prediction of service life from data obtained in accelerated life testing, not only the mathematical treatment but also the testing procedures differ between the deterministic and probabilistic approaches. In this part of the chapter, we will focus on using deterministic mathematical models based on the time-transformation or acceleration factor approach. Such models consist mainly of numerical expressions that relate the change in time of the degradation indicator to the environmental degradation factors or stress factors contributing to the degradation of the materials in the component. The numerical expressions are formulated to model the results from accelerated tests carried out at elevated levels of stress. One problem here is to find an expression that takes into account the influence of all the environmental stress factors contributing to degradation under in-service conditions. Another problem is that the principal mechanism of degradation or the rate-determining step may be different at elevated stress levels than under in-service conditions. In such

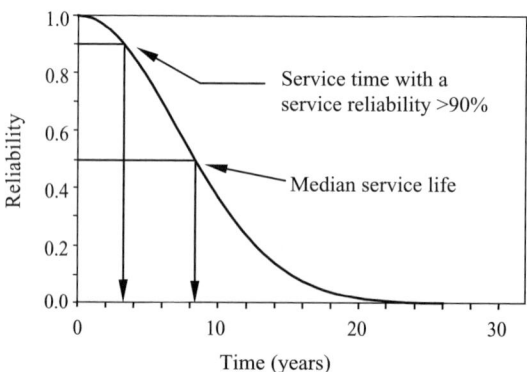

Figure 4.5-1. Deterministic and probabilistic approach in defining service and service reliability. Deterministic approach: requirement on service life in terms of mean or median service life. Probabilistic approach: requirement on minimum service time with a reliability level higher than a selected value (e.g., $\geq 90\%$).

cases, the numerical expressions adopted to model accelerated life data may give erroneous results when used to project a lifetime at in-service conditions.

The probabilistic models that can be used for accelerated life testing are more complex. The problems of finding kinetic models that can be used to transfer accelerated life data to service life data are the same. However, as the variation in component properties and environmental stress is also taken into account, the results obtained will be more realistic and useful for materials selection in component design to match the specific requirements for any specific application. Deterministic models and, in particular, the acceleration factor or time-transformation approach will be considered first.

4.5.2.1 Accelerated Testing and Time-Transformation Functions

To perform an accelerated test, D, means that the intensity of at least one stress factor contributing to degradation is kept at a higher level than actually encountered under in-service conditions. Consequently this means the time to failure in the accelerated test, $\tau_{f,D}$, will be shorter than for the actual service life, τ_s. The ratio between τ_s and $\tau_{f,D}$ is the acceleration factor, A_s.

If the applied stress is constant in time during the accelerated test and during service conditions, the acceleration factor A_s can be expressed as

$$A_s = \tau_s/\tau_{f,D} = g(\underline{S_D})/g(\underline{S_s}) \tag{1}$$

where the expression $g(\underline{S_D})/g(\underline{S_s})$ is called the time-transformation function (Martin, 1982; Carlsson, 1988) or the acceleration factor function.

For equation (1) to be valid, it is assumed that the degradation in performance (P) with time (t) in the interval $\underline{S_D}$ to $\underline{S_s}$ can be described by a product function of the following type: $dP/dt = k(X) \times g(\underline{S})$ where $\underline{X}$ is a vector specifying a set of inherent property parameters of the material, which are of importance for the performance variable P (Carlsson, 1988).

When more than one stress factor has to be taken into account, the function $g(\underline{S})$ can often be approximated to a product function of the following type:

$$g(\underline{S}) = h_1(\underline{S_1}) \times h_2(\underline{S_2}) \cdots \cdots \tag{2}$$

Thus, from equation (2)

$$A_s = [h_1(S_{1,D})/h_1(S_{1,s})] \cdot [h_2(S_{2,D})/h_2(S_{2,s})]. \tag{3}$$

or

$$A_s = a_{S1} \cdot a_{S2} \cdot a_{S3} \cdots \tag{4}$$

where $a_{S1} \cdot a_{S2} \cdot a_{S3}$ are the acceleration factors for the stresses S_1, S_2, and S_3, respectively.

In a series of aging tests, the parameters of the time-transformation functions can be determined varying the intensity of one stress factor and keeping the intensities of the other stress factors constant. In Table 4.5-1, some examples of acceleration factor equations are given for different stress factors.

Accelerating the degradation process by increasing the temperatures is by far the most frequently applied method in accelerated life testing. The results are usually analyzed using the Arrhenius equation (see equation (a) for thermal stress in Table 4.5-1). Examples of materials that can be tested and evaluated this way are elastomers and plastic materials (Thomas and Sinnott, 1969; Bartenev and Zelenenev, 1974; Dixon, 1980). The method has been adopted in connection with solar energy applications to analyze test results, e.g., plastic absorber materials of HDPE (Carlsson, 1988), and selective inorganic absorber coatings (Carlsson et al., 1994). The success of using the Arrhenius equation in chemical reaction kinetics is well documented. Its success in describing the temperature dependence of material degradation processes seems in many cases, however, to be associated with its numerical form. If the degradation mechanism is known, the order of magnitude of the activation energy can often be found in the literature. In the accelerated testing of any material, it is highly recommended that the activation energy be determined for the degradation process from the results of a series of tests performed at several elevated temperatures.

The Eyring equation, equation (b) in Table 4.5-1, contains an additional parameter and can be derived from statistical mechanical principles. It is, however, less frequently used in modeling the temperature dependence of material degradation in lifetime testing. To assess the importance of the initial corrosion of electronic components by accelerated lifetime testing, the Eyring–Flood equation (see equation (c) in Table 4.5-1), in combination with the Arrhenius equation has frequently been used for modeling. The overall acceleration factor is expressed as:

$$A = a_H \cdot a_T = \exp[B(1/RH_D - 1/RH_s)] \cdot \exp[-(E_a/R) \cdot (1/T_D - 1/T_s)] \qquad (5)$$

The two parameters B and E_a are determined first from a series of constant load tests at elevated levels of temperature and relative humidity, and then the expected service life is assessed by using equation (5).

Equation (d) in Table 4.5-1 has also been used in combination with the Arrhenius equation for predicting the in-service loss in optical performance of a nickel-pigmented, anodized aluminum, selective solar absorber surface resulting from hydratization of aluminum oxide. In this case, a series of constant condensation tests

Table 4.5-1. Examples of time-transformation functions for different kinds of environmental stress factors

Stress factor	Time-transformation function	Remarks
Thermal stress (a)	$a_T = \exp\left[-(E_a/R) \cdot (1/T_D - 1/T_s)\right]$ E_a = activation energy R = general gas law constant T_D = temperature of test T_s = temperature in service	(a) is based upon the Arrhenius equation for expressing the temperature dependence of the rate constant of a chemical reaction.
(b)	$a_T = (T_D/T_s)^w \cdot \exp\left[G(1/T_D - 1/T_s)\right]$ G = material-dependent constant w = material-dependent constant T_D = temperature of test T_s = temperature under service conditions	(b) is based upon the Eyring equation for expressing the temperature dependence of the rate constant of a chemical reaction (see, for example, Martin (1982) and Carlsson (1988)).
Humidity stress (c)	$a_H = \exp\left[B(1/RH_D - 1/RH_s)\right]$ B = material-dependent constant RH_D = relative humidity of test RH_s = relative humidity under in-service conditions	(c) is the Eyring–Flood time-transformation function recommended for describing degradation resulting from the initial corrosion of metals in the relative humidity interval 60–95%, (see, for example, Handbook in Environment Engineering (1998)).
Condensed water (d)	$a_{tow} = (\tau_{tow,D}/\tau_{tow,s})$ $\tau_{tow,D}$ = fraction-of-condensation time in test cycle $\tau_{tow,s}$ = fraction-of-condensation time under service cycle	In (d), the assumption is that degradation only occurs during condensation, time-of-wet-ness, conditions (see, for example, Carlsson *et al.*, (1994)).
Photoactive light (e)	$a_I = (I_D/I_s)^p$ p = material-dependent constant I_D = intensity of photoactive light in test I_s = intensity of photoactive light in service	(e), when $p = 1$ the rate of degradation is limited by a single photochemcial reaction (see, for example, Martin (1982) and Carlsson (1988))
Electrical Stress (f)	$a_e = (V_D/V_s)^p$ p = material-dependent constant V_D = applied elevated electrical potential in test V_s = applied electrical potential in service	(f) is based upon Levenbach's equation to describe failure resulting from electrical stress of electrically insulating fluids (see, from example, Nelson (1975))
Mechanical stress (g)	$a_\sigma = \exp\left[b(\sigma_{f,D} - \sigma_{f,s})\right]$ b = material dependent constant $\sigma_{f,D}$ = applied elevated mechanical stress during testing $\sigma_{f,s}$ = mechanical stress in service	(g) is based upon an equation that has been used to model degradation of amorphous polymers resulting from mechanical stress at temperatures below the glass transition point (Martin, 1982).

(Continued)

Table 4.5-1. Continued

Stress factor	Time-transformation function	Remarks
(h)	$a_\sigma = [((\sigma_{f,D} - \sigma_{f,\infty})/(\sigma_{f,s} - \sigma_{f,\infty})]^{1/q}$ q = material-dependent constant $\sigma_{f,\infty}$ = the largest value of mechanical stress such that the lifetime is infinite $\sigma_{f,D}$ = applied elevated mechanical stress during testing $\sigma_{f,s}$ = mechanical stress in-service	(h) is based upon an equation that has been used to model degradation of amorphous polymers resulting from mechanical stress at temperatures above the glass transition point (Martin, 1982).
(i)	$a_\sigma = (\sigma_{f,D}/\sigma_{f,s})^n$ n = material dependent constant $\sigma_{f,D}$ = applied elevated mechanical stress in test $\sigma_{f,s}$ = mechanical stress in service	(i) is based upon an equation that has been used to model degradation of solar mirrors resulting from mechanical stress (Freiman et al., 1984)
Chemical stress (j)	$a_C = (c_{i,D}/c_{i,s})^n \cdot (c_{j,D}/c_{j,s})^m \ldots$ $c_{i,D}$, $c_{j,D}$ = concentrations in test of the chemical compounds i and j $c_{i,s}$, $c_{j,s}$ = concentrations during in-service use of the chemical compounds i and j n, m = reaction order constants	(j) is based upon on the well-known general rate equation in chemical reaction kinetics (see, for example, Carlsson (1988))
Corrosive stress (k)	$a_{Co} = \tau_{M,s}/\tau_{M,D}$ $\tau_{M,s}$ = time to reach a certain extent of corrosion of reference metal during in-service use $\tau_{M,D}$ = time needed to reach the same extent of corrosion of reference metal during testing	(k) is based upon on the principles of comparative testing. The choice of reference metal may be crucial (see, for example, Carlsson et al., 1994)

at varying temperatures was used as the basis for evaluating the service life (Carlsson et al., 1994).

A value of $p = 1$ in equation (e) in Table 4.5-1 results when the turnover in a photochemical reaction is proportional to the light dose absorbed by a photoactive species. When $p \neq 1$, the reaction scheme is more complex and consists most probably of photochemical and thermochemical steps acting in parallel.

The influence of chemical stress, equation (j) in Table 4.5-1, can sometimes be taken into account by applying the general rate equation in chemical reaction kinetics. However, equation (j) is only relevant for those cases when a chemical compound in the environment acts as a reactant or catalyst in a chemical degradation reaction. For other cases of chemical stress, the role played by the chemical compound may not be directly coupled to a chemical reaction. An example of the appearance of chemical stress in different roles is the degradation of carbon steel by way of atmospheric corrosion. The most important degradation factors here are atmospheric oxygen, pollutants such as sulfur dioxide and chlorides, and the degradation factor of moisture. Atmospheric oxygen is the reactant. Sulfur dioxide is a catalyst, which reacts chemically in the corrosion process. Moisture is adsorbed on the surface and forms a liquid film so that the electrochemical process of corrosion occurs. Chlorides in the air, which are also deposited on the surface, increase the electrical conductivity of the adsorbed film when present, that in turn increases the corrosion rate. In addition, deposited chlorides on the surface are hygroscopic, which lowers the critical humidity level for corrosion. This increases the time-of-wetness and thus the time of corrosion.

Atmospheric corrosion of carbon steel, copper, zinc, and aluminum can be described quite well in terms of the degradation factors, i.e., the time-of-wetness, yearly deposited amount of sulfur dioxide, and yearly deposited amount of chloride. Accordingly, the corrosion rate in different climates can be estimated (ISO 9223:ISO-9223:1992). However, the example shows that when chemical stress is involved, detailed knowledge of the degradation mechanisms is necessary before reasonably proceeding with mathematical modeling of the degradation processes.

Atmospheric corrosion involves quite complex degradation mechanisms and therefore the principle of comparative testing using standard metals as references may sometimes be applied in such cases, e.g., equation (k) in Table 4.5-1. This kind of approach has been used in accelerated testing of selective solar absorber surfaces (Carlsson *et al.*, 1994). In accelerated lifetime testing of electronic components, metallic copper is frequently used for this purpose (Eriksson *et al.*, 2001). In developing the standard test method described in Eriksson *et al.* (2001), laboratory tests were designed to reproduce the in-service corrosion behavior of metallic copper with respect to the extent of corrosion, composition of corrosion products, and rate-limiting step in the corrosion process. Sometimes more than one reference metal is used to relate laboratory test conditions to different service conditions. In ISO 16701, a cyclic general accelerated corrosion test is presented in which both zinc and carbon steel are used as reference metals (ISO 16701, 2003).

4.5.2.2 *Varying Levels of Stress with Time during an Aging Process*
Thus far, the acceleration factor approach, as represented by equation (1) can be applied only when the levels of the degradation factors are constant in time and can be assumed to act independently of each other. If this is not true, it may be useful to introduce some approximations. If the stress varies periodically with time during an

aging process and the stress factors can be assumed to act independently of each other, the following equation may be derived,

$$\int_0^{\tau_f} g(\underline{S}) \cdot dt = n \cdot \int_0^{y_p} g(\underline{S}) \cdot dt = n \cdot \int_0^{y_p} (h_1(S_1) \cdot h_2(S_2) \ldots \ldots) \cdot dt \qquad (6)$$

where y_p is the length of the period and n is the number of periods until failure is reached.

If only the variation in one stress factor, S_1, needs to be taken into account, and only the time period at a certain state of environmental influence is important, and not the history of changes in the environmental stress factors with time, when calculating the extent of degradation, equation (6) may be written as

$$\int_0^{\tau_f} g(\underline{S}) \cdot dt = n \cdot C(S_{2,c}, S_{3,c}, \ldots) \cdot \int_0^{y_p} h_1(S_1) \cdot dt$$
$$= n \cdot C(S_{2,c}, S_{3,c}, \ldots) \cdot y_p \cdot \int_{S_{1,min}}^{S_{1,max}} h_1(S_1) \cdot f(S_1) \cdot dS_1 \qquad (7)$$

where $f(S_1)$ is the frequency function for distribution of S_1 within one time-period of length y_p, and $C(S_{2,c}, S_{3,c}, \ldots)$ is a constant that depends on the time-invariant levels of stress of $S_2, S_3, \ldots$

An effective mean value, $S_{1,EFF}$, can also be introduced defined by

$$h_1(S_{1,EFF}) = \int_{S_{1,min}}^{S_{1,max}} h_1(S_1) \cdot f(S_1) \cdot dS_1 \qquad (8)$$

or if the acceleration factor is introduced, equation (8) may be rewritten as

$$a_{S1}(S_{1,EFF}, S_{1,c,D}) = \int_{S_{1,min}}^{S_{1,max}} a_{S1}(S_1, S_{1,c,D}) \cdot f(S_1) \cdot dS_1 \qquad (9)$$

If S_1 denotes absolute temperature and the Arrhenius equation is used as a basis for expressing the time-transformation function, equation (9) becomes

$$\exp[-(E_a/R) \cdot (1/T_{EFF})] = \int_{T_{min}}^{T_{max}} \exp[-(E_a/R) \cdot (1/T)] \cdot f(T) \cdot dT \qquad (10)$$

In equation (10), the effective mean value of a stress factor is material dependent.

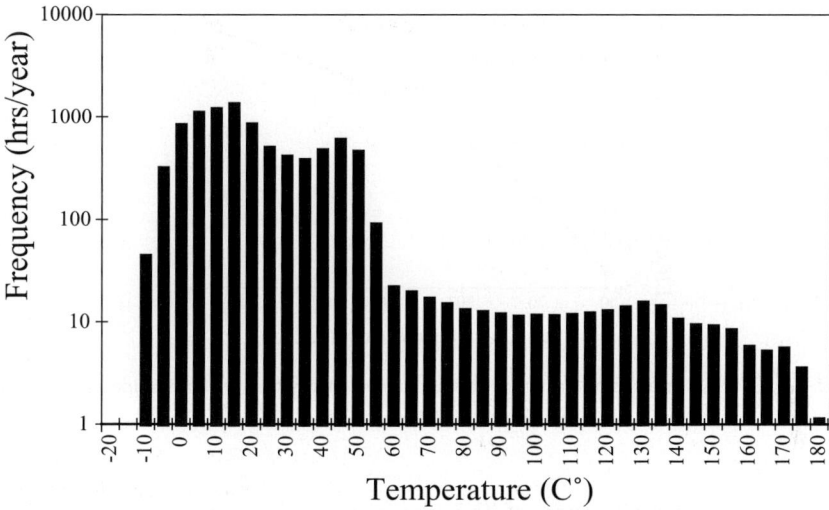

Figure 4.5-2. Absorber temperature frequency function for one year valid for a typical single glazed flat-plate solar collector. The collector was operated at stagnation conditions for one summer month of the year (Carlsson *et al.*, 1994).

The activation energy dependence of the effective mean temperature means that periods with relatively high temperatures are of prime importance. From the study of accelerated lifetime testing of selective solar absorber surfaces (Carlsson *et al.*, 1994), it was concluded that the period when a solar collector is operating at stagnation conditions more or less solely determined the effective mean temperature of the absorber in a solar collector, although it represented only one month of the year (see Figure 4.5-2).

If the variation in more than one stress factor needs to be taken into account, combined time-transformation functions are used. For example, the effective mean temperature during time-of-wetness conditions for a selective solar absorber surface in a solar collector was calculated from measured absorber temperatures and the corresponding air humidity data for one year by using equation (11) (Carlsson *et al.*, 1994).

$$\exp\{-E_{H,T}/(R \cdot T_{H,\text{eff}})\} = \int_{T_{H,\text{min}}}^{T_{H,\text{max}}} \exp\{-E_{H,T}/(R \cdot T_{H,\text{eff}})\} \cdot f_H(T_H) dT_H \qquad (11)$$

where $f_H(T_H)$ is the yearly-based frequency function for the service temperature of the absorber surface in a solar collector when the relative humidity level exceeds 99%. Thus, it is the time fraction of a year when the service temperature is in the interval T to $T + dT$ and the relative humidity level exceeds 99%. $T_{H,\text{max}}$ (in K) is the

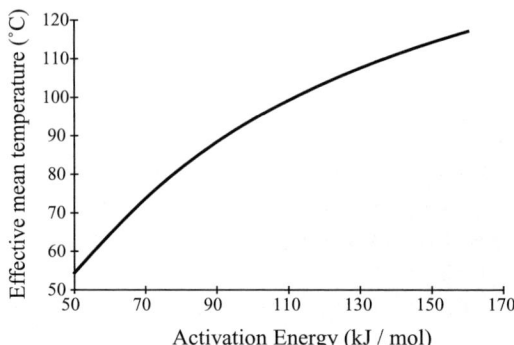

Figure 4.5-3.   Effective mean temperature versus activation energy for the thermal load profile illustrated in Figure 4.5-2.

maximum service temperature of the absorber surface in a collector, when the relative humidity level exceeds 99%. $T_{H,min}$ is 273 K, because below this temperature ice is formed on the surface of the absorber. $E_{H,T}$ is the Arrhenius activation energy and expresses the temperature dependence for a possible degradation reaction at the absorber surface resulting from condensation.

The acceleration factor for a constant load condensation test D, $a_{T,H}$, can therefore be expressed as a combination of equations (a) and (d) in Table 4.5-1, i.e.

$$a_{T,H}^{-1} = \tau_H \cdot \exp\{-(E_{H,T}/R) \cdot (T_{H,eff}^{-1} - T_D^{-1}\} \tag{12}$$

where $T_{H,eff}$ is the effective mean temperature of the absorber surface (in K), defined by equation (11). The variable τ_H is the time fraction of the year, for the time-of-wetness during which the relative humidity in the air gap is equal to or higher than 99%. The Arrhenius activation energy $E_{H,T}$ was determined from a series of constant load condensation tests performed at several different temperatures.

4.5.2.3   *More Than One Degradation Reaction*

In cases where two or more degradation reactions contribute to the deterioration in performance, simplifications need to be introduced, as can be illustrated by the model developed for accelerated life testing of solar absorber surfaces in IEA Task X (Carlsson *et al.*, 1994). In the IEA study, a constant-load accelerated aging test program was formulated to take into account the different environmental stress factors that might result in degradation in the optical performance of selected coatings. Three different kinds of aging tests were included in the program that related to (A) high-temperature degradation, (B) degradation caused by sulfur dioxide at high humidity, and (C) degradation by the action of condensed water. To distinguish between processes induced by test conditions (A) and (B), a certain threshold value for the relative humidity was assumed to exist, RH_c, such that the effect of water vapor on the rate of degradation is negligible for humidity levels below this value. For modeling, it was found reasonable to assume that the different processes of degradation (A), (B), and (C) proceed in parallel and independently of each other and that the contributions of the different processes to the changes in performance were additive.

If n denotes the overall service life in number of years, it then follows that

$$1/n = \sum_{i=A,B,C} \int_0^{y_i} (g_i(S)/g_i(S_{Sc,Di})) \cdot (1/\tau_{f,Di}) dt \tag{13}$$

where y_i is the time period of a year when service conditions according to test category i (A, B, or C) prevail.

Equation (13) may also be expressed as

$$1/n = 1/n_A + 1/n_B + 1/n_C \tag{14}$$

The quantity $1/n_A$ represents testing according to category A and at in-service conditions when the relative humidity is less than RH_c; the quantity $1/n_B$, represents testing according to category B and at in-service conditions when the relative humidity is between RH_c and 100%, and the quantity $1/n_C$, represents testing according to category C and at in-service conditions during condensation.

From the above, it can be concluded that the parameter n_A is the expected service life when only processes of category A contribute to the overall degradation. The parameters n_B and n_C are the corresponding reaction times when processes of categories B and C respectively act alone. The tentative model described enables processes of each category A, B, or C to be evaluated separately. Reference constant-load tests can also be selected for each humidity interval, so that the acceleration factor or time-transformation functions will be as simple as possible.

4.5.2.4 Nonmechanistic Phenomenological Models

Another approach to obtaining correlations between in-use and accelerated exposure results is a nonmechanistic or phenomenological methodology. This procedure is similar to the mechanistic approach in that damage functions are hypothesized, except that they are based upon macroscopic observations and effects rather than microscopic mechanisms. For example, the impact of light exposure may be known to result in degradation of a certain material and so a model is developed that accounts for this stress factor without necessarily expressing an understanding of the detailed reactions that may proceed at a microscopic level (Jorgensen *et al.*, 2001). The first step is to obtain a suitable material-specific damage function model that accurately relates changes in an appropriate response variable to relevant applied environmental stresses. For organic materials susceptible to photochemical degradation the following cumulative damage function (LD) has been proposed (Martin *et al.*, 1994):

$$LD(t) = \int_0^t L(t) \cdot dt \tag{15}$$

where $L(t)$ is the time-dependent incident spectral irradiance. $L(t)$ depends on $I(\lambda,t)$, which is convoluted with the absorption spectra of the material being exposed, $\alpha(\lambda,t)$, and the quantum efficiency, $\phi(\lambda,t)$, of the absorbed photons to propagate reactions that are harmful to the material and is integrated over an appropriate bandwidth (defined by λ_{min} and λ_{max}) throughout which light-induced damage occurs. Thus,

$$L(t) = \int_{\lambda_{min}}^{\lambda_{max}} I(\lambda, t) \cdot \alpha(\lambda, t) \cdot \phi(\lambda, t) \times d\lambda \tag{16}$$

Exposure tests of a back-metallized, polymeric film materials have produced useful results (Jorgensen *et al.*, 1996, 1999) by approximating the absorption spectra and quantum efficiency as constants in equation (16) and defining:

$$I_{UV}(t) = \int_{\lambda_{min}}^{\lambda_{max}} I(\lambda, t) \cdot d\lambda \tag{17}$$

where λ_{min} is 280 nm and λ_{max} is 315 nm (for UV-B) or λ_{min} is 280 nm and λ_{max} is 385 nm (for the total UV). For constant (controlled) irradiance, an approximate generalized cumulative degradation model is obtained in which the loss in performance, ΔP (change in response variable), with time is proportional to a power law expression of the ultraviolet irradiance I_{UV}:

$$\Delta P/\Delta t \propto (I_{UV})^n \tag{18}$$

To account for thermal effects, an Arrhenius term can be included and the change in performance after the ith time interval is:

$$\Delta P_i = A \cdot (I_{UV})^n \cdot \Delta t_i \cdot e^{-E/RT} \tag{19}$$

where T is the temperature (K) experienced by samples during exposure, R is the ideal gas law constant, and E is the Arrhenius activation energy. For constant accelerated stresses, I_{UV} and T are known; this allows equation (19) to be fit to measured values of ΔP_i and subsequent determination of the coefficients A, E, and n. For variable in-service use stresses, the time-dependent form of equation (19) must be used:

$$\Delta P(t) = A \cdot \int_0^t [I_{UV}(t)]^n \cdot e^{-E/RT(t)} \cdot dt \tag{20}$$

From the relevant coefficients determined from accelerated exposure tests performed at constant stresses, equation (20) can be used to compute a predicted loss in performance after some time t where the relevant stresses are monitored during in-use conditions; these predicted values can then be compared with actual measured values.

The approach outlined above may be slightly modified to allow moisture effects to be included in the analysis of metallized polymer reflector materials. A model was developed (Jorgensen *et al.*, 1996) that relates the time-dependent loss in performance (degradation in hemispherical reflectance at 400 nm), $\Delta \rho_{400}(t)$, to relevant environmental stresses, i.e.

$$\Delta \rho_{400}(t) = A \cdot \int_0^t L_{UV\text{-}B}(t) \cdot e^{-E/RT(t)} \cdot e^{-C \cdot RH(t)} \cdot dt \tag{21}$$

where $L_{UV\text{-}B}(t)$ is the time-dependent flux within the UV-B bandwidth (W/m^2), T(t) is the time dependent ambient temperature (K), RH(t) is the time-dependent ambient relative humidity (%), and A, E, and C are material-dependent coefficients to be determined from the measured data. Equations (20) and (21) have been used to demonstrate their usefulness and validity for organic glazing and reflector materials (see Chapter 6.2, this volume).

4.5.2.5 *Probabilistic Models*

The formulation presented in the preceding sections provides a mechanistic/deterministic treatment of life prediction. Sometimes failure mechanisms are not known or multiple mechanisms interact in a complicated manner that makes it difficult and tedious to explicitly deal with them. Another approach involves a probabilistic formalism in which observed failures are fit to life distributions that are characterized by their respective probability density functions (PDF), f(t). This results in an analytical expression for the failure rate, h(t). Life distributions that have proven to be useful in this regard include the exponential, Weibull, and lognormal functions (Martin, 1982; Carlsson, 1988; Tobias and Trindade, 1995). The failure rate can be obtained from the PDF by computing the cumulative distribution function (CDF):

$$F(t) = \int_0^t f(t) \cdot dt \tag{22}$$

The failure rate is then:

$$h(t) = f(t)/[1 - F(t)] \tag{23}$$

A plot of failure rate versus time often resembles a "bathtub" curve (Figure 4.5-4).

Early in a product's life, failures can be quite frequent as deficient parts are weeded out. As the sub-population having inadequate reliability fail ("infant mortality"), the failure rate rapidly decreases and a long period of stability is reached in which failures occur randomly or at a fairly constant and predictable rate. After an extended period of service, parts begin to wear out and the failure rate rapidly increases, marking the end of the useful life of the product. A parametric model allows projections to times much earlier than the time of the first observed failure, when only a small percentile (e.g., 0.001%) of the population may fail. This is particularly important when say 100 samples are being tested and the first observed

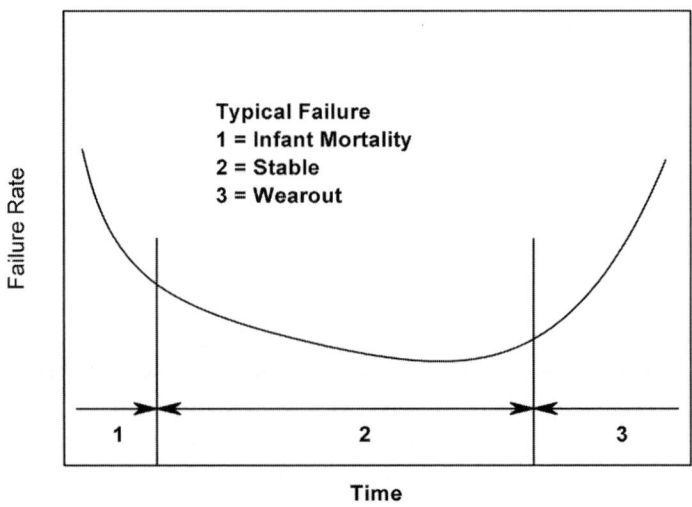

Figure 4.5-4. Bathtub shaped failure rate curve.

failure already represents 1% of the population. Justification of such a probabilistic approach should be based primarily on the fact that it works, that is, the chosen distribution fits the data well and that it results in reasonable projections when extrapolating beyond the range of the data. Ideally, the selection and use of a particular life distribution should be further justified by theoretical considerations. For example, a lognormal distribution can be shown to be theoretically applicable to chemically induced failure mechanisms such as corrosion, diffusion, or electromigration (Tobias and Trindade, 1995). As such, the lognormal distribution typically provides a good representation of the wear-out part of the failure rate curve. Although no single distribution function adequately describes the whole curve, the probability of failure can be represented as a mixture of appropriately weighted terms representing, for example, a small proportion (α) of components susceptible to manufacturing defects (d) and a larger population that follows normal competing failure modes (N):

$$F_{tot}(t) = \alpha F_d(t) + (1 - \alpha) \cdot F_N(t) \tag{24}$$

Then, the overall failure rate will have a bathtub shape (Mann *et al.*, 1974) given by:

$$h_{tot}(t) = [\alpha f_d(t) + (1 - \alpha) \cdot f_N(t)]/[1 - F_{tot}(t)] \tag{25}$$

The functional form of the Weibull distribution is particularly robust in terms of being able to represent various parts of the failure rate curve. The PDF of the Weibull distribution is:

$$f_W(t) = (m/t) \cdot (t/c)^m \exp\{-(t/c)^m\} \tag{26}$$

and the associated failure rate is:

$$h_W(t) = (m/c) \cdot (t/c)^{m-1} \tag{27}$$

where m is the shape factor and c is the characteristic life parameter. For example, Figure 4.5-2 was generated using equation (25) with a Weibull representation for defects (early failures) having $m = 0.75$, and a wearout period characterized by $m = 10$.

The key to using accelerated testing for life prediction purposes is the relation of time at the in-use stress (τ_s) to time at the elevated/accelerated stress ($\tau_{f,D}$) via an acceleration factor A_s or a time-transformation function $g(\underline{S}_D)/g(\underline{S}_s) = A_s$ (see equation (1)).

In addition to transforming the time parameter as in equation (1), the appropriate acceleration factors discussed in Section 4.5.2.1 can be used to transform the failure rate functions as well. Having determined a failure rate from accelerated testing (h_D), the failure rate at in-use stress, h_s, is (Tobias and Trindade, 1995):

$$h_s(t) = h_D(t/A_s)/A_s \qquad (28)$$

Components can often be modeled as a series system (Nelson, 1990) in which the device has a potential time-to-failure for each of M competing failure modes. The life of such a system is the smallest of those M potential times to failure, and the system fails when the first failure mode occurs. If the different times-to-failure associated with the M modes are statistically independent then the failure rate function, h(t), follows the addition model for failure rates:

$$h(t) = \sum_{i=1}^{M} h_i(t) \qquad (29)$$

For example, if a glazing material is susceptible to damage from elevated temperature (T), moisture, UV light, and other environmental stresses then:

$$h_{glazing}(t) = h_T(t) + h_{moisture}(t) + h_{UV\text{-}irradiance}(t) + \cdots\cdots \qquad (30)$$

Once the overall failure rate is known, the time, t, at which $[1 - F(t)]$ units still survive can be calculated (Mann *et al.*, 1974) from

$$\ln[1 - F(t)] = -\int_0^t h(t) \cdot dt \qquad (31)$$

4.5.2.6   *Change of Mechanism with Varying Levels of Stress*

One complication in connection with extrapolation of accelerated test results is that the mechanism of degradation or the rate-determining step in the degradation process is different at elevated stress levels than for in-service conditions. In such cases, however, both the mechanistic and nonmechanistic approaches most probably result in erroneous results. However, an approach based on a physical and chemical model of the degradation process seems preferable, in general, because extrapolation should be more accurate. However, if many competing mechanisms of degradation are involved, it may not be possible to design reasonable mechanistic models and a more general numerical approach may be preferable.

4.5.3 ACCELERATED LIFE TESTING AND ASSESSMENT OF THE EXPECTED SERVICE LIFE

In accelerated life testing, a quantitative assessment is made for the sensitivity of the overall deterioration in performance of the component and its materials to the various degradation factors. Mathematical models are then formulated to characterize the different degradation mechanisms identified. Accelerated life testing, therefore, requires the conduction of a series of tests, which also offers an opportunity to validate that the predominate degradation mechanism or mechanisms identified during screening testing do not change when the level of stress is changed. It is particularly important to validate this using tests with the lowest possible stress levels. The longest available testing time will set a limit for the lowest possible stress levels that can be selected for the tests.

Equations (32), (33), and (34) may be helpful for evaluating the results of the test program. If the accelerated tests truly represent the in-service degradation behavior, the change in performance, ΔP, with service time, y_s, should be described by the same function as the change in performance with testing time y_D for a test with acceleration factor A_s, i.e.,

$$\Delta P = f(y_s) = f(A_s \cdot y_D) \tag{32}$$

By introducing the inverse function f^{-1}, equation (32) may be rewritten as

$$y_D = (1/A_s) \cdot f^{-1}(\Delta P) \tag{33}$$

or

$$\ln(y_D) = -\ln(A_s) + \ln\{f^{-1}(\Delta P)\} \tag{34}$$

Equation (34) forms the basis for determining the parameters in the time-transformation or acceleration-factor equation and also for establishing if the performance versus transformed-time relation is the same, irrespective of the stress level.

There are general problems in applying equation (34) to evaluate the results of tests. For high-stress level tests, the performance versus time curve may be difficult to determine for small ΔP values because the degradation reaction(s) is too fast. However, for low stress level tests the appearance of the performance versus time curve at high ΔP values may be difficult to determine because the testing time may be too long. These restrictions limit the range for which it is possible to establish if the relation between the performance and the transformed time is the same, irrespective of the stress level of test. Table 4.5-2 and Figure 4.5-5 show how equation (34) may

Table 4.5-2. Steps in the evaluation of the service life of the nickel-pigmented anodized aluminum absorber surface with respect to its resistance to hydratization studied in IEA task 10 (Carlsson *et al.*, 1994)

Component/materials	Nickel-pigmented, anodized aluminum absorber surface
Degradation mechanisms	Hydratization of aluminum oxide and electrochemical corrosion of metallic Ni particles by the action of condensed water
Degradation indicator	$PC = \Delta\alpha - 0.25\Delta\varepsilon$ α = solar absorptance, ε = thermal emittance (PC of 0.05 corresponds to a loss in 5% of the solar system performance)
Environmental resistance tests	Exposure tests under constant condensation (the sample surface was cooled 5°C below the surrounding air that was kept at 95% RH) and temperatures ranging from 12 to 30°C
Performance versus time relation	$\ln(\text{testing time}) = \{(E_{H,T}/R)\,(T_D^{-1} - T_s^{-1}) + \ln(\sum_{n=0} d_n \cdot (\Delta PC))^n)$ T_D = testing temperature; $T_s = 20°C$ (see (a) in Table 4.5-1) Performance versus time relations at different temperatures are shown in Figure 4.5-5
Service life prediction	See mechanism (C) in Table 4.5-3

be applied. The example chosen is from the IEA absorber surface case study (Carlsson *et al.*, 1994).

After it has been concluded that the assumed model for degradation is applicable, the service life determined by the pertinent degradation mechanism is estimated by extrapolation to service conditions. If the in-service conditions vary, effective mean values of stress need to be assessed from measured service stress data (see Chapter 4.5, Section 4.5.2.2)

As a result, it may be possible to express the importance of different degradation mechanisms in terms of expected service life, which are valid only under the assumption that each degradation mechanism acts alone, see for example, the results from the absorber surface case study given in Table 4.5-3. To estimate the overall service life when two or more degradation reactions contribute to the degradation in materials performance, some guidance is given in Chapter 4.5, Section 4.5.2.3.

In cases where the degradation process is very complex and the degradation rate is difficult to model mathematically in terms of measurable degradation factors, the use of reference materials or components in life testing may be applied (see, for example, mechanism (B) in Table 4.5-3). Other examples on how reference materials are used

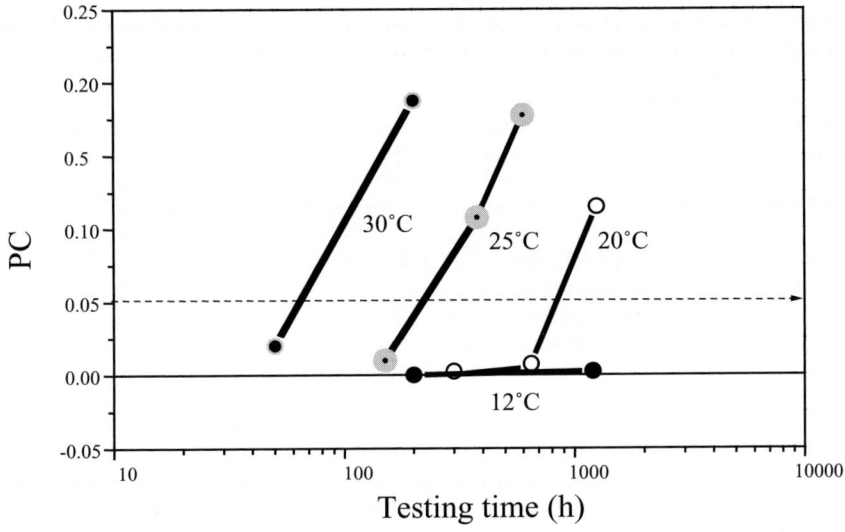

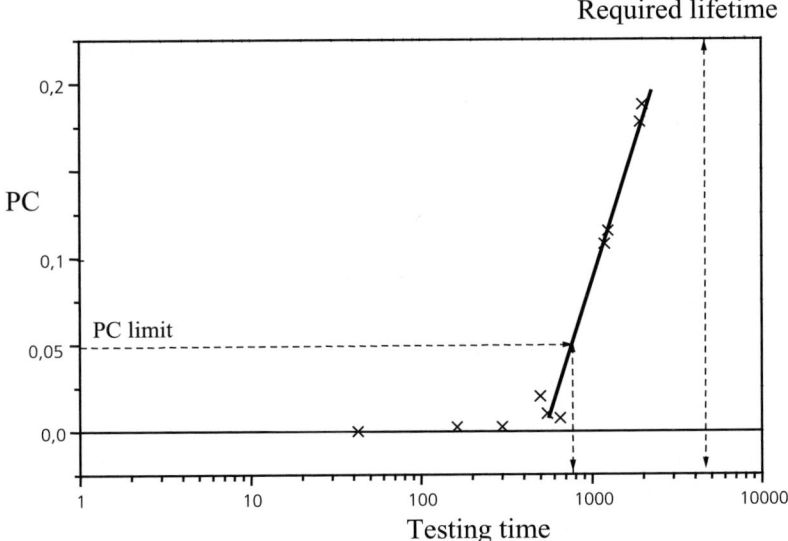

Figure 4.5-5. Example from accelerated testing of a nickel-pigmented anodized aluminum absorber surface with respect to its resistance to the action of condensed water on its surface. Top: Results of a series of constant-load condensation tests at different temperatures. Bottom: All test results from top figure transformed to the 20°C test condition using the performance versus time relation given in Table 4.5-2.

Table 4.5-3. Estimated service life of the nickel-pigmented anodized aluminum absorber surface resulting from different degradation mechanisms and assuming that each degradation mechanism acts alone (Carlsson *et al.*, 1994)

Degradation mechanism	Time-transformation function (see Table 4.5-1)	Estimated service life with $PC = -\Delta\alpha + \Delta\varepsilon < 0.05$[1] (years)
(A) High-temperature oxidation of metallic Ni particles	$a_T = \exp[-(E_a/R) \cdot (1/T_D - 1/T_s)]$ E_a = activation energy R = ideal gas law constant T_D = testing temperature T_s = effective mean temperature during in-service use	$> 10^5$
(B) Electrochemical corrosion of metallic Ni particles at high-humidity levels and in the presence of sulfur dioxide	$a_{Co} = \tau_{M,s}/\tau_{M,D}$ $\tau_{M,s}$ = time to reach a certain extent of corrosion of reference metal during in-service use $\tau_{M,D}$ = time needed to reach the same extent of corrosion of reference metal in test D (zinc was used as the reference metal)	12 (The coating is assumed to be installed in a permeable, strongly ventilated collector) 34 (The coating is assumed to be installed in an airtight collector with controlled ventilation)
(C) Hydratization of aluminum oxide and electrochemical corrosion of metallic Ni particles by the action of condensed water	$a_{T,H}^{-1} = \tau_H \exp\{-(E_{H,T}/R)(T_{H,eff}^{-1} - T_R^{-1})\}$ $T_{H,eff}$ = effective mean temperature of the absorber surface when the relative humidity in the air gap is equal to or higher than 99%. τ_H = the time fraction of the year, time-of-wetness, during which the relative humidity in the collector air gap is equal to or higher than 99%. $E_{H,T}$ = Arrhenius activation energy	9 (The coating is assumed to be installed in a permeable strongly ventilated collector)

[1] $PC = 0.05$ corresponds to a decrease in the solar system performance of 5%.

in connection with atmospheric corrosion testing are available (Eriksson *et al.*, 2001; ISO 16701, 2003).

4.5.4 REASONABILITY ASSESSMENT AND VALIDATION

By using accelerated life testing, potential degradation mechanisms limiting the service life of a component may be identified. However, the service life is determined by the material degradation mechanisms observed in the accelerated tests at relatively high

Table 4.5-4. Some general characteristics of six solar DHW systems from which nickel-pigmented anodized aluminum absorber samples were analyzed after long period of service (Carlsson *et al.*, 2000)

DHM-system and location	Age	Collector	Remark
DK 1 (Copenhagen, Denmark)	12 years	BATEC 22 SEL	
DK 2 (Zealand, Denmark)	11 years	BATEC 22 SEL	Frost burst, one fin replaced
DK 3 (Karlslunde, Denmark)	10 years	BATEC 22 SEL	
DK 4 (Zealand, Denmark)	10 years	BATEC 22 SEL	Frost burst, one fin replaced
CH 1 (Gisikon, Switzerland)	15 years	Mühlemann Einbau	Collector leaky 1993
CH 2 (Frauenfeld, Switzerland)	15 years	SOLTOP Einbau	Plastic cover replaced 1988 Collector leaky 1990

levels of stress. Life-limiting degradation mechanisms may exist that cannot be identified by accelerated life testing, because the knowledge and experience of causes of degradation for a particular material in a component may be too limited.

The best approach for validating an estimated service life from accelerated testing, therefore, is to use the results from the accelerated life tests to predict expected changes in materials properties or component performance versus service time and then validate the estimate by using in-service long-term tests. If the predicted change in performance with time is actually observed the projection was correct, if not, further analysis of the degradation mechanisms is required. The case study on polymeric glazing and reflectors in Chapter 6.2 is an example of a successful validation of a lifetime projection. The advantage with such an approach is, of course, that deviations from predicted performance may provide an early indication that the assumptions made about the life-limiting degradation processes are not completely correct and the result might indicate that some important process of degradation was not anticipated.

The results of validation tests therefore can be used to revise a predicted service life and form the starting point also for improving the component tested with respect to environmental resistance, if so required. It should be remembered that the main objective of accelerated life testing is to try to identify those failures that may result in an unacceptably short service life of a component. In terms of service life, the critical question is the probability that the service life is greater and an economically viable value.

Some general characteristics of six solar DHW systems from which nickel-pigmented anodized aluminum absorber samples were analyzed are listed in Table 4.5-4. The method used to validate the predicted service life of the nickel-pigmented anodized aluminum absorber coating in the absorber surface case study is illustrated in Figures 4.5-6 and 4.5-7. In this case the estimated service life from accelerated life testing for the absorber coating in air tight collectors and permeable

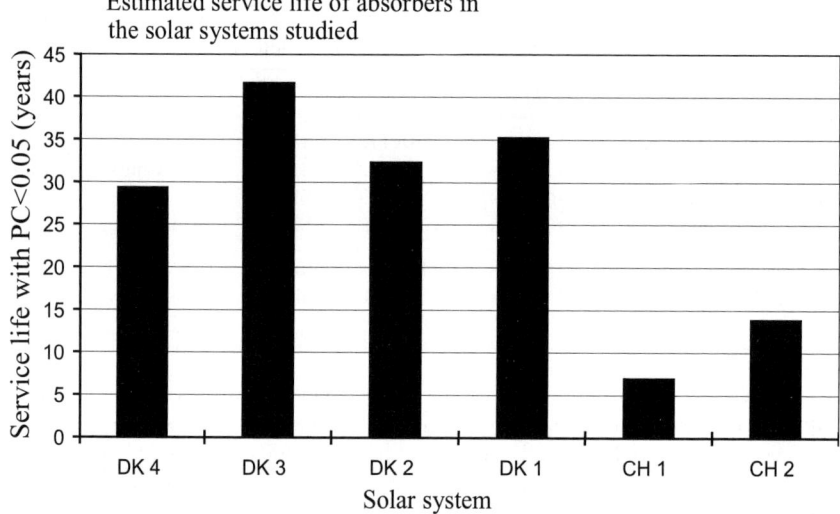

Figure 4.5-6. Service life for the absorber coatings in the different solar DHW systems, see Table 4.5-4, estimated from measurements on samples from the different systems (Carlsson *et al.*, 2000).

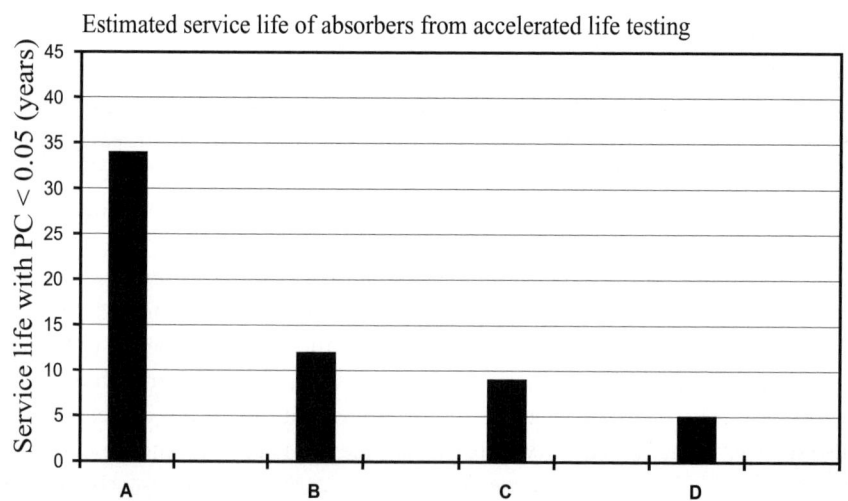

Figure 4.5-7. Service life of absorber coating, as determined from the results of accelerated life testing. A: Corrosion induced by sulfur dioxide (airtight collector); B: Corrosion induced by sulfur dioxide (permeable collector); C: Hydratization (permeable collector); D: Both mechanisms operative (permeable collector) (Carlsson *et al.*, 2000).

collectors, Figure 4.5-7, agreed well with actually measured data, Figure 4.5-6. For the absorber coating in a properly designed solar collector, the service life seems sufficiently good. For the absorber coating in a permeable strongly ventilated solar collector, where probably the glazing or sealant has failed, the humidity level is raised to such high levels that the service life is reduced to an unacceptably low value.

REFERENCES

Bartenev, G.M. & Zelenenev Y.V. (1974) *Relaxation Phenomena in Polymers*, John Wiley & Son, London.

Carlsson, B. (Ed.) (1988) Survey of Service Life Prediction Methods for Materials in Solar Heating and Cooling, *International Energy Agency*, Solar Heating and Cooling Programme Task X: Solar Energy Materials Research and Development, Technical Report, Swedish Council for Building Research Document D16:1988, Stockholm, Sweden.

Carlsson, B., Frei, U., Köhl, M. & Möller, K. (1994) Accelerated Life Testing of Solar Energy Materials - Case Study of Some Selective Solar Absorber Coatings for DHW systems, International Energy Agency, Solar Heating and Cooling Programme Task X: Solar Materials Research and Development) Technical Report, SP- Report 1994:13, SE-50115 Borås, Sweden.

Carlsson, B., Möller, K., Frei, U., Brunold, S. & Köhl, M. (2000) Comparison Between Predicted and Actually Observed in-service Degradation of a Nickel Pigmented Anodized Aluminium Absorber Coating for solar DHW Systems. *Solar Energy Materials and Solar Cells*, **6**, 223.

Dixon, R.K. (1980) Thermal Aging Prediction from an Arrhenius Plot with Only one Data Point; *IEEE Transaction on Elec. Insula.*, E1 – 15, 4 331.

Eriksson, P., Carlsson, B. & Odnevall, Wallinder I. (2001) Design of Accelerated Corrosion Tests for Electronic Components in Automotive Applications, *IEEE Transactions of components and Packaging technologies*, **24**(1), 1521.

Freiman, S.W., Ganzalea, A.C. & Wiederhorn, S.M. (1984) Life Prediction for Solar Glasses, *Ceramic Bulletin*, **63**, 597.

Handbook in Environmental Engineering, Swedish Environmental Engineering Society, 1998 (This book can be ordered from SP Swedish National Testing and Research Institute, P.O. Box 857, SE-51015 Borås, Sweden).

ISO 9223:ISO-9223:1992 Corrosion of metals and alloys – Corrosivity of atmospheres – Classification; International Standardization Organization, CH -1211 Geneva, Switzerland.

ISO 16701 (2003) Accelerated Corrosion Testing under Controlled Conditions of Humidity Cycling and Intermittent Spraying of a Salt Solution; CH-1211 Geneva, Switzerland.

Jorgensen, G., Bingham, C., King, D., Lewandowski, A., Netter, J., Terwilliger, K. & Adamsons, K. (2001) In: Bauer, D.R. & Martin, J.W. (Eds.), *Service Life Prediction Methodology and Metrologies*, p. 396, ACS Symposium Series 805; ISBN 0841236933, American Chemical Society.

Jorgensen, G., Bingham, C., Netter, J., Goggin, R. & Lewandowski, A. (1999) In: Bauer, D.R. & Martin, J.W. (Eds.), *Service Life Prediction of Organic Coatings, A Systems Approach*, pp. 170–185, ACS Symposium Series 722; American Chemical Society, Oxford University Press, Washington, DC.

Jorgensen, G., Kim, H.-M. & Wendelin, T. (1996) In: Herling, R.J. (Ed.), *Durability Testing of Nonmetallic Materials*, pp. 121–135, ASTM STP 1294, American Society for Testing and Materials, Scranton, PA.

Mann, N.R., Schafer, R.E. & Singpurwalla, N.D. (1974) *Methods for Statistical Analysis of Reliability and Life Data*, p. 118, John Wiley and Sons, New York.

Martin, J.W. (1982) Time Transformation Functions Commonly used in Life Testing Analysis, *Durability of Building Materials*, **1**, 175.

Martin, J.W., Saunders, S.C., Floyd, F.L. & Weinberg, J.P. (1994) *Methodologies for Predicting the Service Life of Coatings Systems*, NIST Building Science Series 172, National Institute of Standards and Technology, Gaithersburg, MD, USA (available from National Technical Information Service, order number PB95-146387).

Nelson, W. (1990) *Accelerated Life Testing*, John Wiley and Sons, Inc., New York.

Nelson, W.B. (1975) Analysis of Accelerated Life Test Data – Least Squares Method for the Inverse Power Law Model, *IEEE Transaction on Reliability*, **24**, 103.

Thomas, D.K. & Sinnott, R. (1969) Prediction of Shelf Storage Life from Accelerated Heat Aging Tests, *Journal of IRI*, 163.

Tobias, P.S. & Trindade, D.C. (1995) *Applied Reliability*, Van Nostrand Reinhold, New York.

Chapter 4.6

Nonmechanistic Phenomenological Treatment of Glazings and Reflectors

G. Jorgensen

*National Renewable Energy Laboratory, 1617 Cole Boulevard,
Golden, CO 80401-3393, USA*

Abstract: The purpose of this chapter is to illustrate how a phenomenological approach can be accurately used to interpret the results from a number of highly accelerated exposure tests of organic materials. Two approaches are commonly used to derive correlations between in-service and accelerated exposure test results. When detailed degradation mechanisms are well understood, a deterministic formalism can be applied in which a precise damage function model is utilized. If failure mechanisms are not known or multiple mechanisms interact in a complicated manner that makes it difficult and tedious to treat them explicitly, a probabilistic procedure can be used. Observed failures are fit to appropriate life distributions to obtain expressions for related failure rates. A third approach uses a phenomenological methodology. This procedure is similar to the deterministic approach in which damage functions are hypothesized, except that they are based on macroscopic observations and effects, rather than on microscopic mechanisms. In this chapter, the phenomenological approach is used to describe accurately the results from a number of highly accelerated exposure tests of organic materials. The ensuing damage function models are then used to predict behavior during in-service use. Excellent agreement is demonstrated between these predictions and actual measured data, thereby validating the phenomenological approach and providing a very useful way to estimate service life of organic-based materials.

Keywords: Accelerated   testing,   Outdoor   testing,   Organic   materials, Performance degradation, Service life prediction, Polymeric glazings, Solar reflectors, Weathering

LIST OF ABBREVIATIONS AND ACRONYMS

A	material-dependent constant
AET	Accelerated Exposure Test
C	material-dependent constant
D	cumulative light dosage (MJ/m^2)

E activation energy (kJ/mole)
I spectral irradiance (W/m^2/nm)
I_{UV} ultraviolet irradiance (W/m^2)
k Boltzmann constant
L integrated irradiance (W/m^2)
n material-dependent constant (exponent)
OET   Outdoor Exposure Test
PC   polycarbonate
PMMA   polymethylmethacrylate
PVC   polyvinyl chloride
RH   relative humidity (%)
t time
T temperature
UV   ultraviolet
UV-B   the region of the electromagnetic spectrum between 280 and 315 nm
α absorption spectrum
ϕ quantum efficiency
λ wavelength (nm)
ΔP change in performance or response variable
$\Delta \rho_{400}$ change in hemispherical reflectance at 400 nm

4.6.1 INTRODUCTION

Several service life prediction methods were discussed in Chapter 4.5. Another approach to obtaining correlations between in-use and accelerated exposure test (AET) results is a nonmechanistic or phenomenological methodology. This procedure is similar to the deterministic approach in which damage functions are hypothesized, except that they are based upon macroscopic observations and effects rather than microscopic mechanisms. For example, the impact of light exposure may be known to result in degradation of a certain material and so a model is developed that accounts for this stress factor without necessarily expressing an understanding of the detailed reactions that may proceed at a microscopic level (Jorgensen *et al.*, 2002). The first step is to obtain a suitable material-specific damage function model that accurately relates changes in an appropriate response variable to relevant applied environmental stresses. Organic materials are known to be susceptible to degradation that results from exposure to a cumulative dosage (D) of light (Martin *et al.*, 1994).

$$D(t) \sim \int_0^t L(t)\mathrm{d}t \tag{1}$$

$L(t)$, which is calculated from the equation

$$L(t) = \int_{\lambda_{\min}}^{\lambda_{\max}} I(\lambda,\, t)\alpha(\lambda,\, t)\phi(\lambda,\, t)\mathrm{d}\lambda \tag{2}$$

is the time-dependent incident spectral irradiance in which $I(\lambda,t)$ is convoluted with the absorption spectra of the material being exposed, $\alpha(\lambda,\, t)$ and $\phi(\lambda,\, t)$, the quantum efficiency of the absorbed photons to propagate reactions that are harmful to the material, and integrated over an appropriate bandwidth (defined by $\lambda_{\min}$ and $\lambda_{\max}$) throughout which light-induced damage occurs.

Exposure tests of back-metallized polymeric film organic materials have yielded useful results (Jorgensen *et al.*, 1996, 1999) by approximating the absorption spectra and quantum efficiency as constants in equation (2) and defining:

$$I_{\mathrm{UV}}(t) = \int_{\lambda_{\min}}^{\lambda_{\max}} I(\lambda,\, t)\,\mathrm{d}\lambda \tag{3}$$

with $\lambda_{\min} = 280\,\mathrm{nm}$ and $\lambda_{\max} = 315\,\mathrm{nm}$ for UV-B, or $\lambda_{\min} = 280\,\mathrm{nm}$ and $\lambda_{\max} = 400\,\mathrm{nm}$ for the total ultraviolet (UV) exposure. For a constant (controlled) irradiance, this results in an approximate generalized cumulative dosage model in which the loss in performance, ΔP (change in the response variable), with time is proportional to a power law expression of the UV irradiance I_{UV} (Martin, 1982; Carlsson, 1989):

$$\Delta P / \Delta t \cong (I_{\mathrm{UV}})^n \tag{4}$$

To account for thermal effects, an Arrhenius term can be included, so ΔP after the ith time interval is:

$$\Delta P_i = A(I_{\mathrm{UV}})^n \Delta t_i \exp(-E/kT) \tag{5}$$

where T is the temperature (K) experienced by the samples during exposure, k is Boltzmann's constant, and E is the activation energy. Because I_{UV} and T are known for constant accelerated stresses, equation (5) can be fit to measured values of ΔP_i and the determination of the coefficients A, E, and n can be accomplished. For the

variable stresses encountered during in-service exposures, the time-dependent form of equation (5) must be used:

$$\Delta P(t) = A \int_0^t [I_{UV}(t)]^n \exp[-E/kT(t)] \, dt \tag{6}$$

Because the relevant coefficients can be determined from accelerated exposure tests performed at constant stresses, equation (6) can be used to compute a predicted loss in performance after some time t where the relevant stresses are monitored during in-use conditions; these predicted values can then be compared with actual measured values. This nonmechanistic phenomenological approach has been usefully demonstrated and validated for organic glazing and reflector materials.

4.6.2   PREDICTED DEGRADATION OF ORGANIC GLAZINGS

For predicting the performance losses in organic glazings, values of ΔP_i were measured for samples of glazing materials exposed outdoors while the relevant time dependent stress variables were monitored (Chapter 4.7). Thus, equation (6) could be used to predict ΔP_i for comparison with the measured data. To validate the methodology presented above, two types of sheet (0.32-cm thick) glazing materials that were exposed at a variety of outdoor test sites were considered. These included polyvinyl chloride (PVC) and a UV-stabilized polycarbonate (PC). These materials were also exposed (ASTM G155-00ae1, 2003; ASTM D2565-99, 2003) in an Atlas Ci-5000 WeatherOmeter® at a UV intensity of about twice that of the typical outdoor terrestrial level, and at 50 times and 100 times using a UV concentrator (Jorgensen *et al.*, 2002). The response variable was chosen to be the hemispherical transmittance between 400 and 500 nm because, in general, that is the spectral region most sensitive to the stress exposure induced loss in performance (see Chapter 2.5, Figure 2.5-1). The same damage functions expressed in equations (5) and (6) were assumed. Data from the Atlas Ci-5000 and the UV concentrator exposures were used to fit equation (5) and to obtain the model coefficients; the results are given in Table 4.6-1. Values of the activation energies derived are reasonable for mechanisms

Table 4.6-1. Coefficients derived from equation (5) for representative clear polymer sheet samples

Polymer Sheet	A	n	E (kJ/mole)
Polyvinyl Chloride	2892	0.669	35.33
UV-stabilized polycarbonate	5.497	1.093	28.00

resulting from photothermal degradation processes. A value of $n \sim 2/3$ for PVC implies that exposure to 50 to 100 times the in-service light intensity had a net effect of only 15 to 25 times. To explain the lower net effect, we suspect some shielding or rate limiting reaction(s) occur that precludes all the photons from participating in the degradation mechanism(s). For the UV-stabilized PC samples, a value of $n = 1$ supports our conclusion that exposure of this material follows strict reciprocity even up to 100 times the exposure during in-service. Thus, all incident photons fully contribute to degradation reactions that proceed at twice the rate undergone at 50 times the exposure and 50 times the rate experienced at 2 times the exposure.

Using the coefficients from Table 4.6-1 and the time-monitored values of sample temperature and UV irradiance, the loss in performance was predicted using equation (6) for PVC and PC sheets exposed outdoors in Golden, CO, and Phoenix, AZ. The predicted values were then compared with actual measured data for these materials exposed at these sites. The results are presented in Figure 4.6-1. Time-dependent changes in weathering variables produce the irregular shapes of the predicted curves. Excellent agreement is evident between the measured and predicted data, thereby validating the phenomenological approach, i.e., using accelerated test

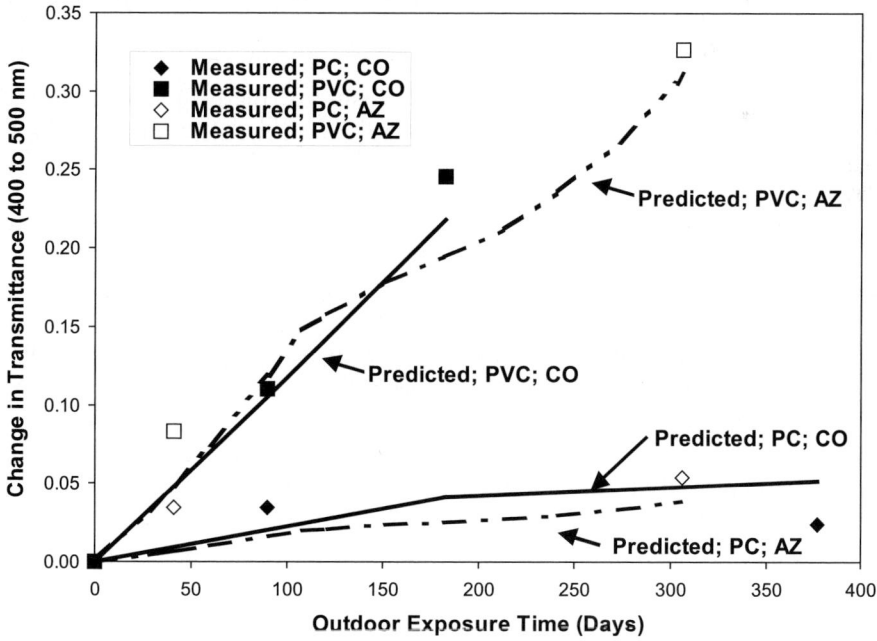

Figure 4.6-1.   Measured versus predicted change in hemispherical transmittance between 400 and 500 nm for sheets of PVC and PC exposed at outdoor test sites in Phoenix, AZ and Golden, CO.

results to obtain model coefficients, and then using these coefficients to predict the time-dependent in-service degradation, and the assumed damage function model.

4.6.3 PREDICTING DEGRADATION OF METALLIZED POLYMER REFLECTORS

For predicting performance losses of metallized polymer reflector materials, the approach described in Section 4.6.2 was slightly modified to allow moisture effects to be included in the analysis of the exposed specimens. The model, developed in reference Jorgensen *et al.* (1996), relates the time-dependent loss in performance, which was the change in hemispherical reflectance at 400 nm, $\Delta\rho_{400}(t)$, to the relevant environmental stresses, i.e.,

$$\Delta\rho_{400}(t) = \int_0^t AL_{\text{UV-B}}(t)\exp[-E/kT(t)]\exp[C*\text{RH}(t)]\mathrm{d}t \tag{7}$$

where $L_{\text{UV-B}}(t)$ is the time-dependent flux within the UV-B bandwidth (W/m^2), $T(t)$ is the time-dependent ambient temperature (K), $\text{RH}(t)$ is the time-dependent ambient relative humidity (%), and A, E, C are the material-dependent coefficients to be determined from measured data.

The hemispherical reflectance at 400 nm was chosen as the performance parameter because it provides a particularly sensitive measure of the degradation (Jorgensen *et al.*, 1996). $L_{\text{UV-B}}(t)$ is obtained by integrating the spectral irradiance in the UV-B bandwidth ($\lambda = 280$ to 315 nm):

$$L_{\text{UV-B}}(t) = \int_{\lambda=\text{UV-B}} I(\lambda, t)\,\mathrm{d}\lambda \tag{8}$$

where $I(\lambda, t)$ is the time-dependent terrestrial solar spectral irradiance (W/m^2/nm). The UV-B bandwidth is emphasized because light in this spectral range has been found to be the most damaging to the solar reflector materials of interest. The values of $L_{\text{UV-B}}$ are the data recorded and reported by an EKO Model 210-W UV-B pyranometer deployed at our outdoor test sites (see Chapter 4.7). Temperature and relative humidity data are also measured and monitored at these sites. For discrete time intervals, the degradation in performance will be the sum of the coincident temporal variations of the appropriate environmental stresses:

$$\Delta\rho_{400}(t) = \sum_t AL_{\text{UV-B}}(t)\exp[-E/kT(t)]\exp[C*\text{RH}(t)]\Delta t \tag{9}$$

Equation (9) is also applicable for tests of materials exposed to laboratory-controlled conditions. If the various relevant environmental stresses are controlled at constant levels, the loss in performance after some cumulative exposure time, Δt, will be:

$$\Delta \rho_{400}(t) = A L_{UV\text{-}B} \exp[-E/kT] \exp[C * RH] \Delta t \tag{10}$$

By performing a series of laboratory experiments for which equation (10) is valid, the coefficients A, C, and E were obtained. The procedure is to apply constant levels of stresses ($L_{UV\text{-}B}$, T, and RH) to materials of interest and then to measure the loss in performance, $\Delta \rho_{400}(t)$, at periodic intervals, Δt. The reflectance, $\Delta \rho_{400}(t)$, must be measured at several intensities of each stress for the coefficient associated with that stress to be determined. For example, if only one temperature is used, then the effect of the $\exp[-E/kT]$ term will be constant and can be combined with the coefficient A. The best approach is to use a factorial designed experiment in which at least two intensities of each stress of interest (for example, T_{high} and T_{low}) are used. Sample exposure chambers that allow two different temperatures and relative humidities to be concurrently applied to samples during exposure to a common level of light intensity have been designed and used for exposure testing (Figure 4.6-2). This exposure chamber allows a four-fold increase in experimental throughput.

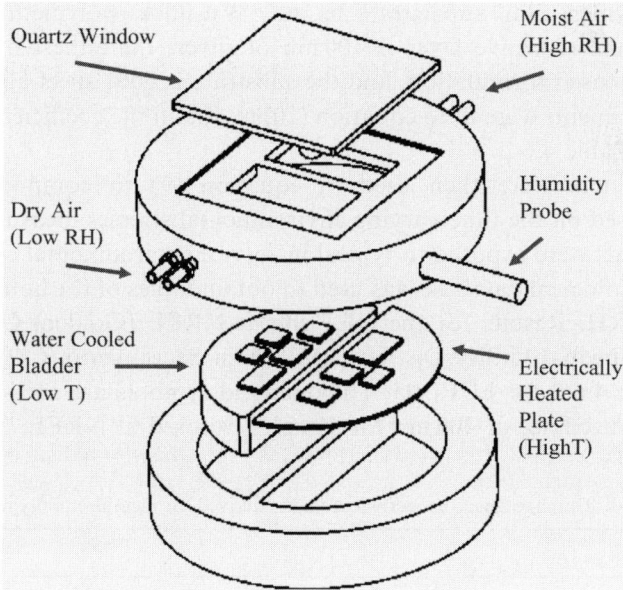

Figure 4.6-2. Exposure chamber allowing concurrent exposure to light, two different temperatures, and two different relative humidities.

Furthermore, by using elevated intensities of the applied stresses and with the assumption that the stresses do not accelerate the reactions to the extent that unrealistic failure mechanisms are introduced, measurable degradation can be achieved in highly abbreviated times, allowing estimates of relevant coefficients to be rapidly determined. Once the coefficients are known, they can be used in equation (9), along with meteorological data monitored at the geographical locations of interest, to compute performance loss as a function of time. By defining failure as the loss in performance below some desired or preferred value, the associated service lifetime can be predicted. Alternatively, if measured data are available for a given site at which meteorological data is available, equation (9) can be used to validate the methodology for predicting material lifetimes.

Silvered polymer reflectors, which were also being exposed at outdoor exposure test (OET) sites, were subjected to ultra-accelerated levels of natural sunlight in a High Flux Solar Furnace at the National Renewable Energy Laboratory (NREL, Golden, CO), with L_{UV-B} equal to 50, 75, and 100 suns (Jorgensen *et al.*, 1999). The silvered polymer reflectors, ECP-305, were made by the 3M Company. The reflector has the construction:

Polymer Film Superstrate/Reflective Layer/Adhesive/Substrate

where the polymer film superstrate is an 89 μ-thick polymethylmethacrylate (PMMA) film, the reflective layer is 100 nm of silver, the adhesive is a pressure-sensitive acrylic-based formulation, and the substrate is 6061 sheet aluminum. Data from these experiments were fit to equation (10) to obtain the coefficients A, C, and E as compiled in Table 4.6-2.

These parameters were then used in equation (9) to compute the loss in performance based on the time-varying environmental stresses measured at the sites at which materials were exposed to typical in-service environmental conditions. The OET site meteorological database was used to obtain tables of the hourly averages of L_{UV-B}, T, and RH. Results for the OET site at NREL (Golden, CO) and at the Sacramento Municipal Utility District (SMUD, in Sacramento, CA) OET site are shown in Figure 4.6-3 for ECP-305. The diamond symbols are measured values of hemispherical reflectance at 400 nm for samples exposed at NREL, and the square

Table 4.6-2. Damage function coefficient values derived for metallized polymer films

Coefficient	Value
A	2.27×10^{-5}
C	0.0073
E(kJ/mole)	11.47

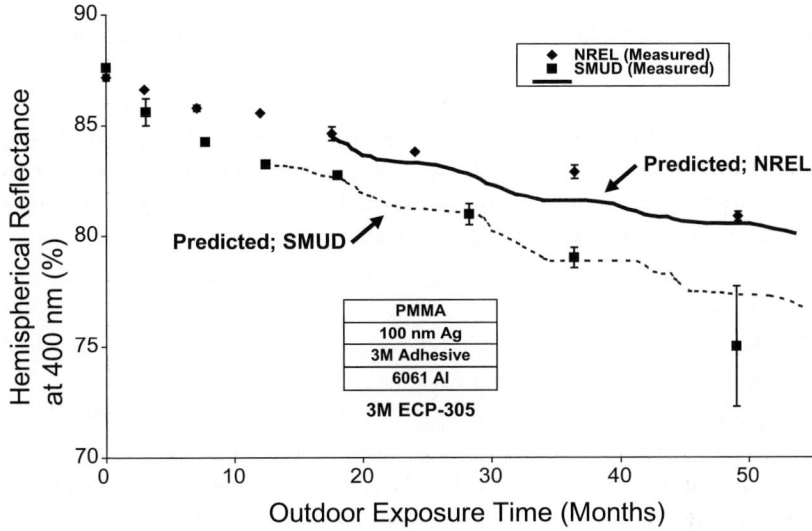

Figure 4.6-3. Measured versus predicted hemispherical reflectance at 400 nm for metallized polymer samples (3M ECP-305) exposed at NREL (Golden, CO) and SMUD (Sacramento, CA). The inset shows a cross section of ECP-305.

symbols are measured values for samples exposed at SMUD. The bar range is for plus or minus one standard deviation. The solid or dashed lines are the predicted reflectance calculated from equation (9).

As indicated in Figure 4.6-3, exposure of ECP-305 samples was initiated before acquisition of meteorological data was begun. Therefore, the predicted data were renormalized to the first measurement made after meteorological data became available. Excellent agreement between the predicted and measured data is seen for ECP-305 exposed at both sites. As with the results in Section 4.6.2, this again demonstrates the usefulness of the phenomenological methodology, in this case, for predicting the loss in performance of metallized polymer reflectors.

4.6.4   CONCLUSIONS

A practical approach to obtain correlations between accelerated and outdoor exposure test results has been demonstrated and validated for several optical elements comprised of representative organic materials. The procedure is based on phenomenological principles, but docs not rcly on a dctailcd understanding and analysis of microscopic mechanisms that lead to degradation. However, it does rely on knowledge of the macroscopic relationship between applied stresses and their effects on relevant response variables. In this regard, this approach may be more palpable than probabilistic treatments and more readily allow product improvements to be made. A straightforward method has been outlined that allows determination of coefficients that relate applied stresses to response variables in the damage function. Once validated, the fully developed damage function can then be used with monitored or statistically generated real-world variable weather data to allow loss in performance to be predicted with confidence.

REFERENCES

ASTM D2565-99, Standard Practice for Xenon Arc Exposure of Plastics Intended for Outdoor Applications. *Annual Book of ASTM Standards 2003*, **8.02** American Society for Testing and Materials, West Conshohocken, PA.

ASTM G155-00ae1, Standard Practice for Operating Xenon Arc Light Apparatus for Exposure of Non-Metallic Materials, *Annual Book of ASTM Standards 2003* **14.04**, American Society for Testing and Materials, West Conshohocken, PA.

Carlsson, B. (1989) *Solar Materials Research and Development; Survey of Service Life Prediction Methods for Materials in Solar Heating and Cooling; International Energy Agency Solar Heating and Cooling Program, Report # BFR-D-16-1989 (DE90 748556), ISBN 91-540-5063-4*, Vol. 27, Swedish Council for Building Research, Stockholm, Sweden, 27.

Jorgensen, G., Bingham, C., King, D., Lewandowski, A., Netter, J., Terwilliger, K. & Adamsons, K. (2002) Use of Uniformly Distributed Concentrated Sunlight for Highly Accelerated Testing of Coatings, In: Martin, J.W. & Bauer, D.R. (Eds.), *Service Life Prediction Methodology and Metrologies, ACS Symposium Series 805*, pp. 100–118, American Chemical Society, Oxford University Press, Washington, DC.

Jorgensen, G., Bingham, C., Netter, J., Goggin, R. & Lewandowski, A. (1999) A Unique Facility for Ultra-Accelerated Natural Sunlight Exposure Testing of Materials, In: Bauer, D.R. & Martin, J.W. (Eds.), *Service Life Prediction of Organic Coatings, A Systems Approach, ACS Symposium Series 722*, pp. 170–185, American Chemical Society, Oxford University Press, Washington, DC.

Jorgensen, G., Kim H-M. & Wendelin, T. (1996) Durability Studies of Solar Reflector Materials Exposed to Environmental Stresses, In: Herling, R.J. (Ed.), *Durability Testing of Nonmetallic Materials ASTM STP 1294*, pp. 121–135, American Society for Testing and Materials: Scranton, PA.

Martin, J.W. (1982) Time Transformation Functions Commonly Used in Life Testing Analysis, *Durability of Building Materials*, **1**, 175–194.

Martin, J.W., Saunders, S.C., Floyd, F.L. & Weinberg, J.P. (1994) *Methodologies for Predicting the Service Life of Coatings Systems, NIST Building Science Series 172*, National Institute of Standards and Technology, Gaithersburg, MD.

Chapter 4.7
Outdoor Exposure Testing

G. Jorgensen

National Renewable Energy Laboratory, 1617 Cole Boulevard,
Golden, CO 80401-3393, USA

Abstract: The purposes of outdoor exposure testing (OET) include: (1) obtaining data on material durability and lifetimes at a number of representative locations; (2) understanding why materials degrade at different rates at different test sites; (3) correlating in-service measurements with accelerated life testing results and subsequently making quantitative predictions of material service lifetimes; and (4) validating damage function models derived from accelerated exposure tests. To allow validation, the detailed exposure conditions that candidate sample materials experience must be monitored. For this, test sites must be fully instrumented in terms of monitoring meteorological conditions and solar irradiance. Candidate materials are optically characterized before being subjected to exposure at OET sites. Optical durability is quantified by periodically remeasuring an appropriate response variable (Chapters 2.5, 4.5, and 4.6) as a function of exposure time. By closely monitoring the site- and time-dependent environmental stress conditions experienced by the samples, the actual site dependent loss of performance is obtained. In addition, parallel accelerated exposure testing of these materials in laboratory-controlled conditions permits correlations to be made with in-service results and subsequent prediction of service lifetimes.

Keywords: Outdoor exposure testing, Environmental stresses, Meteorological and radiometric monitoring, Weathering, Durability, Service lifetime prediction

LIST OF ABBREVIATIONS AND ACRONYMS

$a_{air}(T)$	temperature-dependent thermal diffusivity of air (m^2/s)
$C_{air}(T)$	temperature-dependent specific heat of air ($J/kg\,K$)
$D(t)$	time-dependent cumulative dose (J/m^2)
$h(t)$	time-dependent convective heat transfer coefficient ($W/m^2\,K$)
$I(t, \beta, \lambda)$	time-dependent spectral irradiance in plane at tilt angle β ($W/m^2/nm$)
k_T	sky clearness index
L	characteristic length of a sample (m)

n	day of the year
N	cloud cover
OET	Outdoor Exposure Testing
$Q_g(t, \beta, \text{TS})$	average global TS irradiance at time t and tilt angle β (W/m^2)
$Q_g(t, \beta, \text{UV})$	average global UV irradiance at time t and tilt angle β (W/m^2)
$Q_g(t, \beta, \text{UV-B})$	average global UV-B irradiance at time t and tilt angle β (W/m^2)
Q_g	incident global horizontal radiation (W/m^2)
Q_d	diffuse component of Q_g (W/m^2)
Q_r	ground-reflected component of Q_g (W/m^2)
Q_s	specular component of Q_g (W/m^2)
Q_{sc}	solar constant (1367 W/m^2)
Q_0	extraterrestrial irradiance on a horizontal surface (W/m^2)
RH	relative humidity (%)
t	time
t_i	ith time of day (h) $= 0 \rightarrow 24$
T	temperature
T_{amb}	ambient temperature (°C)
T_{back}	background temperature (K)
T_{dew}	dew point temperature (°C)
T_{grnd}	ground temperature (°C)
T_{rack}	back-of-rack temperature (°C)
T_{samp}	sample temperature (°C)
T_{sky}	sky temperature (K)
TMY	typical meteorological year
TOW	time of wetness (h/y)
TS	total solar (integrated over terrestrial solar wavelengths, $\lambda = 305\text{–}2500 \, \text{nm}$)
UV	ultraviolet light or radiation
UV-B	the region of the electromagnetic spectrum between 280–315 nm
v_w	wind speed (m/s)
Δt	time increment
α	solar absorptance of the sample
ε	thermal emittance of the sample
σ	Stephan Boltzmann constant: $5.67 \times 10^{-8} \, \text{W/m}^2 \, \text{K}^4$
β	tilt angle of exposure rack relative to horizontal (°)
φ	latitude angle (°)
δ	declination angle (°) $= 23.45 \times \sin[(284 + n) \times 360°/365]$
ω_i	ith hour angle (°) $= 15° \times (t_i - 12)$

λ	wavelength (nm)
θ	the angle between the wind direction and the sample plane (°)
$\rho_{air}(T)$	temperature dependent density of air (kg/m^3)
$v_{air}(T)$	temperature dependent kinetic viscosity of air (m^2/s)

4.7.1 INTRODUCTION

Careful planning and proper execution of in-service exposure testing are required to enable an understanding of why materials degrade at different rates at geographically diverse test sites. For example, Figure 4.7-1 shows vastly different losses in performance of the same reflectors (3M ECP-305) exposed at Miami, FL, Golden, CO, and Phoenix, AZ in the United States that are representative of three different types of climates. Samples exposed at these three locations experience dissimilar weather conditions (environmental stresses) that result in varying rates of degradation. Samples exposed at Golden, CO (a dry, high solar UV radiation, high-altitude climate) exhibit little loss in performance after 2 years of exposure. Samples exposed at Miami, FL (a hot, moist, high solar UV radiation, subtropical climate) demonstrate a fairly constant rate of degradation. Samples exposed at Phoenix, AZ (a hot, dry, very high solar UV radiation, desert climate) exhibit an induction period during which little degradation occurs, but after which a rapid/catastrophic loss in performance is evident. We know we need to understand which stress factors are important and what controlling role each plays in contributing to

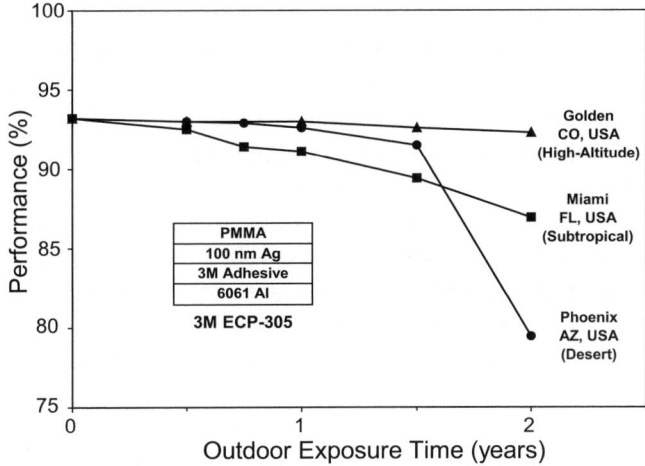

Figure 4.7-1. Variability in loss in performance of 3M ECP-305 reflectors at three outdoor exposure sites.

the performance loss. However, because of the highly variable nature of weather conditions, it is difficult to isolate the separate and synergistic effects of multiple stresses. The most straightforward approach is to perform laboratory exposure tests to identify and then control the significant stress factors experienced by candidate samples. As discussed in Chapters 4.3 and 4.5, factorial experiments can be carried out to determine the effect of specific environmental stress factors upon the critical properties of materials. By complementing the outdoor test activities with parallel accelerated laboratory testing of the same materials, correlation of these results may allow quantitative prediction of the service lifetime of the materials in the samples, constructions, or systems. In this way, convincing estimates of the optical durability can be made for samples deployed at arbitrary locations (given the meteorological and radiometric characteristics of that site) and for new candidate materials based upon accelerated test results only. This in turn will greatly facilitate and support commercialization of solar technologies by providing greater confidence in life-cycle cost estimates and less uncertainty in warranty projections.

Considerable compilations of data for weathered materials are available (Gilligan, 1980; Gilligan *et al.*, 1980; Zerlaut and Anderson, 1989; SPF-Info thermische Sonnenenergie, 1998) but these are generally inadequate to derive correlations with accelerated exposure test results because an accurate record of the in-use environmental stresses experienced by the materials during their weathering are not available. Outdoor weather conditions are too highly variable to allow long-term totals or averages to be useful in describing weathering effects. Temporal links between synergistic weather factors cannot be properly described by average values. For example, incident photons will do more damage to materials at elevated temperatures than at lower temperatures. Meeker *et al.* (2002) have shown the ineffectiveness of using average weather conditions rather than environmental time series data, e.g., actual measured meteorological and radiometric data. Use of an average temperature can greatly underestimate the amount of degradation compared with stochastic sample paths that simulate the effects of variable temperature environments (see also, Chapter 4, Section 4.4).

Extensive effort has been devoted to outdoor weathering of organic materials such as coatings (Jacques, 2000) and polymers (Wypych, 1995; Masters and Bond, 1999). The details regarding the exposure methods, sample mounting, sample characterization (measurement of performance indicators), schedules, and the number of replicated samples discussed (Wypych, 1995; Masters and Bond, 1999; Jacques, 2000) are directly applicable to other materials intended for outdoor use and should be incorporated into any OET program. In addition, standard test methods have been developed for "near service conditions," i.e., 1-sun light intensity and temperatures close to operational values, OET of plastics (ISO 877:1994; ASTM

D1435-99, 2003), and other nonmetallic materials (ASTM G7–97, 2003). These procedures should also be followed, including the details of exposure rack construction and configuration. For example, radiometers should measure the global irradiance incident on the samples by being mounted in the same plane in which the samples are exposed. To maximize the annual solar exposure for stationary (nontracking) exposure racks in the northern (southern) hemisphere, racks should be installed and positioned facing due south (north) and have an angle with the horizontal equal to the latitude angle (ISO 877:1994; Wypych, 1995; ASTM D1435-99, 2003; ASTM G7-97, 2003). Typical exposure rack dimensions are 1.6 m × 3.6 m. Samples can be exposed "unbacked" or with plywood backing that provides some insulation and convective loss protection for the backside of the samples being exposed. Use of backing sheets can result in sample exposure temperatures that are elevated by up to 10°C.

Materials can also be subjected to accelerated natural exposure testing. New advanced materials are being developed at a rapid pace to meet increasingly demanding service lifetime requirements. To enhance commercialization efforts and to achieve and maintain competitiveness in international markets, industry must be able to determine accurately the expected lifetime of their products that use the new materials. A service lifetime requirement for polymeric materials of up to 30 years, which is a significant increase in existing lifetimes, is anticipated. Advanced materials and devices in many technologies are being developed so rapidly, and the competition is so intense, that companies cannot afford to wait for in-service outdoor test results. To date, the most widely used accelerated testing equipment are the EMMAQUA devices that use natural sunlight concentrated by Fresnel reflector elements (Robbins, 1994). These units provide a light acceleration factor of 5 to 6 times, and are therefore capable of providing direct assurance of 10 year lifetimes after about 2 years of testing. Standard test procedures for accelerated OET of plastics (ISO 877:1994; ASTM D4364-94, 2003) and other nonmetallic materials (ASTM G90-98, 2003) have been developed.

The National Renewable Energy Laboratory (NREL) at Golden, CO, USA has exposed organic materials, e.g., back-metallized polymeric films, up to 100 times the intensity of solar UV (Jorgensen *et al.*, 1999). More recently, a new UV concentrator system has been developed that is less complicated and more amenable to future commercialization (Jorgensen *et al.*, 2002). Using this device, samples, e.g., paint coatings and polymeric sheets and films, have been exposed at closely controlled sample exposure temperatures and to uniform intensity levels of 50 to 100 times the terrestrial solar UV contained between 290 and 385 nm. At an exposure of "100 suns," this device could provide a 10-year equivalent exposure in only 10 weeks of testing. Obviously, for a valid interpretation, the degradation mechanism must be the same at "1 sun" as at "100 suns."

4.7.2   RELEVANT ENVIRONMENTAL STRESSES

Natural weathering involves environmental stresses that adversely effect the essential performance properties of materials otherwise expected to function for extended lifetimes during usually harsh outdoor service conditions. Meteorological and radiometric factors that can induce physical and/or chemical damage include sunlight, especially the UV part of the solar spectrum, temperature, moisture, pollution, and wind. Solar irradiance can result in harmful photolytic reactions, as well as contributing to thermally-induced degradation by directly heating exposed materials. Defects can be formed at elevated temperatures, and typically the reaction rates of other degradation mechanisms are increased at higher temperatures. Cyclic temperature variations can also result in thermo-mechanical stress failures, e.g., by parts having mismatched coefficients of thermal expansion properties, and from cyclic freeze-thaw damage. Moisture in the form of humidity, precipitation, condensed water, etc. can accelerate hydrolytic reactions and may change the morphological structure of materials. Cyclic swelling can also result from sorption and desorption processes that can, for example, weaken adhesive bonds and result in interfacial delamination. Potentially harmful pollutants are NO_x, SO_x, ozone, particulates, salts, and acid rain. Any of these pollutants can participate in materials degradation via corrosive acid or base reactions, oxidation, and the coating and/or etching of surfaces. Wind can change the sample exposure temperature by convective heat transfer and change the concentration level of moisture and/or pollutants experienced by a material (see, for example, Chapter 3.4). Wind can also apply mechanical loading, abrade surfaces with sand or particulates, and deposit dust and dirt (Wypych, 1995). In addition to these individual stresses, synergistic interactions can also occur that result in multifactor-induced degradation processes.

   A precise and detailed knowledge of the specific environmental stress conditions experienced by weathered samples is needed to understand site-specific performance losses and permit service-lifetime prediction of candidate solar materials. Consequently, operational exposure sites must be fully equipped with appropriate meteorological and radiometric instrumentation and data-logging capability. Figure 4.7-2 shows an example of the hardware associated with a typical exposure site. Each data channel should be sampled at ~10-s intervals and 5 to 10-min averages should be recorded. Standard specifications (ISO 877:1994; ASTM D1435-99, 2003; ASTM G7-97, 2003) for outdoor exposure testing should be used, with the following minimum information routinely monitored: (a) the average global total solar (TS) irradiance at time t, $Q_g(t, \beta, TS)$, in watts per square meter, measured with the instrument oriented due south and tilted relative to horizontal by an angle, β, equal to the latitude, φ, of the site; (b) the average global total UV solar irradiance at

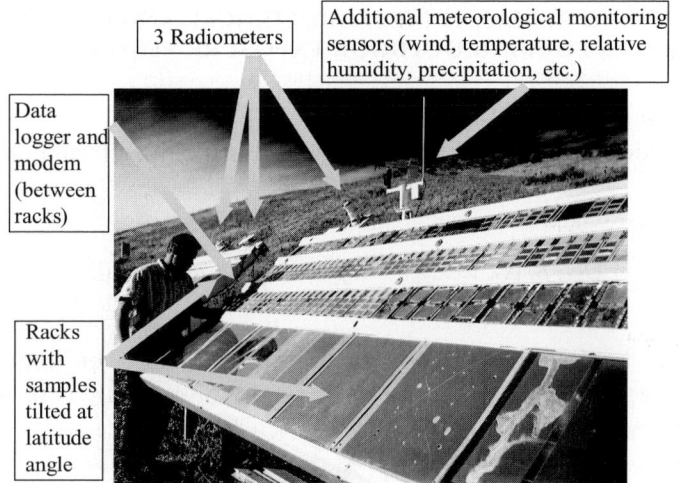

Data logger and modem (between racks)

3 Radiometers

Additional meteorological monitoring sensors (wind, temperature, relative humidity, precipitation, etc.)

Racks with samples tilted at latitude angle

Figure 4.7-2. View of a typical outdoor exposure test site.

time t, $Q_g(t, \beta, \mathrm{UV})$, in watts per square meter, measured with the instrument oriented due south and tilted relative to horizontal by an angle equal to the latitude of the site ($\beta = \varphi$); (c) the average global narrowband UV-B solar irradiance at time t, $Q_g(t, \beta, \mathrm{UV\text{-}B})$, in watts per square meter, measured with the instrument (e.g., an EKO 210W) oriented due south and tilted relative to horizontal by an angle equal to the latitude of the site ($\beta = \varphi$); (d) the average ambient air temperature (T_{amb}) in degrees Celsius; (e) the average temperature measured on the backside of the sample exposure rack (T_{rack}) in degrees Celsius; (f) the average temperature measured approximately 2.5 cm below the ground surface (T_{grnd}) underneath the sample exposure rack in degrees Celsius; (g) the average relative humidity (RH) in percent; (h) the precipitation (rain, snow, sleet, hail, or ice) measured in millimeters per time interval; and (i) the average wind speed (km/h) and wind direction that conforms to the "wind rose" convention to provide a way of resolving ambiguities associated with averaging 1° with 359° and calculating the correct result 360°, instead of 180° for wind direction.

Real-time monitoring of atmospheric pollutants is very expensive. In some cases, pollution data may be available from monitoring stations located geographically close to the exposure sites. However, such data may be irregular or variable. To provide some consistent assessment of the pollutant stresses, SO_2 levels can be monitored using a sulfation plate technique as specified by ISO 9225:1992 and ASTM G91-97 (2003). Correlations between parallel measurements, e.g., sulfation plates and nearby stations, can be performed to quantify the usefulness of the data.

4.7.3 MONITORING RELEVANT METEOROLOGICAL AND RADIOMETRIC PARAMETERS DURING OUTDOOR EXPOSURE TESTING

4.7.3.1 *Sample Irradiance*

The time-dependent integrated irradiance, Q, on a plane tilted from horizontal by an angle, β, is:

$$Q(t, \beta, \Delta\lambda = \lambda_{max} - \lambda_{min}) = \int_{\lambda_{min}}^{\lambda_{max}} I(t, \beta, \lambda)\,d\lambda \qquad (1)$$

where $I(t, \beta, \lambda)$ is the time-dependent spectral irradiance on that plane. The cumulative dose that materials receive during outdoor exposure is then:

$$D(t) = \int_0^t Q(t', \beta, \Delta\lambda)\,dt' \qquad (2)$$

The total solar irradiance is $Q(t, \beta, \Delta\lambda = 2500$ to $305\,\text{nm}) \equiv Q(t, \beta, \text{TS})$. Materials exposed or in-use outdoors are typically most susceptible to damage resulting from the most energetic (UV) portion of the solar spectrum. UV radiation can have a particularly deleterious effect on organic-based materials because the photon energies are sufficient to break important bonds in this spectral range (Table 4.7-1). A specific bond can be adversely affected by photons with wavelengths less than or equal to the corresponding bond energy. Only the O–H and C–F bonds listed in Table 4.7-1 are not readily prone to rupture by photons within the terrestrial solar spectrum.

For terrestrial sunlight, the Commission Internationale de l'Eclairage (CIE) Committee E-2.1.2 defines the UV spectral region as UV-A ($\lambda = 315$–$400\,\text{nm}$) and UV-B ($\lambda = 280$–$315\,\text{nm}$). The response bandwidth of commercial UV radiometers

Table 4.7-1. Bond energies in organic materials and associated photon energy

Chemical Bond	Bond Energy (kJ/mole)	Wavelength of associated photon energy (nm)
O–H	463.0	259
C–F	441.2	272
C–H	413.6	290
N–H	391.0	306
C–O	351.6	340
C–C	347.8	342
C–Cl	328.6	364
C–N	290.9	410

(pyranometers) is typically from 315 to 385 nm for UV-A ("total UV") and from 295 to 315 nm for UV-B. Such pyranometers measure total hemispherical (global) irradiance. Recommendations for selection and use of pyranometers are provided (ISO 9060:1990; ISO/TR 9901:1990) and standard procedures for calibrating these devices are described (ISO 9059:1990; ISO 9847:1992; ISO 9846:1993; ASTM E816–95, 2003; ASTM G167-00, 2003; ASTM E824-94, 2003). The initial deployment of UV-B pyranometers (Zerlaut and Miyake, 2002) demonstrated that UV-B is much more sensitive than UV-A radiation to atmospheric conditions, particularly the moisture vapor content of the sky and dynamic changes in the ozone layer. Thus, although $Q(t, \beta, \text{TS})$ associated with a particular terrestrial solar irradiance spectrum may be fairly constant in both time and location, the UV-A and especially the UV-B components of solar irradiance, $Q(t, \beta, \text{UV})$ and $Q(t, \beta, \text{UV-B})$, may exhibit considerable variation. Furthermore, because these bandwidths represent a small fraction of the entire solar spectrum, such deviations may not be detectable by broadband solar radiometers. Figure 4.7-3 shows a standard terrestrial global solar spectrum and the different spectral regions of interest to materials weathering are identified (ISO 9845-1:1992; ASTM G159-98:2003). The total UV bandwidth represents about 4.5% of the total solar spectrum; the UV-B range encompasses less than 0.1% of the full spectrum. Variations in the intensity of solar radiation are most uncertain at the wavelengths that are most damaging to materials. Therefore, it is crucial to have appropriate radiometric instrumentation operational during OET of materials, samples, and constructions.

Samples exposed outdoors experience global solar irradiance (Q_g) comprised of radiation arriving from the total hemispherical sky dome. In general, this includes

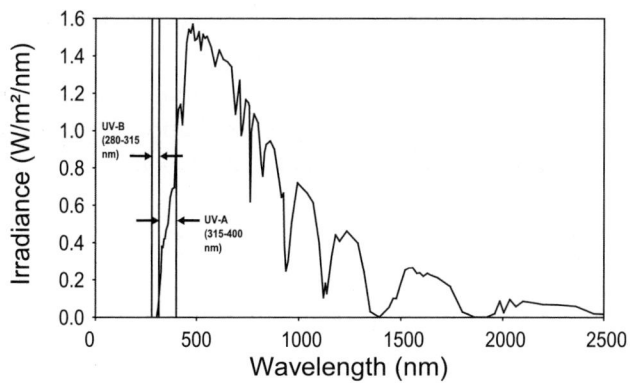

Figure 4.7-3. Typical global terrestrial solar spectrum (ISO 9845-1, 1992; ASTM G/59–98: 2003) at airmass 1.5. The UV-A and UV-B intervals are based on definitions chosen by the CIE Committee E-2.1.2 (see text).

light incident directly from the solar disk (specular), light incident after scattering through the earth's atmosphere (diffuse), and (for $\beta \neq 0$) light incident after reflecting off the ground:

$$Q_g(t, \beta, \Delta\lambda) = Q_s(t, \beta, \Delta\lambda) + Q_d(t, \beta, \Delta\lambda) + Q_r(t, \beta, \Delta\lambda) \tag{3}$$

where Q_s is the specular component, Q_d is the diffuse component, and Q_r is the ground-reflected component.

To quantify the effect of outdoor exposure, an accurate time history is needed for the UV and UV-B exposure experienced by the samples being tested. Chapter 4.6 provides a detailed example of how accurately monitored weather data allows correlations between outdoor and accelerated test results to be derived and validated. The best way to ensure that proper irradiance data are available is to mount the appropriate pyranometers, that have the required spectral sensitivity, directly onto the exposure rack, i.e., in the same plane as the samples, so that they concomitantly measure the global irradiance experienced by the materials.

However, the only data frequently available are for the global total solar irradiance on a horizontal plane, $Q_g(t, \beta = 0, \text{TS})$. These generic data may be useful in simulating environmental exposures; "typical" meteorological year weather data files are contained in a TMY2 data base (Marion and Urban 1995). These data files provide 30-years average hourly weather data for more than 200 geographical locations in the USA. TMY2DATA parameters include ambient temperature (T_{amb}), dewpoint temperature (T_{dew}), ambient relative humidity (RH), wind speed (v_w), precipitation, cloud cover (N), and incident global horizontal radiation (Q_g). If Q_g is measured or otherwise known, a method for obtaining the requisite spectral portion of irradiance incident on samples exposed at a given tilt is needed. Perez and Stewart (1985) presented an algorithm to compute global irradiance values on tilted planes from horizontal specular and diffuse irradiance. To obtain the specular and diffuse components from the measured global horizontal value, Duffie and Beckman (1991) provided a correlation for the fraction of irradiance on a horizontal plane that is diffuse, Q_d/Q_g, as a function of sky clearness index, k_T:

$$\frac{Q_d(t, \beta = 0, \text{TS})}{Q_g(t, \beta = 0, \text{TS})} = 1.0 - 0.09k_T \quad \text{for } k_T \leq 0.22$$

$$\frac{Q_d(t, \beta = 0, \text{TS})}{Q_g(t, \beta = 0, \text{TS})} = 0.9511 - 0.1604\,k_T + 4.388\,k_T^2$$

$$-6.638\,k_T^3 + 12.336\,k_T^4 \quad \text{for } 0.22 < k_T \leq 0.80 \tag{4}$$

$$\frac{Q_d(t, \beta = 0, \text{TS})}{Q_g(t, \beta = 0, \text{TS})} = 0.165 \quad \text{for } k_T > 0.80$$

When $k_T = 1$, the sky is perfectly clear and Q_g is comprised of ~16.5% diffuse radiation and ~83.5% specular radiation; when $k_T = 0$, the sky completely scatters the radiation and all of the global radiation is diffuse. The clearness index is defined (Duffie and Beckman, 1991) as:

$$k_T = Q_g(t, \beta = 0, \text{TS})/Q_0(t, \beta = 0, \text{TS}) \tag{5}$$

where Q_0 is the extraterrestrial irradiance on a horizontal surface during the time interval $\Delta t = t_2 - t_1$, and is also given (Duffie and Beckman, 1991) by:

$$Q_0(t, \beta = 0, \text{TS}) = \left\{(12 \times 3600)/\pi\right\} Q_{sc}[1 + 0.033 \cos(360n/365)]$$
$$\times \left\{\cos\varphi \cos\delta (\sin\omega_2 - \sin\omega_1) + [\pi(\omega_2 - \omega_1)/180] \sin\varphi \sin\delta\right\} \tag{6}$$

where Q_{sc} is the solar constant (1367 W/m^2); n is the day of the year; φ is the latitude angle (°); δ is the declination angle (°) and is given by $23.45 \times \sin[(284 + n) \times 360°/365]$; ω_i is the ith hour angle (°) and is given by $15° \times (t_i - 12)$; and t_i is the ith time of day (h) and is given by $0 \rightarrow 24$.

Then, the specular irradiance on a horizontal plane can be obtained from the measured $Q_g(t, \beta = 0, \text{TS})$ and equation (4) as:

$$Q_s(t, \beta = 0, \text{TS}) = Q_g(t, \beta = 0, \text{TS}) \times \left[1 - \frac{Q_d(t, \beta = 0, \text{TS})}{Q_g(t, \beta = 0, \text{TS})}\right] \tag{7}$$

Once the specular and diffuse components are obtained from equations (4) and (7), $Q_g(t, \beta, \text{TS})$ can be calculated using equation (3), in which the necessary expressions for the various irradiance components on a tilted surface have been given (Perez and Stewart, 1985).

As discussed above, it is generally more important to monitor UV irradiance than $Q_g(t, \beta, \text{TS})$. Zerlaut and Miyake (2002) have measured $Q_g(t, \beta, \text{UV-B})$ and $Q_g(t, \beta, \text{UV})$ as well as $Q_g(t, \beta, \text{TS})$ at several sites representative of different types of climates (Zerlaut and Miyake, 2002). This has allowed preliminary relationships between UV-B, total UV, and total solar irradiance to be derived. Their approach may be used when such data are not available, although this is less preferable to having actual measured data during OET.

4.7.3.2 *Sample Temperature and Wind*
Ideally, the temperature of every sample being exposed should be directly and independently monitored. Identical samples mounted at different locations on an

exposure rack will experience different temperatures. Fischer *et al.* (1991) found that depending upon sample mounting configuration, i.e., open-backed or plywood-backed, and windspeed, average temperatures of 50 to 60°C were obtained with temperature differentials across the rack of 5 to 11°C. Furthermore, these thermal differences were shown to result in different material failure rates as characterized by gloss retention of the exposed PVC film. To minimize these effects, the use of control samples, adequate replicates, and elimination of sample placement near the edge of the racks (to significantly reduce temperature variations) were recommended (Fischer *et al.*, 1991). The criterion that is to be used for the selection and characterization of a weathering reference material intended for use in monitoring operating conditions to establish the consistency of exposure tests, is described in ASTM G156-97 (2003).

Usually, too many samples are being exposed to allow concurrent measurement of individual sample temperatures. In such cases, sample temperatures, T_{samp}, can be derived from other monitored quantities (Saunders *et al.*, 1990; Myers, 1999). For samples mounted horizontally, T_{samp} can be obtained at any time from an energy-balance equation (Saunders *et al.*, 1990), i.e.,

$$\alpha Q_g(t, \beta = 0, TS) = 2h(t)[T_{samp}(t) - T_{amb}(t)] - \varepsilon\sigma[T^4_{sky}(t) - T^4_{samp}(t)]$$
$$- \varepsilon\sigma[T^4_{back}(t) - T^4_{samp}(t)] \tag{8}$$

where α is the solar absorptance of the sample, h is the convective heat transfer coefficient ($W/m^2\,K$), T_{amb} is the ambient temperature (K), ε is the thermal emittance of the sample, σ is the Stephan Boltzmann constant given by $5.67 \times 10^{-8}\,W/m^2\,K^4$, T_{sky} is the sky temperature (K), and T_{back} is the background temperature (K).

Saunders *et al.* (1990) further suggests that $T_{back} \approx T_{amb}$. In general, the convection coefficient will be a complicated function of the monitored wind speed (v_w) and direction, attack angle, θ, which is the angle between the wind direction and the sample plane, and the characteristic length, L, of the sample, which is defined as four times the area divided by the perimeter. For a wide range of θ, Sparrow *et al.*, (1979) presented a global correlation equation for the convection coefficient, i.e.,

$$h = \frac{0.86\rho_{air}(T)C_{air}(T)\sqrt{v_w}}{(v_{air}(T)/a_{air}(T))^{2/3}\sqrt{L/v_{air}(T)}} \tag{9}$$

where $\rho_{air}(T)$ is the temperature-dependent density of air, $C_{air}(T)$ is the temperature-dependent specific heat of air, $v_{air}(T)$ is the temperature-dependent kinetic viscosity of air, $a_{air}(T)$ is the temperature-dependent thermal diffusivity of air, and temporal variations, $h(t)$, arise from the time-dependent wind speed and ambient temperature.

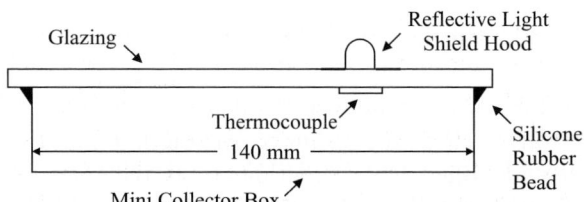

Figure 4.7-4. Cross section of a "mini collector box" used for outdoor exposure of glazing materials.

Values for these temperature-dependent properties of air are available (Eckert and Drake, 1959). Equation (9) agrees quite well with measured results for wind from the east through the south to the west, but winds from the north resulted in heat loss 10 to 20% less than that associated with wind from the southern quadrants (Onur and Hewitt, 1980).

The sky temperature needed in equation (8) can be obtained as a function of the dew point temperature, T_{dew} (Martin and Berdahl, 1984a,b), and T_{dew} can be calculated from the measured T_{amb} and relative humidity, RH (1993 ASHRAE Handbook of Fundamentals, 1993). Thus, T_{samp} can be obtained from the readily monitored parameters: $Q_g(t, \beta, TS)$, v_w, T_{amb}, and RH.

The approach outlined above is applicable to samples of materials exposed as separate articles. Sometimes more complicated configurations are required. For example, glazing materials used in solar thermal flat plate collectors should be affixed to "mini collector boxes" to simulate the elevated temperature the collector covers experience during in-service use (ASTM E881-92, 1996, 2003; ASTM E782-95, 2001, 2003). The "mini collectors" are made of a solar selective coated stainless steel. In this case, the sample temperature must be measured directly. An appropriate thermocouple can be attached to the glazing sample as shown in Figure 4.7-4. Reflective tape should be used to provide a light shielding hood to prevent direct sunlight from unrealistically heating the thermocouple.

4.7.3.3 Moisture and Pollutants

Monitoring meteorological and radiometric conditions provide information about the macroclimate that materials experience during outdoor exposure. Degradation often occurs because of reactions that result from complex microclimate environmental effects. Therefore, it must be possible to relate the actual microclimate conditions experienced by a material being tested to macroclimate weather parameters. The U.S. National Institute of Standards and Technology (NIST) has developed a moisture and heat transfer model that allows prediction of the temperature and relative humidity experienced by polymer coatings during outdoor exposure (Burch and Martin, 1999). Other approaches for other material constructions are discussed in Chapters 3.1–3.4.

For materials with metal-containing components (metallized reflectors, absorber-coated metal substrates, etc.), the primary weathering factors for atmospheric corrosion have been identified from experiments as the time of wetness (TOW), temperature, relative humidity, and UV intensity. However, the presence of pollutants, such as sulfur dioxide, nitrogen oxides, ozone, particulates, hydrogen sulfide, and chlorides greatly accelerates atmospheric corrosion. The type and rate of atmospheric corrosion depend on the formation of electrolytes at the surface or interface of the material being exposed. These, in turn, are related to the level and type of pollutants in the atmosphere and to the length of time they are allowed to interact. An overview of procedures for measuring, modeling, and mapping air pollution and aerosols has been presented (Haagenrud, 1999). A number of specifications have been developed to quantify the effects of atmospheric corrosion of exposed materials; these include both ISO and ASTM standards described below.

The primary factors involved in atmospheric corrosion are TOW, pollution by sulfur dioxide, and air-borne salinity (ISO 9223:1992). Five categories of atmospheric corrosivity are defined, which are broadly described as being very low, low, medium, high, and very high. These categories are related to the corrosion rates (g/m^2/y) of four types of metals: carbon steel, copper, aluminum, and zinc. These rates, in turn, are related to the TOW (h/y) and the deposition rate (mg/m^2/d) of SO$_2$ and air-borne chlorides onto the samples being exposed. Guidelines for corrosion values and characteristics of corrosion for these categories have been provided (ISO 9224:1992). Three methods for measuring the deposition rates of SO$_2$ and Cl$^-$, including methods for characterization of test site corrosivity, have been specified (ISO 9225:1992). Suggested procedures for the characterization of atmospheric test sites have also been given (ASTM G92-86, 1997). A description of how to measure mass loss to determine the corrosion rates of standard metal specimens is available (ISO 9226:1992). By measuring the corrosion rates of standard metal test specimens, the concentrations of SO$_2$ and Cl$^-$ can be inferred, which can then be correlated with the extent and rate of degradation experienced by test samples of interest. The effective SO$_2$ content of the atmosphere can also be determined by using the sulfation plate method (ASTM G91-97, 2003). The results of this practice correlate approximately with volumetric SO$_2$ concentrations, although the presence of dew or condensed moisture can enhance the capture of SO$_2$ by the plate.

In the International Standards Organization standards, the TOW is calculated as the fraction of time in which relative humidity is greater than 80% at an air temperature greater than 0°C. Alternatively, samples will experience condensed moisture whenever $T_{samp} \leq T_{dew}$, where these temperatures can be measured or obtained as discussed above. The experimental TOW can also be directly determined by various instruments that should be located close to the samples being exposed (Sereda *et al.*, 1980). Methods for measuring surface moisture and the TOW have been reviewed (Norberg, 2002). A standard technique for monitoring the TOW on surfaces exposed to cyclic atmospheric conditions that produce depositions of moisture has been specified (ASTM G84-89, 1999) and discussed (Baker and Lee, 1980).

4.7.4 CONCLUSIONS

The environmental conditions that materials encounter during outdoor service and exposure are highly variable, and the effects of weathering on materials performance depend on the detailed time-dependent stresses they experience. Consequently, it is important to quantify environmental stress levels experienced by materials subjected to OET by monitoring the relevant meteorological and radiometric variables. A means of relating macroclimate measurements to microclimate conditions experienced by materials being tested is also required. Samples should be exposed outdoors in a configuration that is representative of their intended in-service use. For example, candidate glazing materials should be attached to mini-collector boxes to provide elevated exposure temperatures. Measurement instrumentation should be matched to the types of materials being tested. Monitoring narrow-band UV-B may not be appropriate for the OET of inorganic materials. Conversely, determination of the detailed spectral content of light incident on organic materials may be essential for understanding the degradation processes that are photo- or photothermally activated. Detailed accurate monitoring is required to permit making correlations with accelerated test results.

REFERENCES

1993 ASHRAE Handbook of Fundamentals (1993) American Society of Heating, Refrigerating and Air Conditioning Engineers, Inc., Atlanta, GA. 6.7–6.9.

ASTM D1435-99 (2003) Standard Practice for Outdoor Weathering of Plastics, *Annual book of ASTM Standards 2003* **08.01**, American Society for Testing and Materials, West Conshohocken, PA.

ASTM D4364-94 (2003) Standard Practice for Performing Outdoor Accelerated Weathering Tests of Plastics Using Concentrated Sunlight, *Annual book of ASTM Standards 2003* **08.03**, American Society for Testing and Materials, West Conshohocken, PA.

ASTM E782-95 (2001) (2003) Standard Practice for Exposure of Cover Materials for Solar Collectors to Natural Weathering Under Conditions Simulating Operational Mode *Annual book of ASTM Standards 2003* **12.02**, American Society for Testing and Materials, West Conshohocken, PA.

ASTM E816-95 (2003) Standard Test Method for Calibration of Pyrheliometers by Comparison to Reference Pyrheliometers, *Annual book of ASTM Standards 2003*, **14.04**, American Society for Testing and Materials, West Conshohocken, PA.

ASTM E824-94 (2003) Standard Test Method for Transfer of Calibration from Reference to Field Radiometers, *Annual book of ASTM Standards 2003*, **14.04**, American Society for Testing and Materials, West Conshohocken, PA.

ASTM E881-92 (1996) (2003) Standard Practice for Exposure of Solar Collector Cover Materials to Natural Weathering Under Conditions Simulating Stagnation Mode. *Annual book of ASTM Standards 2003* **12.02**, American Society for Testing and Materials, West Conshohocken, PA.

ASTM G156-97 (2003) Standard Practice for Selecting and Characterizing Weathering Reference Materials Used to Monitor Consistency of Conditions in an Exposure Test. *Annual book of ASTM Standards 2003*, **14.04**, American Society for Testing and Materials, West Conshohocken, PA.

ASTM G159-98 (2003) Standard Tables for References Solar Spectral Irradiance at Air Mass 1.5: Direct Normal and Hemispherical for a 37° Tilted Surface, *Annual book of ASTM Standards 2003*, **14.04**, American Society for Testing and Materials, West Conshohocken, PA.

ASTM G167-00 (2003) Standard Test Method for Calibration of a Pyranometer Using a Pyrheliometer, *Annual book of ASTM Standards 2003*, **14.04**, American Society for Testing and Materials, West Conshohocken, PA.

ASTM G7-97 (2003) Standard Practice for Atmospheric Environmental Exposure Testing of Nonmetallic Materials. *Annual book of ASTM Standards 2003* **14.04**, American Society for Testing and Materials, West Conshohocken, PA.

ASTM G84-89 (1999) & (2003) Standard Practice for Measurement of Time-of-Wetness on Surfaces Exposed to Wetting Conditions as in Atmospheric Corrosion Testing. *Annual book of ASTM Standards 2003* **03.02**, American Society for Testing and Materials, West Conshohocken, PA.

ASTM G90-98 (2003) Standard Practice for Performing Accelerated Outdoor Weathering of Nonmetallic Materials Using Concentrated Natural Sunlight, *Annual book of ASTM Standards 2003* **14.04**, American Society for Testing and Materials, West Conshohocken, PA.

ASTM G91-97 (2003) Standard Practice for Monitoring Atmospheric SO_2 Using the Sulfation Plate Technique, *Annual book of ASTM Standards 2003* **03.02**, American Society for Testing and Materials, West Conshohocken, PA.

ASTM G92-86 (1997) & (2003) Standard Practice for Characterization of Atmospheric Test Sites. *Annual book of ASTM Standards 2003* **03.02**, American Society for Testing and Materials, West Conshohocken, PA.

Baker, E.A. & Lee, T.S. (1980) Calibration of Atmospheric Corrosion Test Sites, In: Dean, Jr. S.W. & Rhea, E.C. (Eds.), *Atmospheric Corrosion of Metals*, *ASTM STP 767*, pp. 250–266, American Society for Testing and Materials, Philadelphia.

Burch, D.M. & Martin, J.W. (1999) Predicting the Temperature and Relative Humidity of Polymer Coatings in the Field, In: Bauer, D.R. & Martin, J.W. (Eds.), *Service Life Prediction of Organic Coatings, A Systems Approach*, *ACS Symposium Series 722*, pp. 85–107, American Chemical Society, Oxford University Press, Washington, DC.

Duffie, J.A. & Beckman, W.A. (1991) *Solar Engineering of Thermal Processes*, 2nd Edition, John Wiley & Sons, Inc., New York.

Eckert, E.R.G. & Drake, R.M. (1959) *Heat and Mass Transfer*, 2nd Edition, McGraw-Hill Book Company, New York

Fischer, R.M., Murray, W.P. & Ketola, W.D. (1991) Thermal Variability in Outdoor Exposure Tests, *Progress in Organic Coatings*, **19**, 151–163.

Gilligan, J.E. (1980) *Exposure Testing and Evaluation of Solar Utilization Materials*, DOE/CH/90034-T1, IIT Research Institute, Chicago, IL.

Gilligan, J.E., Brzuskiewicz, J., Brzuskiewicz, J.E. & Mell, R. (1980) *Handbook of Materials for Solar Energy Utilization (Low Temperature Applications)*, DOE/CH/90034-T1 (Handbook), IIT Research Institute, Chicago, IL.

Haagenrud, S.E. (1999) Monitoring and Characterizing Air Pollutants and Aerosols, In: Bauer, D.R. & Martin, J.W. (Eds.), *Service Life Prediction of Organic Coatings, A Systems Approach, ACS Symposium Series 722*, pp. 108–129, American Chemical Society, Oxford University Press, Washington, DC.

ISO 877:1994, Plastics – Methods of exposure to direct weathering, to weathering using glass-filtered daylight, and to intensified weathering by daylight using Fresnel mirrors. International Organization for Standardization, Geneva, Switzerland.

ISO 9059:1990, Solar energy – Calibration of Field Pyrheliometers by Comparison to a Reference Pyrheliometer, International Organization for Standardization, Geneva, Switzerland.

ISO 9060:1990, Solar energy – Specification and Classification of Instruments for Measuring Hemispherical Solar and Direct Solar Radiation, International Organization for Standardization, Geneva, Switzerland.

ISO 9223:1992, Corrosion of metals and alloys – Corrosivity of atmospheres – Classification. International Organization for Standardization, Geneva, Switzerland.

ISO 9224:1992, Corrosion of metals and alloys – Corrosivity of atmospheres – Guiding values for the corrosivity categories. International Organization for Standardization, Geneva, Switzerland.

ISO 9225:1992, Corrosion of Metals and Alloys – Corrosivity of Atmospheres – Measurement of Pollution. International Organization for Standardization, Geneva, Switzerland.

ISO 9226:1992, Corrosion of metals and alloys – Corrosivity of atmospheres – Determination of corrosion rate of standard specimens for the evaluation of corrosivity, International Organization for Standardization, Geneva, Switzerland.

ISO 9845-1:1992, Solar energy – Reference Solar Spectral Irradiance at the Ground at Different Receiving Conditions – Part 1: Direct Normal and Hemispherical Solar Irradiance for Air Mass 1.5, International Organization for Standardization, Geneva, Switzerland.

ISO 9846:1993, Solar energy – Calibration of a Pyranometer using a Pyrheliometer, International Organization for Standardization, Geneva, Switzerland.

ISO 9847:1992, Solar energy – Calibration of Field Pyranometers by Comparison to a Reference Pyranometer, International Organization for Standardization, Geneva, Switzerland.

ISO/TR 9901:1990, Solar energy – Field Pyranometers – Recommended Practice for Use, International Organization for Standardization, Geneva, Switzerland

Jacques, L.F.E. (2000) Accelerated and Outdoor/Natural Exposure Testing of Coatings, *Prog. Polym. Sci.*, **25**, 1337–1362.

Jorgensen, G., Bingham, C., King, D., Lewandowski, A., Netter, J., Terwilliger, K. & Adamsons, K. (2002) Use of Uniformly Distributed Concentrated Sunlight for Highly Accelerated Testing of Coatings, Martin, J.W. & Bauer, D.R. (Eds.), *Service Life Prediction Methodology and Metrologies*, *ACS Symposium Series 805*, pp. 100–118, American Chemical Society, Oxford University Press, Washington, DC.

Jorgensen, G., Bingham, C., Netter, J., Goggin, R. & Lewandowski, A. (1999) A Unique Facility for Ultra-Accelerated Natural Sunlight Exposure Testing of Materials, In: Bauer, D.R. & Martin, J.W. (Eds.), *Service Life Prediction of Organic Coatings, A Systems Approach*, *ACS Symposium Series 722*, pp. 170–185, American Chemical Society, Oxford University Press, Washington, DC.

Marion, W. & Urban, K. (1995) *User's Manual for TMY2s Typical Meteorological Years: Derived from the 1961–1990 National Solar Radiation Data Base*, NREL/SP-463-7668, National Renewable Energy Laboratory, Golden, CO.

Martin, M. & Berdahl, P. (1984a) Summary of Results from the Spectral and Angular Sky Radiation Measurement Program, *Solar Energy*, **33**(3/4), 241–252.

Martin, M. & Berdahl, P. (1984b) Characteristics of Infrared Sky Radiation in the United States, *Solar Energy*, **33**(3/4), 321–336.

Masters, L.W. & Bond, L.F. (1999) Choices in the Design of Outdoor Weathering Tests, Wypych, G. (Ed.), *Weathering of Plastics, Testing to Mirror Real Life Performance*, pp. 15–27, Plastics Design Library, Norwich, NY.

Meeker, W.Q., Escobar, L.A. & Chan, V. (2002) Using Accelerated Tests to Predict Service Life in Highly Variable Environments, In: Martin, J.W. & Bauer, D.R. (Eds.), *Service Life Prediction Methodology and Metrologies*, *ACS Symposium Series 805*, pp. 396–413. American Chemical Society, Oxford University Press, Washington, DC.

Myers, D.R. (1999) Predicting Temperatures of Exposure Panels: Models and Empirical Data, In: Bauer, D.R. & Martin, J.W. (Eds.), *Service Life Prediction of Organic Coatings, A Systems Approach*, *ACS Symposium Series 722*, pp. 71–84, American Chemical Society, Oxford University Press, Washington, DC.

Norberg, P. (2002) Surface Moisture and Time of Wetness Measurements, In: Martin, J.W. & Bauer, D.R. (Eds.), *Service Life Prediction Methodology and Metrologies*, *ACS Symposium Series 805*, pp. 23–36, American Chemical Society, Oxford University Press, Washington, DC.

Onur, N. & Hewitt, J.C. Jr. (1980) A Study of Wind Effects on Collector Performance, *ASME Century 2 Solar Energy Conference*, pp. 1–6, San Francisco, CA.

Perez, R. & Stewart, R. (1985) Solar Irradiance Conversion Model. *Proceedings of the Photovoltaics and Measurement Workshop, June 30–July 3, 1985, Vail, CO*, SERI/CP-215-2773, pp. 43–52, Solar Energy Research Institute, Golden, CO.

Robbins, J.S. III. (1994) A Review of Recent Developments in the Use of ASTM Standard Practice G 90 for the Testing of Nonmetallic Materials, In: Ketola, W.D. & Grossman, D. (Eds.), *Accelerated and Outdoor Durability Testing of Organic Materials*, *ASTM STP 1202*, pp. 169–182, American Society for Testing and Materials, Philadelphia, PA.

Saunders, S.C., Jensen, M.A. & Martin, J.W. (1990) *A Study of Meteorological Processes Important in the Degradation of Materials Through Surface Temperature*, *NIST Technical Note 1275*, National Institute of Standards and Technology, Washington DC.

Sereda, P.J., Croll, S.G. & Slade, H.F. (1980) Measurement of the Time-of-Wetness by Moisture Sensors and Their Calibration, In: Dean, Jr. S.W. & Rhea, E.C. (Eds.),

Atmospheric Corrosion of Metals, *ASTM STP 767*, pp. 167–283, American Society for Testing and Materials, Philadelphia.

Sparrow, E.M., Ramsey, J.W. & Mass, E.A. (1979) Effect of Finite Width on Heat Transfer and Fluid Flow about an Inclined Rectangular Plate, *J. Heat Transfer*, **101**, 199–204.

SPF-Info thermische Sonnenenergie 1998, Solartechnik Prüfung Forschung, Rapperswil, Switzerland, CD-ROM.

Wypych, G., (1995) *Handbook of Material Weathering*, 2nd Edition, ChemTec Publishing, Toronto-Scarborough.

Zerlaut, G. & Anderson, T. (1989) *Commercial Solar Materials Exposure Studies*, DSET-R2658, DSET Laboratories, Inc., Phoenix, AZ.

Zerlaut, G.A. & Miyake, Y. (2002) Relationships between Daily UV-A, UV-B, and Hemispherical Solar Radiation, In: James, R. Slusser, Jay, R. Herman & Wei Gao (Eds.), *Proc. SPIE Vol. 4482, Ultraviolet Ground- and Space-based Measurements, Models, and Effects*, pp. 127–141.

Chapter 4.8

Analytical Techniques for Studying Solar Materials Degradation Processes

K. Möller

Swedish National Testing and Research Institute,
P. O Box 857, SE-501 15, Borås, Sweden

Abstract: The purpose of this chapter is to provide a brief overview of some modern analytical techniques that are selected because of their general-purpose and powerful qualities in obtaining elemental and molecular information. To assess the long-term performance and durability of materials, it is necessary not only to measure the deterioration of macroscopic physical properties, but also to gain information about degradation processes taking place at a molecular level. The techniques described are also chosen because they represent robust analytical tools and are part of most well-equipped materials laboratories. For these reasons, techniques such as Proton Induced X-ray Emission and Mössbauer Spectroscopy are excluded in this overview. For analyzing solar energy materials and interfaces, bulk as well as surface properties are essential for obtaining the understanding desired. Therefore, the instruments discussed range from having state-of-the art surface sensitivities to those that are more bulk oriented.

Keywords: Degradation, Aging, Bulk analysis, Surface sensitive, Surface analysis, Routine, Topography, Molecular information, Nondestructive, Depth profile

LIST OF ACRONYMS AND ABBREVIATIONS

AES	Auger Electron Spectroscopy
AFM	Atomic Force Microscopy
ATR	Attenuated Total Reflection
CHA	Concentric Hemispherical Analyzer
CMA	Cylindrical Mirror Analyzer
CRT	Cathode Ray Tube
DRIFT	Diffuse Reflectance Spectroscopy
DSC	Differential Scanning Calorimetry
EDX	Energy Dispersive X-ray

ESCA	Electron Spectroscopy for Chemical Analysis
FTIR	Fourier Transform Infrared
IR	Infrared
MIR	Multiple Internal Reflection
RAS	Reflection Absorption Spectroscopy
SEM	Scanning Electron Microscopy
SIMS	Secondary Ion Mass Spectrometry
SNMS	Secondary Neutral Mass Spectrometry
STM	Scanning Tunneling Microscopy
TOF	Time of Flight
UHV	Ultra High Vacuum
UV-VIS-NIR	Ultraviolet-Visible-Near Infrared
WDX	Wavelength Dispersive X-ray
XPS	X-ray Photoelectron Spectroscopy

4.8.1 INTRODUCTION

Progress in our understanding of how materials degrade during natural as well as artificial accelerated aging depends to a large extent on the proper choice and use of analytical tools that can extract valuable information related to the degradation processes. In the past, aging behavior of materials has been investigated to a great extent by examining changes in the performance properties. For example, the aging of solar absorber coatings has been evaluated by changes in solar absorptance and thermal emittance and glazing materials by changes in a mechanical property such as impact strength. Although of great importance, changes in macroscopic physical properties do not provide insight into the molecular nature of the aging of materials. In the last three decades, a number of advanced analytical techniques have become commercially available and are now available in materials research laboratories. A very brief overview is given here. Many of these techniques are based on the bombardment of the sample by particles (the primary beam), for example electrons, photons, or ions, and the subsequent analysis of the energy, intensity, mass, or direction of the particles that are emitted, reflected, or transmitted as a result of the bombardment (the secondary beam).

Degradation and aging processes frequently initiate at the surface or in the near-surface region of a material. However, the bulk of a material can also be affected by aging processes. The latter is definitely true for polymeric materials (see Chapters 4.6, 6.1, and 6.2). Consequently, it is important to have access to analytical tools that yield information about degradation at the surface as well as in the bulk of a material. Compared with charged particles, photons interact the least with the

Table 4.8-1. Analytical techniques described and acronyms used

	Acronym
Photon Absorption Spectroscopies	
Fourier Transform Infrared Spectroscopy	FT-IR
Ultraviolet-Visible-Near Infrared Spectroscopy	UV-VIS-NIR
Thermal Analysis	
Thechniques for imaging surface topography and "bulk" analysis	
Scanning Electron Microscopy	SEM
Scanning Tunneling Microscopy	STM
Atomic Force Microscopy	AFM
X-ray Spectroscopy	EDX, WDX
Electron Spectroscopies for Surface Analysis	
Auger Electron Spectroscopy	AES
X-ray Photoelectron Spectroscopy or	XPS
Electron Spectroscopy for Chemical Analysis	ESCA
Secondary Ion Mass Spectrometry	SIMS
Dynamic SIMS	
Static SIMS or Time of Flight (TOF)-SIMS	

medium they penetrate. Hence, analytical techniques that use photons as a probe will have a much larger information depth than techniques using ions or electrons. For example, FTIR spectroscopy can be regarded as a "bulk" technique, while the electron spectroscopies and SIMS techniques are surface sensitive. The techniques that will be discussed are listed in Table 4.8-1.

4.8.2   PHOTON ABSORPTION SPECTROSCOPIES

Photon absorption spectroscopies are based on the absorption of electromagnetic radiation by matter at specific wavelengths. The wavelength region of interest usually extends from 200 nm to 50 µm. The lower limit is set by absorption because of oxygen and nitrogen in the atmosphere. Absorption at wavelengths longer than 50 µm results from excitation of rotational states. Rotation of molecules is not possible in solid materials so the quantum mechanically allowed transitions between these energy levels cannot occur. Both UV-VIS-NIR and FTIR spectroscopy is of essential importance for studying degradation of solar thermal materials because these techniques provide data for the determination of performance properties such as solar absorptance, solar transmittance, and thermal emittance.

4.8.2.1   Infrared and Fourier Transform Infrared Spectroscopy
Infrared (IR) spectroscopy is one of the oldest and most commonly used spectroscopic techniques for molecular level characterization of materials. It is

perhaps the most important tool in the investigation of oxidation and photo-oxidation of polymeric materials. The infrared region of the electromagnetic spectrum extends from the red end of the visible spectrum to the microwave region. The spectral range mostly used is the mid-infrared region between 4000 and 200 cm^{-1} (2.5 to 50 μm). Infrared spectroscopy depends on the twisting, bending, and stretching vibrational motion of atoms in a molecule. Upon interaction with infrared radiation, portions of the incident radiation are absorbed at specific wavelengths resulting in a transition between the vibrational energy levels in the molecules. The multiplicity of vibrations occurring simultaneously in a molecule produces a complex absorption spectrum that is uniquely characteristic of the functional groups that make up the molecule and of the whole molecule as well.

The mid-infrared region can be divided into the group frequency region from 4000 to 1300 cm^{-1}, and the finger print region, 1300–600 cm^{-1}. In the group frequency region, the absorption bands are assigned to vibrational units consisting of two atoms of a molecule, i.e., the absorption is more or less dependent on only the functional groups and not on the complete molecule. The major fraction of absorption bands in the spectrum between 1300 and 600 cm^{-1}, on the other hand, results from vibrations involving the whole molecule (skeletal frequencies). This is the finger print region. Individual bands cannot be correlated to specific functional groups, but collectively the absorption bands in this region help in identifying the material. Gases, liquids, and solids can be examined. The development of a number of accessories or attachments has turned IR-spectroscopy into a very powerful technique when characterizing solid materials.

Two widely known techniques can be used to obtain IR absorption spectra. In the older one, dispersion of light by a prism or grating is utilized. However, a fundamental drawback of prism or grating spectrometers is in the monochromator. Narrow slits are situated at both the entrance and the exit of the monochromator, which limits the frequency range of the radiation that can reach the detector to a narrow band. As the width of the entrance and exit slits are narrowed, the resolution of the spectrometer becomes higher. Because of the narrow slits, only a small fraction of the radiation entering the monochromator actually reaches the detector. For example, in a dispersive infrared spectrophotometer operating at a resolution of 4 cm^{-1}, only about 0.1% of the radiation reaches the detector.

In Fourier transform instruments, a Michelson interferometer is used to replace the energy wasting monochromator. The interferometer consists of a beam splitter, and a fixed and moving mirror. Radiation from a broad band source is divided at the beam splitter. Half the beam passes to the fixed mirror and half is reflected to the oscillatory moving mirror. After reflection at the mirrors, the two beams recombine at the beam splitter. For any particular wavelength, constructive or destructive interference occurs, depending on the difference in optical paths (retardation) between the two arms of the interferometer. The resulting signal from the detector as a function of the retardation is known as an interferogram. The interferogram contains all the information required to reconstruct the broad band emission spectrum from the source through a process called Fourier transformation. The transformation, which is very comprehensive, has to be carried out by a computer that is an essential part of the spectrometer. At present, almost all infrared instruments produced are of the interferometer type (FTIR).

In attenuated total reflection (ATR) or multiple internal reflection (MIR) spectroscopies, a beam of IR radiation enters a crystal plate surrounded by the sample. If the angle of incidence is greater than a critical angle, which is a function of reflective index relation between the crystal and the sample, total reflection occurs. The beam appears, however, to penetrate slightly beyond the interface before it returns to the crystal. The depth of penetration into the sample is approximately one wavelength. Because of this penetration before returning, the beam loses energy at those wavelengths where the sample absorbs energy. By using these techniques infrared absorption spectra are easily obtained from most solid materials without the need for grinding, or making a mull. Because of the small penetration depth this method is also quite surface sensitive. Of greatest importance for acquiring high-quality spectra, the optical contact must be very good, which requires plane samples with smooth surfaces. In micro-ATR these requirements have to be met only over a small area (a few square millimetres).

Another IR reflectance method is reflection absorption spectroscopy (RAS). In this external reflection method, the radiation is collected at an angle, which is the same as the angle of incidence. At high angles of incidence, i.e., when the beam is nearly parallel to the surface, there is a high gain in intensity for the measurements of thin surface films (0.5 nm) and absorbed molecules on substrates such as metals.

A third IR reflectance technique is diffuse reflectance spectroscopy (DRIFT). This attachment is based on two ellipsoidal mirrors; the first is used to condense the beam and focus it onto a small spot and the second one collects a significant portion of the diffusely reflected radiation. Most of the specular part is rejected. This accessory is very useful for analyzing powder samples or diffusely reflecting surfaces.

In an IR microscope, which is a combination of an ordinary optical microscope, IR-beam condensers, and a variable aperture, samples can be analyzed at sizes of the order of 10 μm. An IR-microscope can be operated both in the transmittance and reflectance mode. In the latter case, the sensitivity is quite limited.

Using an integrating sphere, a hemispherical reflectance spectrum is obtained. From such a spectrum, the emittance of a surface can be calculated. Further details are available (Griffiths and De Haseth, 1986; Hobart et al., 1988).

4.8.2.2   *Ultraviolet-Visible-Near Infrared Spectroscopy (UV-VIS-NIR)*

UV-VIS spectroscopy is very closely related to IR-spectroscopy. The principal difference is the spectral region investigated. In contrast to IR-spectroscopy, where transitions between vibrational energy levels are induced with infrared light, these spectra are associated with transitions between electronic energy levels of the molecules. NIR spectroscopy is an extension into the near infrared region, which is a part of the full infrared region. In the near infrared region, many materials experience a "vibrational overtone absorption" where the fundamental absorption is in the IR region. In recent years, NIR-spectroscopy has attracted considerable interest. Gases, liquids, and solids can be examined. Both types of UV-VIS-NIR spectrometers are manufactured. An integrating sphere attachment can be used for hemispherical reflectance or transmittance measurements that may be used for calculating solar absorptance of absorber coatings or the solar transmittance of transparent materials, respectively (see Chapter 2.1). Further details are available (Rao, 1975; Bower and Maddams, 1992).

4.8.3   THERMAL ANALYSIS

Thermal analysis includes a number of analytical techniques in which some parameter of the sample, such as its mass or dimension is determined as a function of temperature. The temperature may be held constant or varied at a controlled rate. Thermogravimetry is one of the most common techniques of thermal analysis. It is based on the change of sample mass as a function of temperature or time that results from physical and/or chemical reactions induced by the temperature change. Typical applications include identifying the phases in degradation products of metal and plastics and evaluating the amount of sorbed compounds like moister in exposed materials. The sensitivity for detecting specific compounds by thermogravimetry ranges from 0.01 to 1% of the total mass. Thermal analysis instruments are simple to operate and the evaluation of the measured results is generally straightforward

Of the various types of thermal analysis, differential scanning calorimetry (DSC) is the most important type, at least for analysing the degradation and stability of polymers. In DSC, heat production resulting from exothermic reactions or heat consumption because of endothermic reactions are measured as a function of temperature or time at a constant elevated temperature. DSC can give information about the degree of crystallinity in semicrystalline materials, degree of curing in two-component adhesives or thermosetting plastics, and melting and glass transition temperatures.

The stability of a polymeric material or the stabilising power of a stabiliser package can be examined by exposing a sample to a temperature ramp in the presence of oxygen. In an oxygen induction temperature measurement, the temperature at which a rapid exothermic reaction initiates is a measure of the stability of the exposed material. In an alternative procedure, the temperature is maintained at a constant value and the induction time is measured. Both techniques are very commonly used in investigations of polymeric material degradation and aging. Further details are available (Wendlandt, 1974; Haines, 2002).

4.8.4 TECHNIQUES FOR IMAGING THE SURFACE TOPOGRAPHY AND ANALYSIS OF THE NEAR-SURFACE REGION

4.8.4.1 Scanning Electron Microscopy
Scanning electron microscopy (SEM) is an imaging technique for studying all kinds of solid materials such as metals, ceramics, glasses, polymers, and wood in addition to biological materials. It is now a routinely used technique; the data can be easily interpreted by those with little experience. The greatest advantages of SEM measurements are a resolution of about 10 nm and large depth of field. The latter is perhaps the most important. In addition, three-dimensional imaging is possible.

In principle, a scanning electron microscope consists of a high-energy electron gun, lenses for focusing the primary electron beam on the object of interest, and an electron detector, all inside a high vacuum system. The amplified signal from the detector modulates the brightness of a cathode ray tube (CRT). An image is obtained on the CRT screen by scanning the primary electron beam and the CRT electron beams in a raster pattern in synchronism across the analyzed surface and the CRT screen. The magnification is determined by the ratio of the deflection of the CRT beam and the deflection of the primary electron beam. Usually, detection of secondary emitted electrons yields the best resolution and imaging. However, backscattered primary electrons provide topographical as well as some chemical information (different modes).

In an environmental scanning electron microscope, imaging of the sample is obtained at a pressure of greater than 20 mbar. In this environment, moisture is retained on the material surface, thereby permitting in situ analysis at pressures close to atmospheric conditions.

In addition to the electrons used for imaging, a number of other particles are emitted, e.g., X-ray photons and Auger electrons. These can be used for obtaining information of which elements are present in the sample. Both techniques will be described below. Further details are available (Watt, 1985).

4.8.4.2 Scanning Tunneling Microscopy

In the scanning tunneling microscopy (STM) technique, the topography of a surface is monitored by scanning a sharp tip over it. The distance from the tip to the surface is measured from the analysis of the tunneling current between the tip and the surface. The technique has a resolution of 1 nm or better, thus permitting individual atoms to be imaged. The sample has to be a good conductor that obstructs analysis of insulating materials or thicker layers of insulating or even semiconducting compounds on top of conducting materials like metals. As with atomic force microscopy (AFM, discussed below), an STM can be operated under vacuum conditions or in environments at atmospheric pressure. Therefore, the technique has the potential capability for in situ studies of the initial stages of degradation of materials before the degradation products formed hamper the tunneling current between tip and sample. Further details are available (Loretto, 1984).

4.8.4.3 Atomic Force Microscopy

In atomic force microscopy (AFM), the topography of a surface is obtained by monitoring the force between a cantilever and the surface to be analyzed. When the distance between the tip of the cantilever and the surface is close enough, an attractive force results in a bending of the cantilever. By measuring the deflection as a function of position over the surface an image of the surface is obtained with a resolution of the order of 1 nm or better. As opposed to STM, one can study conducting as well as nonconducting materials that makes this technique more versatile than STM. Further details are available. (Loretto, 1984).

4.8.4.4 X-ray Spectroscopy

Two classes of X-ray photons are emitted from atoms exposed to high-energy electrons in a scanning electron microscope. The first class, characteristic X-rays, have well-defined energies that are characteristic of the atoms in the sample. These X-rays form sharp peaks in the X-ray energy spectrum and contain information about the elements present in the specimen. Atoms in the specimen emit X-rays as they return to their ground state, after an inner shell electron has been removed by an interaction with a high-energy electron beam. The wavelength or energy of the photons is different for different elements. The second class consists of continuum X-rays that have a wide range of energies and are the background in the X-ray spectrum with no useful information. They are produced when scattering near the atomic nucleus slows incident high-energy electrons. This type of X-ray radiation is also called brake radiation.

Two different types of detectors are used to measure X-ray intensities as a function of energy or wavelength. In an energy dispersive X-ray spectrometer (EDX), X-rays generated from the specimen enter a solid state, lithium-drifted silicon detector. Because of their interaction, electron–hole pairs are formed and results in a pulse of current through the detector circuit. The number of electron–hole pairs produced by an X-ray photon is proportional to its energy. The pulses produced are amplified, sorted by size with a multichannel analyzer, and displayed as an energy spectrum.

In a wavelength dispersive X-ray spectrometer (WDX), the X-rays are incident on a bent crystal and are reflected only if Bragg's law is satisfied. The crystal is set to focus X-rays of one specific wavelength onto a detector and rotates to scan the wavelength detected. Only one wavelength at a time can be detected. WDX spectra have better spectral resolution than EDX spectra but poorer spatial resolution. The analysis is also slower for WDX. In current SEM instruments, EDX is almost always used.

Often it is useful to show the concentration of a specific element as a function of the position on the specimen, which is done by elemental mapping. The X-ray emission from a specific element is used to modulate the intensity of the CRT electron beam as the primary electron beam scans the specimen and forms the map. The X-ray signals are produced from almost the entire interaction volume of the primary electron beam in the specimen. The diameter of this volume is generally of the order of 1 μm, which determines the spatial resolution of the EDX technique. This diameter of the interaction volume is generally also much greater than the primary electron beam diameter, which determines the resolution of secondary electrons. Secondary electrons formed deep in the interaction volume will not reach the surface and will not contribute to the SEM image. Consequently, an image produced by an X-ray peak has a much poorer resolution than an ordinary SEM image. Further details are available (Loretto, 1984; Watt, 1985).

4.8.5 ELECTRON SPECTROSCOPIES FOR SURFACE ANALYSIS

Auger electron spectroscopy (AES) and X-ray photoelectron spectroscopy (XPS) are the two most widely used surface analytical techniques. The two techniques are largely complementary. Both can provide elemental identification at comparable sensitivities. However, XPS provides important chemical bonding information resulting from shifts in the binding energy for different valence states. AES is much faster, can provide analyses at spatial resolutions of about 100 times that of current XPS instruments, and is more easily used with sputter depth profiling for in-depth analyses.

In both AES and XPS instruments, the electron energy analyzer or electron spectrometer is the crucial component because both techniques are based on the energy analysis of ejected electrons from surfaces. There exist a number of different types of electron energy analyzers. The most commonly used analyzers are the Cylindrical Mirror Analyser (CMA) and the Concentric Hemispherical Analyzer (CHA). One advantage of CMA is that it has a very high acceptance angle that provides both a high sensitivity and reduces the dependence of the signal on specimen topography, i.e., rough surfaces can be analyzed. Another advantage is the relatively simple design and the possibility of incorporating an electron gun into the analyzer for a very compact unit suitable for AES. The principal advantage of the CHA is a much higher electron energy resolution. A CHA is also smaller than a CMA. The size is an important property because space is often limited. In the past, the CHA was most often used in XPS instruments, whereas the CMA was used in AES instruments. In current equipment, CHA is preferred especially for combined XPS/AES instruments

4.8.5.1 *Auger Electron Spectroscopy*
In Auger Electron Spectrocopy (AES), the Auger effect is the deexcitation of an atom by a nonradiative process. When a high-energy electron beam impinges upon a specimen, inner shell electrons will be ejected from the atoms in the specimen. The resultant electron vacancy in an atom is soon filled by an electron from one of the outer shells. The energy released may appear either as an X-ray photon (low probability), used in X-ray spectroscopy, or be transferred to another electron (high probability), which is ejected from the atom with a very well-defined energy specific for each element. The collision cross section, i.e., the disposition to collide with surrounding electrons and thereby losing energy, is very high for electrons in condensed matter. Therefore, only electrons ejected very close to the surface may escape from the surface with little or no energy loss. An electron energy analyzer as described above is used to measure the energies of the ejected electrons.

The escape depth, i.e., the depth from which Auger electrons can be detected, determines the surface sensitivity of the technique. The escape depth is close to the minimum of about two monolayers at electron energies in the region of about 60 eV. The upper energy limit for ejected Auger electrons is about 3 keV, which corresponds to an escape depth of about ten monolayers. Here, the Auger electron peaks are very weak. When an electron beam has produced the primary ionization, peaks resulting from the Auger electrons appear superimposed on a background of backscattered electrons. Backscattered as well as low-energy secondary electrons may be used to form an image in AES in the same manner as in a scanning electron microscope (SEM). In fact, an AES instrument can be regarded as a SEM instrument equipped with an electron energy analyzer. Normally, the vacuum in a conventional SEM is lower, e.g., about 10^{-4} Pa than in an Auger instrument, e.g., about 10^{-8} Pa. Because the Auger technique is a surface sensitive technique, an ultra high vacuum (UHV) must be maintained to avoid detecting reaction products between the surfaced and residual gases in the vacuum.

The basic components of an AES system are an electron gun, an electron energy analyzer – electron spectrometer, an electron detector to record secondary electron images (SEM), and an ion gun for sample cleaning and sputter depth profiling.

On a sample of unknown composition, the elements present at the surface may be determined from the presence of their characteristic Auger peaks. Standard Auger spectra for the majority of elements have been assembled in handbooks and in computer files. There are usually little problems in identifying the major elements present. Small peaks may be difficult to assign unequivocally, particular at lower energies, where many elements have peaks. Interpretation may be further complicated by peak shifts resulting from chemical bonding effects or surface charging.

In general, AES instruments are relatively easy to operate. The simplest investigation is point analysis from areas of interest selected from the SEM image. Ion bombardment may be used to sputter the surface slowly to remove contamination. A more detailed investigation involves the identification of elements present and a subsequent determination of their spatial distribution by elemental mapping. AES is more rapid and has about 100 times greater spatial resolution than XPS. The higher spatial resolution is achieved because, in AES, the excitation results from an electron beam and not an X-ray beam. An electron beam can be focused into a narrower diameter. In modern AES instruments, a spatial resolution of about 10 nm can be achieved. The electron beam can also be scanned in a raster pattern across the sample surface in the same way as in a SEM. This can be used for elemental mapping in the same way as for EDX.

Sputter depth profiling is one of the most important applications of AES as it provides a convenient way of analyzing the in-depth composition of thin surface layers. If a depth profile is required, the elements of interest are first determined and their Auger intensities recorded as the surface material is sputtered by an ion beam. The sputtering rate is determined by ion beam intensity, ion energy, and surface properties. From time, sputtering rate, and depth resolution considerations, practical sputtering depths are limited to a few micrometers. For analysis deeper into the material, cross sectioning may be considered.

Auger electron spectroscopy is a very powerful surface analytical technique that has applications in many fields of science and technology, and in industrial applications like metallurgy, corrosion, and semiconductor science. In many solar materials, thin metal and oxide films are used. AES is a most powerful tool for the investigation of these materials and their degradation and aging behavior. However, some disadvantages exist. The electron beam can damage the surface (normally no serious problem). Charging can make the analysis of insulators difficult or impossible. The sensitivity is quite moderate (0.1–1 atom%). Further details are available (Briggs and Grant, 2003; Watts and Wolstenholme, 2003).

4.8.5.2 *X-ray Photoelectron Spectroscopy*

X-ray photoelectron spectroscopy also known as electron spectroscopy for chemical analysis (ESCA) is the most commonly used surface analytical technique today. In XPS, the investigated surface is irradiated by a source of "monochromatic" X-rays. The X-rays result in photoionization of atoms in the specimen and photoemission of electrons is detected by measuring the energy spectrum of the emitted photoelectrons from the specimen. In addition to the valence electrons, that are involved in the bonding of the system, each atom possesses core electrons (except hydrogen). The binding energy (E_b) of each core electron is characteristic of the individual atom to which it is bound. When using XPS equipment, the sample surface is typically irradiated by 1253.6 eV or 1486.6 eV X-rays under ultra high vacuum (UHV) conditions. The resulting photoelectrons have a kinetic energy (E_k), which is related to the X-ray energy ($h\nu$) by the Einstein relation, $E_k = h\nu - E_b - \Phi$. Photoemission occurs when the kinetic energy of the electrons is sufficient to overcome the work function barrier, Φ, at the surface. The entire process is called the photoelectric effect. Because the energy levels of electrons in an atom are quantized, the photoelectrons have kinetic energy distribution that consists of discrete bands. The intensity of photo-emitted electrons is monitored as a function of their kinetic energy. For electrons with kinetic energies of less than 2 keV, the information originates from about 0.5 to 3-nm deep, corresponding to 1 to 5 monolayers of solid compounds. Because each element has a characteristic spectrum of emitted electrons, the overall spectrum provides an elemental analysis of the investigated surface region. All elements except hydrogen can be detected. In XPS, the binding energies of the core-level electrons in atoms are sufficiently affected by their chemical environment to cause a detectable shift in the measured photoelectron energy. Therefore, chemical information is obtained from XPS.

The measured area is normally a few square millimeters, but instruments with a spatial resolution of ~10 µm are available. The detection limit is about 0.1% for most elements, but is reduced considerable if the surface is contaminated by hydrocarbons during sample preparation and handling. As with AES, data can be quantified with an uncertainty of 10 to 20% of the absolute value without standards, and to 1 to 2% by using reference standards. Further details are available

(Czanderna and Hercules, 1991; Briggs, 1998; Czanderna *et al.*, 1998; Briggs and Grant, 2003; Watts and Wolstenholme, 2003).

4.8.6 SECONDARY ION MASS SPECTROMETRY

In Secondary Ion Mass Spectrometry (SIMS), a solid surface is bombarded by primary ions with energies that typically range from 5 to 25 keV. The primary ion energy is transferred to target atoms in the sample via atomic collisions and a collision cascade is generated. Part of the energy is transferred to the surface and near surface atoms and results in the ejection of atoms and molecular compounds and fragments that can overcome the surface binding energy. The interaction during the collision cascade with surface molecules is weak enough to allow even large and nonvolatile molecules with masses up to 10,000 amu to escape without or with little fragmentation. Most of the emitted molecular particles are neutrals, but a small fraction is also positively or negatively charged. The subsequent mass analysis of the emitted ions provides detailed information on the elemental and molecular composition of the surface.

SIMS can be a monolayer surface sensitive technique because the emitted particles originate from the uppermost one or two monolayers. The volume dimension of the collision cascade is rather small and all particles are emitted within an area of a few nanometers in diameter. Hence, SIMS can also be used for microanalysis with very high-lateral resolution, provided that the primary ion beams can be focused with diameters of 100 nm or even 10 nm in custom-made instruments.

Most of the emitted particles in SIMS are neutral in charge. Postionization of the ejected neutral particles by electrons, plasma, or photons allows mass spectrometric analysis of these particles. This technique is called Secondary Neutral Mass Spectrometry (SNMS). One of the most efficient ways to ionize the emitted neutrals is by using Laser Postionization (Laser-SNMS). This technique is becoming increasingly used for the analysis of extremely small volumes because the fraction of the neutrals detected can be as high as 10%. However, laser-SNMS is, at present, quite complicated and expensive.

4.8.6.1 Dynamic SIMS

Secondary ion mass spectrometry (SIMS) is destructive in nature because particles are removed from the surface. This can be used to erode the solid in a controlled manner to obtain information on the in-depth distribution of elements. This dynamic SIMS mode is widely applied to analyze thin films, layer structures, and dopant profiles in semiconductors. However, organic materials like polymers cannot be analyzed by dynamic SIMS because the sputtering process more or less carbonizes the material and consequently no valuable information can be extracted.

4.8.6.2   *Static SIMS and Time of Flight-SIMS*

To receive chemical information on the original undamaged surface, the primary ion dose density must be kept low enough ($< 10^{13}\, cm^{-2}$) to prevent a surface area from being hit more than once. This static SIMS mode is widely used for the characterization of molecular surfaces. Practically, static SIMS can be regarded as "nondestructive." To fulfil the requirements of nondestructiveness, highly sensitive detection systems have to be used. Modern high-transmission mass spectrometric techniques and pulse counting equipment fulfil these requirements. In this respect, the most powerful technique is Time-of-Flight (TOF) mass spectrometry. TOF-MS is possible because ions with the same energy but different masses travel with different velocities. In its simplest setup, ions formed by a short ionization event are accelerated by an electrostatic field to a common energy and travel over a certain drift path to the detector. Measuring the flight time for each arriving ion allows the determination of its mass. The TOF spectrum can be repeated with a certain repetition rate depending on the flight time of the highest mass to be recorded.

In a more sophisticated design, the TOF analyzer corrects for small differences in initial energy and angle to achieve high-mass resolution. Combinations of linear drift paths and electrostatic sectors or ion mirrors are used and results with mass resolutions, M/dM, above 10,000 can be achieved. Major advantages of this approach over quadrupole and magnetic sector type analyzers are the extremely high transmission, the parallel detection of all masses, and the unlimited mass range.

In TOF-SIMS, a start time of all secondary ions can be defined by using a pulsed primary ion beam. Extremely short ion pulses with a duration below 1 ns are applied in a high-mass resolution analysis. These ion pulses are formed from a continuous beam by a pulsing unit and can be compressed in time by electro-dynamic fields (Bunching). The pulsed beam can be focused to a small spot (Ion Microprobe) to select a small area of interest and can be rastered to determine the lateral distribution of elements and molecules (Imaging SIMS).

During the drift time of the secondary ions, the extraction field can be switched off and low-energy electrons can compensate for any surface charging resulting from primary or secondary particles (charge compensation or electron flooding). As a matter of fact most of the charges particles that leave the surface are electrons. Consequently, the surface is always positively charged. By electron flooding, all types of samples can be analyzed. Even bulk insulators can be analyzed without any problems.

4.8.6.3 *Brief Summary of the Most Commonly Used Operating Modes of TOF-SIMS*

For *Surface Spectroscopy*, the purpose of a static SIMS investigation is to analyze the original, nonmodified surface for its composition. Because SIMS is a destructive technique, this means that the contribution of those secondary ions to the spectrum originating from already bombarded surface areas must be negligible. This quasi nondestructive surface analysis can be achieved by the application of very low primary ion dose densities as discussed above for static and TOF-SIMS. For *Surface Imaging*, a finely focused ion beam is rastered over the surface, like an electron beam in an electron microprobe; mass resolved secondary ion images (chemical maps) can be obtained simultaneously. For *Sputter Depth Profiling*, the time when the extraction field is switched off can also be used to apply low-energy ion beams for sample erosion. In this case, the low-energy beam forms a sputter crater, the center of which is analyzed by the pulsed beam (Dual Beam Mode). For the analysis of the sputtered neutrals, pulsed laser beams can be used and compare favorably with dynamic SIMS. For *Scanning Electron Microscopy*, electrons are the most commonly charged particles leaving a surface bombarded by primary ions. These electrons can be used to form an image in the same way as in an ordinary scanning electron microscope.

TOF-SIMS is a most powerful analytical technique. All kinds of solid materials can be analyzed without any restrictions except from sputter depth profiling of organic materials, as was discussed above. In fact, after cooling with liquid nitrogen even liquids can be analyzed. The sensitivity is very high compared with AES and XPS. During analysis, a modern TOF-SIMS spectrometer produces an enormous amount of data that can only be handled by a computer, which is an essential part of a TOF-SIMS system. The software package not only contains software for data handling, but also powerful tools for evaluating spectra. However, some drawbacks do exist. The major disadvantage is that the technique is not quantitative unless standards are used, and appropriate standards are frequently not available. Calibration procedures must be performed using samples with known concentrations. Moreover, a TOF-SIMS instrument is not easy to operate and the evaluation of data is often not straightforward. Further details are available (Czanderna and Hercules, 1991; Czanderna *et al.*, 1998; Vickerman and Briggs, 2001).

REFERENCES

Bower, D.I. & Maddams, W.F. (1992) *The Vibrational Spectroscopy of Polymers*, Cambridge University Press, Cambridge.

Briggs, D. & Grant, J.T. (2003) *Surface Analysis by Auger and X-ray Photoelectron Spectroscopy*, IM Publications and Surface Spectra Limited, Manchester.

Briggs, D. (1998) *Surface Analysis of Polymers by XPS and Static SIMS*, Cambridge University Press, Cambridge.

Czanderna, A.W. & Hercules, D.M. (Eds.) (1991) *Ion Spectroscopies for Surface Analysis*, Plenum, NY.

Czanderna, A.W. Madey, T.E. & Powell, C.J. (Eds.) (1998) *Beam Effects, Surface Topography, and Depth Profiling in Surface Analysis*, Plenum, NY.

Griffiths, P.R. & De Haseth (1986) *Fourier Transform Infrared Spectrometry*, John Wiley & Sons, New York.

Haines, P.J. (2002) *Principles of Thermal Analysis and Calorimetry*, RSC Books, Cambridge.

Hobart, H.W. *et al.* (1988) *Instrumental Methods of Analysis*, 7th edition, Wadsworth Publishing Company, Belmont, California.

Loretto, M.H. (1984) *Electron Beam Analysis of Materials*, Chapman and Hall, London.

Rao, C.N.R. (1975) *Ultra-Violet and Visible Spectroscopy*, Butterworths.

Vickerman, J.C. & Briggs, D. (2001) *ToF-SIMS Surface Analysis by Mass Spectrometry*, IM Publications and Surface Spectra Limited, Manchester.

Watt, I.M. (1985) *The Principles and Practice of Electron Microscopy*, Cambridge University Press, Cambridge.

Watts, J.F. & Wolstenholme, J. (2003) *An Introduction to Surface Analysis by XPS and AES*, John Wiley & Sons, New York.

Wendlandt, W.W. (1974) *Thermal Methods of Analysis*, John Wiley & Sons, New York.

Part 5

Methods for Reducing Environmental Stress in Solar Collectors

Part 5 consists of five chapters that are concerned with minimizing the moisture problem in flat-plate solar collectors. For example, the rain tightness of the solar collector box is crucial for avoiding severe problems that result from a humid microclimate. In the first chapter, a rain penetration test is described that can be used to assess the extent to which the box of a solar thermal collector is substantially resistant to rain penetration. In the second chapter, a summary is given for optimizing the ventilation and ventilation rate on the amount of moisture inside the box of flat-plate solar collectors. It is shown that the microclimate in collectors depends strongly on the collector type for three principle reasons, i.e., the type of and amount of air ventilation, insulation materials and framing materials, and changes by aging over time. In the third chapter, design guidelines are presented for optimising the humidity in the microclimate inside a solar collector to improve the service lifetime. Simulation data for the influence of the ventilation rate on the annual hours that areal condensation exceeds certain amounts are presented for a traditional solar thermal collector, collectors with a vapour barrier, and for collectors with different insulation thicknesses. In chapter four, detailed procedures are discussed for modeling the microclimate inside a collector, and in the fifth chapter, a standard test procedure is presented for assessing the tightness of ventilated solar-thermal collectors.

Chapter 5.1

Rain Tightness

O. Holck and S. Svendsen

Department of Civil Engineering, Building 118,
Technical University of Denmark, DK-2800 Kgs. Lyngby, Denmark

Abstract: The purpose of this chapter is to summarize a rain penetration test that is intended to assess the extent to which the box of a solar thermal collector is substantially resistant to rain penetration. Testing for rain penetration is the highest priority over all other tests that result in performance degradation of the solar collector. The rain tightness of the solar collector box is crucial for avoiding severe problems that result from a humid microclimate, which will have a strong negative effect on the durability of the collector. The risk of water penetration is increased when the collector is exposed to both rain and gusts.

Keywords: Rain penetration, Wind gusts, Deflections, Degradation, Microclimate, Collector box, Apparatus

5.1.1 INTRODUCTION

Rain must be prevented from entering the solar collector box to avoid severe problems with a humid microclimate, which has a strong negative effect on the durability of a collector. The risk of water penetration is increased when the collector is exposed to both rain and wind gusts. Gusts may deflect the cover and result in the movement of air into or out of the collector box.

5.1.2 RAIN PENETRATION TESTING

Testing for rain penetration is the highest priority over all other tests that may result in degradation of the solar collector. The collector box must be sealed to prevent water from entering and to assure an acceptable microclimate. The collector box must not permit the entry of either free-falling rain or driving rain. Water is not allowed to come into contact with parts that should be dry under all circumstances. Water penetration may result from leaks in the collector box from capillary action through cracks and apertures. Any penetration of water can only be permitted in the parts of the collector designed to drain off water.

Because of special circumstances, a rain penetration test may not be appropriate in all cases. For example, the collector box construction may differ from normal practice, the mounting construction may be unusual, the collector may have a curved glazing or is unglazed, or the back of the collector is open. In addition to reporting the amount of water penetration and where the water penetrated, information must be obtained about deflections of the cover and the back of the collector. This information is useful for the characterization of the collector for ventilation resulting from wind gusts.

The rain penetration test is included in the EN 12975-2 (EN 12975-2, 2001). Equipment for performing rain penetration tests is available in many testing laboratories and research institutes, e.g., SPF, Institut für Solartechnik, Rapperswil, Switzerland or SP Swedish National Testing and Research Institute, Sweden.

5.1.3   DEFLECTION OF THE COLLECTOR COVER

Because of wind gusts, the collector cover may deflect, result in a positive or negative pressure in the collector box, and air may flow out of or into the collector box. If a flow of air into the collector box takes place, the risk of rainwater penetration of the collector box is increased. In addition, the movement of the cover because of wind gusts may, in some constructions, result in a negative influence on the water tightness of the sealing systems. Negative pressure in the collector box and deflection of the cover may be simulated in four ways. Firstly, a vacuum pump may be connected to the collector box; secondly, the collector may be mounted in an opening in the box in which the pressure can be varied; thirdly a mechanical load may be applied to the cover; and finally, the collector box may be placed in a wind tunnel.

Using a vacuum pump is not recommended because the test is not realistic. Neither the pressure nor the airflow into the collector box results from a deflection of the cover. Approximately 15 Pa is the realistic pressure difference as the driving force for air ventilation between the outside and inside of the collector box, and the pressure resulting in deflection of the cover with a vacuum pump is 600 Pa. This test is only suitable for detecting where leakage may take place. The principal advantage of this method is the low cost for equipment because only a vacuum pump is needed.

In the "pressure box method," a differential pressure is established between the cover and the back of the collector. The collector is mounted in the opening in the box. This method is realistic for collectors built into the roof of a building because deflection of the glass results from the additional pressure over the collector surface. The advantage of this method is that it is the most realistic system test for collectors integrated into the roof because the pressure is distributed uniformly over the whole area. The principal disadvantage is that the apparatus takes up a lot of space because the pressurizing system should be inside the box. This method is not suitable for collectors mounted above the roof or collectors mounted on a rack. This is because the pressure over the collector surface will be the same as the pressure under the collector, and if the collector is inside the pressurized box the disadvantage is similar to the vacuum pump method. Finally, this method is not similar to in-service use because air does not flow over the collector surface.

In the third method, an apparatus is used with which a number of suction cups are placed on the collector cover. The cover is deflected inwards or outwards as it responds to changes in the applied test pressure. Depending on the deflection of the cover and the air tightness of the collector box, a differential pressure will result between the inside and the outside of the collector box and airflow into or out of the box will take place. As an outstanding advantage, this is a realistic method for simulating the effect of wind gusts either for collectors built into the roof or for collectors mounted on top of the roof or on an open frame. The apparatus, including the pressurizing system, is easy to build and can easily be placed at outdoor test facilities. The disadvantage is the relatively high cost of the apparatus. For collectors integrated into the roof, water must not enter from the bottom of the collector, so a simulated roof has to be made for the collector. Because of the adhesion of the suction cups, this method is not appropriate for testing unglazed solar collectors or collectors with a curved glazing.

In the fourth method, the collector box may be mounted inside a wind tunnel, and a pulsating airflow is used. If a wind tunnel is not already available, this method is too expensive.

5.1.4 DIFFERENCES IN COLLECTOR TYPES AND MOUNTING METHODS

In Europe, except for The Netherlands, collectors are usually mounted on top of the roof or on an open frame. Clearly, the collector box must be rain tight if it is mounted in this way, and the tests described in Section 5.1.3 are appropriate. In the Netherlands, tile roofs are predominantly used, so most collectors are mounted directly on the timbers, and a gutter system is used to drain excess rainwater. The back of the collector is not constructed to withstand large amounts of water because this will never be needed during in-service use. Without exception, these are open, ventilating-type collectors. However, a direct ventilation connection through the roof construction should not be used to prevent air with high humidity from condensing inside the collector box.

5.1.5   PROCEDURE FOR CARRYING OUT A RAIN PENETRATION TEST

5.1.5.1   *Objective and Expectations of Collector Boxes*

This test method is intended to assess the extent to which collector boxes are essentially resistant to rain penetration with or without wind and wind gusts. Collector boxes should normally prevent the entry of both free-falling and driving rain. Collector boxes may have vents and drainage holes, but these must not permit the entry of rain.

5.1.5.2   *Apparatus and Procedure*

The collector shall have its fluid inlet and outlet sealed, and be placed in the test equipment at the shallowest angle to the horizontal recommended by the manufacturer. If this angle is not specified, the collector shall be placed at a tilt of 30 degrees to the horizontal or less. Collectors that are designed to be integrated into a roof structure shall be mounted in a simulated roof and have their underneath surface protected. Other collectors shall be mounted in a conventional manner on an open frame or on a simulated roof.

To simulate the effect of wind gusts, the cover of the collector should be deflected mechanically, i.e., by using suction cups and pneumatic cylinders as described by the third method in Section 5.1.3. The load may be applied in different ways but the equipment should be able to apply both negative and positive loads on the cover in a way that allows the cover to move in a flexible way. The collector shall subsequently be sprayed on all sides using nozzles or showers for a test period of four hours without deflection of the cover. An appropriate flow rate shall be used to simulate rain, e.g., > 0.051 s/m^2 of collector aperture. The collector shall also be sprayed with water while exposed to cyclic deflection of the cover.

5.1.5.3   *Detection of Water Ingress*

The collector shall be mounted and sprayed as explained above while the absorber in the collector is kept warm. This can be done either by circulating hot water at about 50°C through the absorber tubes or by exposing the collector to real or simulated solar radiation. The penetration of water into the collector shall be determined by inspection, e.g., by determining if water droplets or condensation form on the inside of the cover glass. The collector should be heated to the test temperature before beginning the water spray, to ensure that the collector box is dry before testing is begun.

5.1.5.4 Test Conditions

The collector shall be sprayed with water at a temperature of less than 30°C with a flow rate of more than 0.05 l/s per square meter of collector aperture. The duration of the test without deflection of the cover shall be 4 h. Cyclic mechanical deflections of the cover shall take place with a positive or negative load of more than 600 Pa. The maximum load shall be maintained for about 3 s and the total of one cycle should not be more than 20 s.

5.1.5.5 Water Tightness Requirements

Water tightness means that penetrating water is not allowed to come into contact with parts that must be dry under all circumstances; glass panes and interior parts must not become wet, unless the parts are designed to drain water. The pass or fail criteria for the part of the collector test without deflection of the cover are: (1) the area with visible evidence of rain penetration must not exceed one part per thousand of the aperture area, e.g., ≤ 1000 mm^2/m^2; and (2) evidence of any rain penetration shall disappear after exposure to rain in 1 h or less. The pass or fail criteria for the part of the test with the collector exposed to cyclic deflection of the cover are: (1) the area with visible evidence of rain penetration must not exceed one percent of the aperture area, e.g., 100 cm^2/m^2, and (2) the evidence of rain penetration shall disappear after exposure to rain in 1 h or less.

5.1.5.6 Report of Results

The collector box shall be inspected for water penetration after each part of the test. The results of the inspection, i.e., the amount of water penetration and the places where the water penetrated, shall be reported. The extent of the movement of the collector cover and collector back from mechanical deflection shall be reported. The amount of water penetration is reported as a percentage of the aperture area.

5.1.6 CONCLUSIONS

The collector box may not permit the entry of either free-falling rain or driving rain under any circumstances. Applying a pulsating air pressure at the maximum range expected during in-service use to the cover of the collector is necessary. The maximum pressure differential of 600 Pa corresponds to a wind speed of 30 m/s and corresponds to the usual practice for windows and building components. The pass or fail criteria for the part of the collector test with deflection of the cover are less severe because of the extreme conditions.

REFERENCES

EN 12975-2 (2001). Thermal Solar Systems and Components – Solar Collectors – Part 2: Test Methods, CEN publication date 2001-06-06, Europe, B-1050, Brussels, Belgium.

Chapter 5.2
Optimization of the Ventilation Rate in Flat-Plate Collectors

M. Heck, V. Kübler, and M. Köhl

Fraunhofer-Institut für Solare Energiesysteme, ISE, D-79110, Freiburg, Germany

Abstract: The primary purpose of this chapter is to summarize the optimization of ventilation and ventilation rate on the amount of moisture inside the box of flat-plate collectors. Most flat-plate collectors are not air tight, which is in marked contrast to evacuated tubular collectors and hermetically sealed flat-plate collectors, and humidity can enter or leave by ventilation. It is shown by measurements of the ventilation rate and of the transient behavior of the humidity in collectors after rapid changes of the ambient humidity, that the microclimate in collectors depends strongly on the collector type for three principle reasons, i.e., the type of and amount of air ventilation, insulation materials and framing materials, and changes by aging over time.

Keywords: Collectors, Ventilation rate, Moisture exchange, Microclimate, Insulation materials

LIST OF DEFINITIONS

x	absolute humidity in g/kg
ρ	density of humid air in kg/m^3
r	radius of ventilation hole in m
v	average air speed in m/s

5.2.1 INTRODUCTION

The microclimate in ventilated solar thermal collectors is a critical factor for the service lifetime of collectors. High humidity in the collector is likely to result in corrosion of moisture-sensitive materials and components. A humid microclimate can be recognized from the condensation on the glazing after radiative cooling into a cold, clear sky (see Figure 5.2-1). Most flat-plate collectors are not air tight, which is

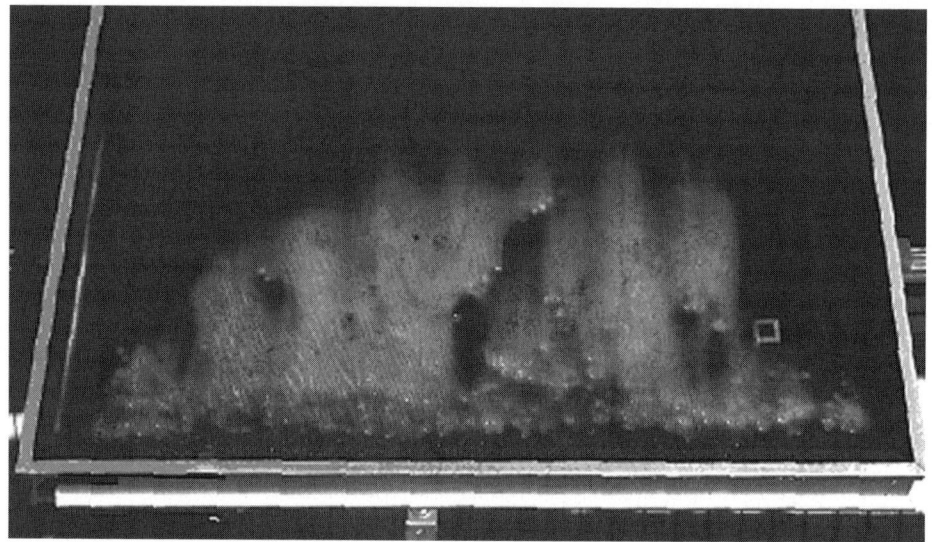

Figure 5.2-1.   Condensation on the glass cover of a flat-plate solar collector.

in marked contrast to evacuated tubular collectors and hermetically sealed flat-plate collectors, and humidity can enter or leave by ventilation. Experience shows that the microclimate depends strongly on the collector type for three principle reasons, i.e., the type of and amount of air ventilation, insulation materials, and framing materials.

5.2.2   VENTILATION

Ventilation results from unintended openings and superfluous vents. Natural ventilation results from buoyancy, as can be seen from Figure 5.2-2, where heating during the day results in an upward-oriented flow and the opposite effect occurs during cooling at night. The ventilation rate is defined as the number of air-exchanges of the total collector volume per hour, e.g., a ventilation rate of 10 means that 10 times the volume of air in the collector flows through the collector per hour. The ventilation rate depends on the absorber temperature, i.e., buoyancy effects; geometrical properties of the collector; kind, dimension, and location of the vents; and pressure differences between the inside and outside of the collector, which principally result from wind. An excessively high ventilation rate might reduce the concentration of moisture in the collector, but may also result in thermal losses and permit entry of dust and insects. Therefore, the ventilation rate should be measured and optimized.

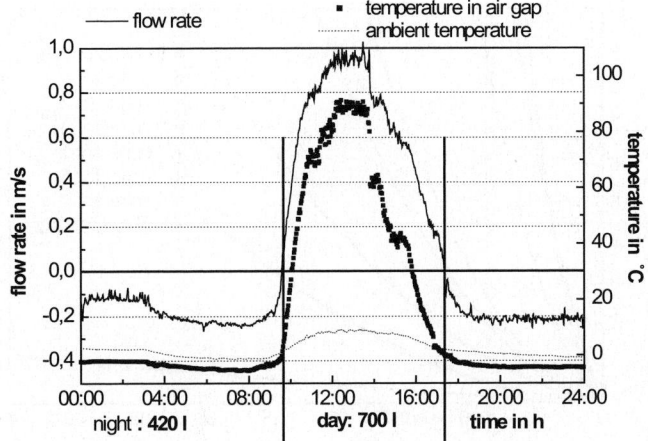

Figure 5.2-2. Vent air speed measured in a collector during stagnation conditions. The negative flow rate means inversion of the flow direction resulted from radiative cooling of the glazing.

5.2.3 VENTILATION RATE MEASUREMENTS

To measure the air ventilation rate, the collector was mounted at an angle of $45°$. Positive and negative pressure differences between the collector interior and the surroundings were applied through an extra opening at the back by using a pump and a flow controller. The resulting pressure difference between the collector and the ambient air was measured by a sensor with a range of 0–50 Pa. The correlation between the flow rate and the pressure difference is a characteristic for the ventilation rate of a collector (see Figure 5.2-3). The test procedure could be simplified because no significant difference was found between an over-pressure or an underpressure in the collector.

During normal operating conditions, i.e., an absorber temperature between $40°C$ and $60°C$ the pressure difference, resulting from thermal buoyancy is about 0.5 Pa to 1 Pa. Therefore, these pressures are appropriate for comparison (see Figure 5.2-4).

Note that the real ventilation is lower by at least a factor of two, since half of the openings are needed for inlet and the other half for outlet. The description of a recommended test and the results of round robin testing for comparison and validation are presented in the appended Chapter 5.5.

5.2.4 VENTILATION RATE OPTIMIZATION

A collector, which did reveal condensation in the morning, was considered to be optimally ventilated. The ventilation rate per collector area of other collector was

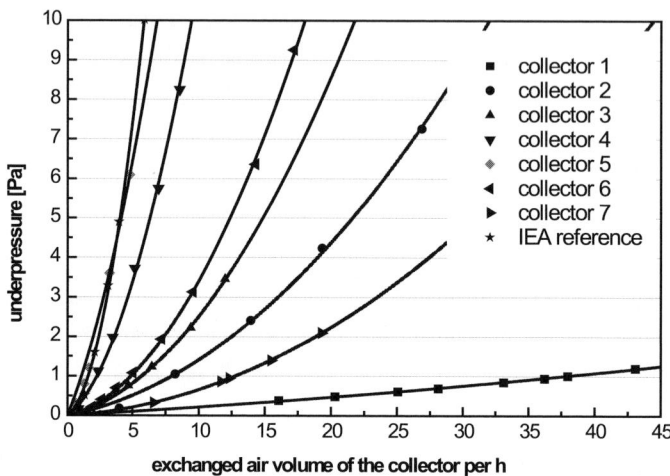

Figure 5.2-3. Dependence of the ventilation rate on pressure difference for seven different collectors (* marks the IEA reference collector).

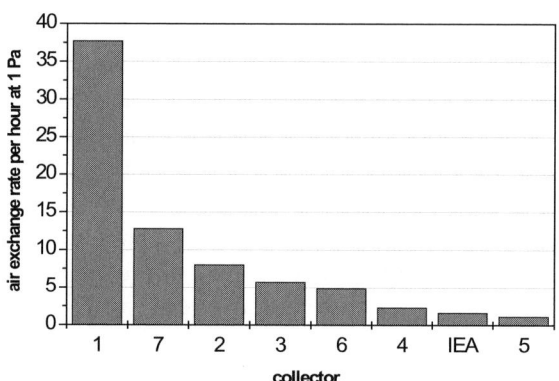

Figure 5.2-4. Ventilation rate for eight different collectors at a pressure difference of 1 Pa.

optimized by using the same flow rate as for the IEA collector (see Figure 5.2-5). Usually, larger vents are necessary to increase the ventilation rate. Aging of frames and sealant caused an increase in the leakage of the collectors, which in general resulted in a higher ventilation rate and especially for collectors with wooden frames (see Figure 5.2-6). A rough estimate of the thermal losses resulting from ventilation yielded a maximum of 1.5% for stagnation conditions, when no energy is used and the absorber temperature could reach more than 200°C. Thus, the losses are essentially negligible under normal operating conditions.

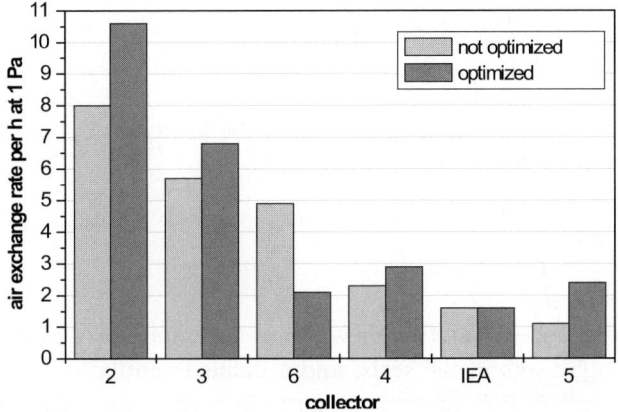

Figure 5.2-5. Ventilation rate for six different collectors at a pressure difference of 1 Pa before and after optimization.

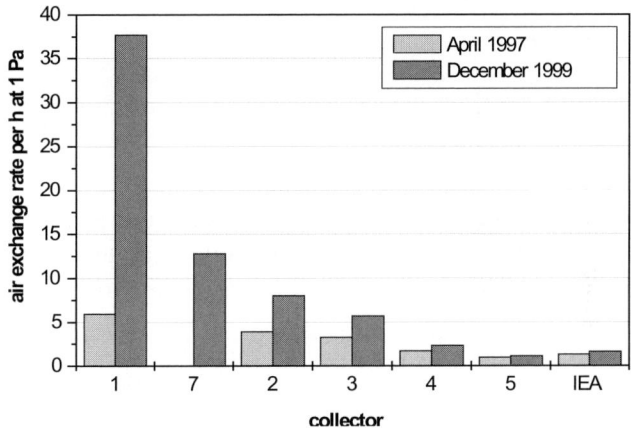

Figure 5.2-6. Ventilation rate for seven different collectors at a pressure difference of 1 Pa before and after natural aging.

5.2.5 MOISTURE EXCHANGE

Drying the inside of the collector box is the principal result from ventilation. Relative humidity in the collector may also be influenced by using materials that can store moisture, e.g., the insulation and the wooden frames. Therefore, the final test for the transient moisture behavior in a collector is the response of the internal humidity to rapid changes of the ambient humidity. Such tests can be carried out in a climate chamber at a high relative humidity, in which collectors are introduced quickly, i.e., a "wet-in test." The alternative is to measure the drying behavior of a wet collector in dry ambient conditions, i.e., a "dry-out test," which turned out to be less feasible. In both cases the collector has to be heated to operating conditions to achieve ventilation by buoyancy.

5.2.5.1 Dry-out Tests

Collectors with a well-sealed frame, which is essential for avoiding uncontrolled ventilation through leaks in the seals, and a defined ventilation rate were operated at an absorber temperature of 40°C or 60°C in a dry, controlled ambient climate, i.e., 23°C and a RH of 30%. Humidity sensors were mounted near the vents at the top and the bottom of the collector and the air speed in the vent was measured (Figure 5.2-7). Transient calculations of the rate of humidity loss (Figure 5.2-8) were made after a defined amount of water, which depends on the collector area, had been injected into the air gap, i.e.,

$$\dot{x}_{\text{output}} = \left(\rho_{\text{air top hole}} \times x_{\text{air top hole}} - \rho_{\text{ambient}} \times x_{\text{ambient}} \right) \times \pi \times r^2 \times v$$

in which x is the absolute humidity, ρ is the density of humid air, r is the radius of the vent holes, and v is the average air speed. The total integrated humidity loss agreed well with the quantity of injected water. However, comparisons between different

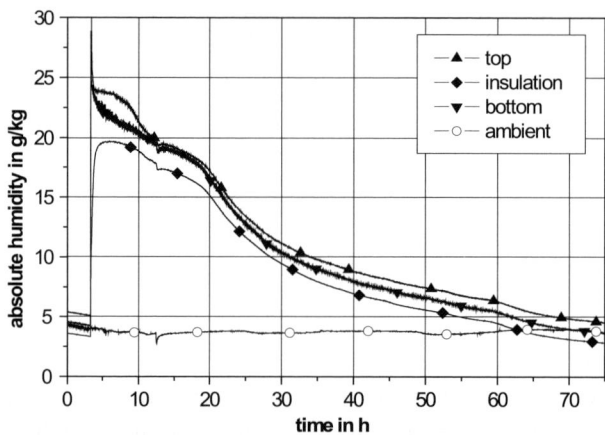

Figure 5.2-7.   Absolute humidity in the ambient air, in the collector near the vents, and in the rear insulation of a solar collector.

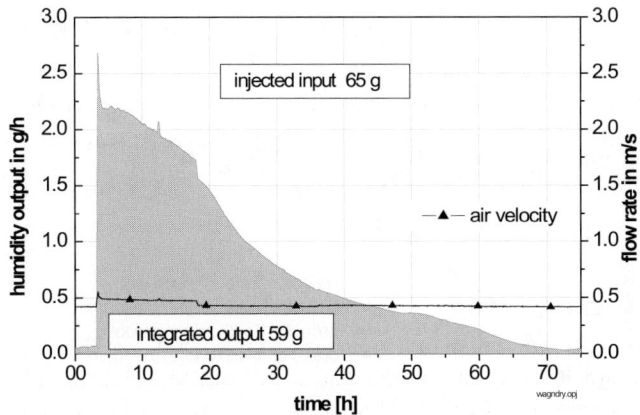

Figure 5.2-8. Absolute humidity near the bottom vent in a solar collector and the air speed through the hole.

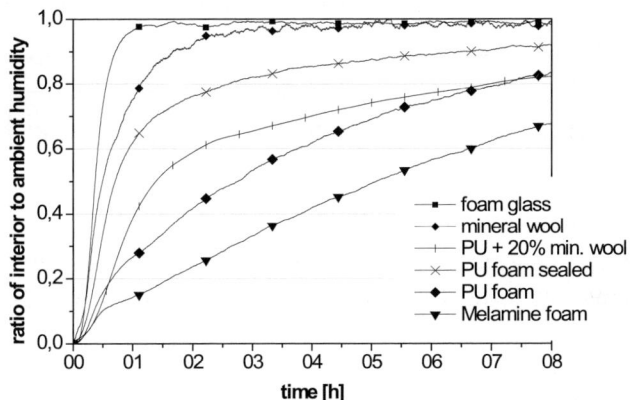

Figure 5.2-9. Humidity ratios from wet-in tests of seven collectors with different insulation materials.

collectors were not satisfying, because sometimes the humidity condensed on the glass cover resulting in delayed drying.

5.2.5.2   Wet-in Tests

In the wet-in test, the opposite procedure is used in which a dry collector is placed into a climate chamber with a controlled high humidity. The response of the humidity inside the collector yields information about the ventilation rate and the sorption behavior of the insulation and construction materials.

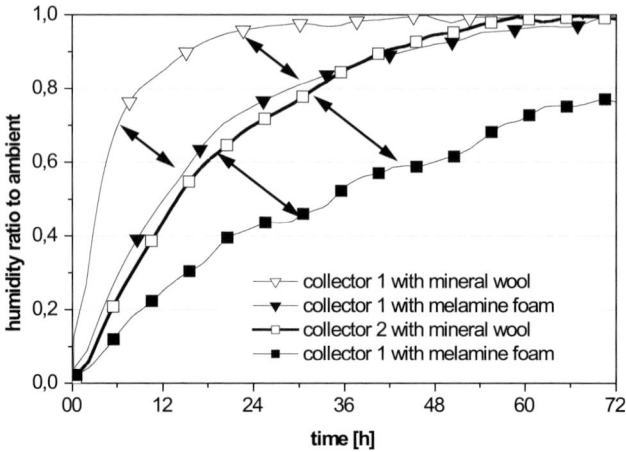

Figure 5.2-10.   Humidity ratios from wet-in tests of two different collectors, each with two different insulation materials.

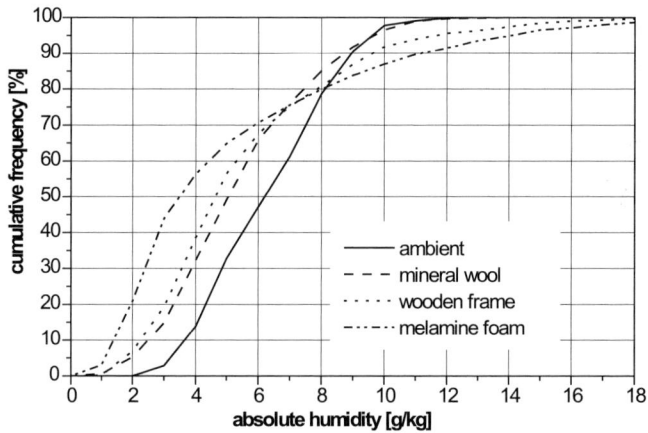

Figure 5.2-11.   Cumulative frequency of the occurrence of increasing amounts of absolute humidity in collectors with two different insulation materials and a collector with a wooden frame during outdoor exposure.

Tests with different collector types show clear differences that depends on the sorption uptake by the insulation materials (Figure 5.2-9). The fastest response was obtained with foam glass material, which sorbs the least amount of moisture. Figure 5.2-10 more clearly shows the influence of two different insulation materials with different water sorption capacities, because the same collectors were used in the tests by simply changing the insulation. The highest absolute humidity was found in collectors with melamine foam (Figure 5.2-11), even though condensation was never observed on the glazing. Wooden frames show the same phenomenon. The reason might probably be that desorption from the media with a high capacity of sorbed water occurs at high collector temperatures, and permits high *absolute* humidity values without reaching a high *relative* humidity or even the dew-point.

Chapter 5.3

Guidelines for Limiting Environmental Stress Factors in Glazed, Ventilated Solar Collectors

O. Holck and S. Svendsen

Department of Civil Engineering, Building 118,
Technical University of Denmark, DK-2800 Kgs. Lyngby, Denmark

Abstract: The purpose of this chapter is to provide design guidelines for optimizing the humidity in the microclimate inside a solar collector to improve the service lifetime. Humidity inside a collector is the major factor that must be minimized to provide the most favorable microclimatic condition for preventing the corrosion of the internal materials of the collector. This microclimate inside the collector is important for determining the service lifetime of an absorber coating. During the design of the collector, the location and size of ventilation holes, properties of the insulation materials, and dimension of the solar collector box are parameters that have to be taken into account for the optimization to achieve the most favorable microclimate to prevent corrosion. Simulation of the microclimate in solar thermal collectors can be a valuable tool for optimization of the ventilation rate for the collector. Manufacturers of collectors can be advised if their solar collectors need to be tighter, if more or fewer ventilation openings are needed, and what influence the insulation material has on the microclimate inside the collector. Guidelines for collector designers are proposed that are suggestions to be considered during the design of solar collectors. Simulation data for the influence of the ventilation rate on the annual hours that areal condensation exceeds certain amounts are presented for a traditional International Energy Agency (IEA) collector, collectors with a vapor barrier, and for collectors with 10 mm of insulation instead of 50 mm. Based on calculations, the optimum ventilation rate is 60 l/h at a pressure difference of 1 Pa.

Keywords: Optimization, Microclimate, Ventilation rate, Lifetime, Design, Air-gap, Collector box, Ventilation holes, Vapor barrier, Insulation thickness, Hygroscopic material.

5.3.1 INTRODUCTION

To select the most cost-effective collector, the performance, cost, and durability should be considered (Chapter 1, Section 1.2). Materials inside the collector box principally degrade because of corrosion resulting from high humidity and temperature. For a longer service life, the microclimate inside the collector box must be optimized to obtain the most favorable environmental conditions for the materials. This optimization will lengthen the service life for the collector and concomitantly improve the financial benefit. The air-gap is the space between absorber and cover in the collector. The collector box encloses the frame, the cover, and the back plate of the collector, protecting the inside components against the outdoor weather. Moisture in the air gap of the collector box can result in degradation of the absorber and other materials in the collector box. The transport of air and humidity into the collector results from air ingress through holes or other leaks in the collector box. The ventilation rate varies with weather conditions and the air temperature inside the air space in the collector box, which was discussed in Chapter 5.2. The wind velocity and the operating temperatures of the collector influence the ventilation rate.

Numerous considerations influence the final collector design. The types of collectors can vary and not all kinds will fit into the recommendations given in this chapter. Because it is not possible to prescribe the optimum collector design, we provide a summary of durability-related problems related to collector design. The summary, i.e., design guidelines, is addressed principally to manufacturers of solar collectors, but can also be valuable for practising architects and engineers involved in designing building-integrated solar collectors.

Two principal research activities are our sources for producing the guidelines. These are annualized simulations of the microclimate inside the collector box by using a balance model (Chapter 3.4.2), and measurements made in many European countries of the actual microclimates in traditional collectors that are supplemented with detailed investigations of relevant special issues.

5.3.2 DESIGN OF COLLECTORS WITH RESPECT TO MOISTURE PROBLEMS

It is recommended that the designer consider the criteria in this section for optimizing solar collectors with respect to moisture problems. First, the application has to be considered, i.e., where will the collector system be used and the purpose for its use. A special design may have to be considered. Special investigations must be carried out in cases where the operating conditions differ from those typical for solar collectors. The operating conditions for solar collectors can differ because of the way they are used. For some systems, the collectors are supposed to operate at low temperatures, e.g., for preheating in hot water applications, and some collectors are supposed to operate at high temperatures. The distribution of internal flow of the heat transfer medium in the absorber also influences the surface temperature of the absorber. The surface temperature, in turn, influences the flow pattern of air inside the collector box. Thus, the absorber used may change the ventilation rate of the collector box. The proposed location of the installed collector will influence the design. Some collectors are intended to be attached above the roof and some are to be integrated in the building.

Secondly, the collector box must be sealed to prevent ingress of moisture from rain, which was discussed in Chapter 5.1. Otherwise, the collector will not have a favorable microclimate. In constructing the collector, the sealant around the cover is important and areas where water can collect have to be avoided. Rain tightness must be validated by using appropriate tests that were discussed in Chapter 5.1. Rain tightness of the collector is indispensable and should be documented carefully. The test for rain tightness should involve using variable pressure differences including the effect from movement of the cover because of wind load and the influence of aging of the sealant materials and application with time. An accelerated test may be used to include degradation changes in the rain tightness (see Chapters 4.3 and 4.4). Rain tightness can be measured using the procedure described in Chapter 5.1, especially using the test methods described (ISO 9806-2, 1995; EN 12975-2, 2001). If a particular design does not pass these tests, the collector design and/or materials used must be reconsidered.

Thirdly, with respect to the ventilation rate and air tightness of the collector box, a relatively high ventilation rate is desired during the day when the collector is warm because this helps dry the inside of the collector. On the other hand, a rapid ventilation rate results in energy losses that should be minimized. Ventilation also has to be minimized at night because the amount of moisture in the collector box will be increased when the collector is cool. The designer has to optimize the size and location of the ventilation holes and also the draining systems for certain applications. All possible sources of leaks in the construction need to be considered for ventilation of the collector box, of course, these include the designed ventilation and drain openings, but cracks and apertures in the rest of the construction may result in undesired ventilation. The high ventilation rate during the day is increased from an increased difference in the elevation between the ventilation holes. Thus, the position of the ventilation holes has to be close to the low and top end of the collector; preferably, these two holes are diagonally opposite each other. When designing the ventilation openings, we know moisture can accumulate in the insulation if it is in contact with the incoming air at an opening. Properly designed channels through the insulation materials to the rear side of the collector can prevent the incoming air from having contact with the insulation materials.

The optimized ventilation rate at a pressure difference of 1 Pa has been shown by calculations to be about 60 l/h (see, Section 5.3.3). The computer program MOMIC can be used to simulate the condensation of moisture on the inside of the cover of any given solar collector design (Holck, 1999). Simulations with this tool can be used for the optimization of the ventilation rate. By controlling the humidity increase from the insulation materials and optimizing the ventilation rate, the risk of wetness caused by water can decrease from the cover to the insulation. As another benefit from this optimization, bad visual impressions are avoided from a dewy collector.

The leakiness of the collector box is determined by using a pressure difference between the inside and outside of the collector box at 1 Pa and by measuring the airflow through the collector box. Air tightness can be measured using the procedure described in Chapter 5.5. Testing must validate the air tightness and drying of collector boxes. Tests can be performed to characterize the ventilation rate in the collector box. The air tightness test of collector boxes is essential for assuring that an optimized microclimate can be achieved. An accelerated test can demonstrate some changes in the air tightness.

The design of the collector can be accepted if an optimized ventilation rate is achieved and if ventilation will dry the collector during a drying-out test; passing this test when combined with the rain tightness test assures that the design has passed two critical tests. The insulation materials or other hygroscopic materials influence the rate of accumulation of moisture and its distribution as a condensate on the cover. The influence of the hygroscopic material on operation of the complete collector must be examined. The computer program MOMIC is a tool that has been useful for improving collector designs. In most cases, the action of the hygroscopic materials will result in an increased amount of condensation in a short time period, and in extreme cases produce visible condensation on the cover. Large concentrations of sorbed moisture in the insulation materials are a factor that may result in a visible condensate. Moisture in the insulation is released into the air inside the collector box in a short period of time and results in the condensate forming on the cover. A high ventilation rate has been shown to result in increasing the high concentration of moisture in the insulation during the night.

Wet-in and dry-out testing of the collector in a climate chamber can be used to investigate the behavior of hygroscopic materials used as a moisture buffer. A buffer effect from the insulation materials will lower the humidity in the air inside the collector box and reduce the risk of forming condensation. The reduced amount of moisture will help reduce the rate of degradation in collector materials. A vapor barrier will not permit buffering of the humidity in the collector box by the rear insulation. The barrier also eliminates the effect of moisture outgassing from the insulation, so it reduces the amount of condensation on the cover. A vapor barrier can be used to separate the air inside the air-gap from the air in the insulation. The consequences of eliminating the buffering effect from the insulation on the collector operation should be considered, or alternative methods of introducing a hygroscopic buffer may be needed. For constructing a vapor barrier, the sealant between the barrier and the collector frame has to ensure the air tightness of the vapor barrier.

The vapor barrier should be made of an ultraviolet light stabilized material that can withstand the elevated collector temperatures that can occur and without photothermal degradation resulting from the incident UV radiation. Finally, no leaks between the vapor barrier and the frame of the collector box can be permitted because of perforations made for the pipe used to transport the heat transfer fluid.

Lifetime estimation tests can be made by investigating the real-time durability of the collector and by comparing the performance degradation from accelerated tests. As stated in Chapters 4.4 and 4.5, accelerated aging tests of component materials in solar collectors can be used to determine the influence of their degradation on the long-term collector performance. The accelerated tests need to be combined with long time outdoor exposure tests to estimate the service lifetime of any particular collector design. As weaknesses in design are discovered, changes can be introduced that will improve the design and the projected service lifetime.

5.3.3 IDENTIFYING THE OPTIMIZED VALUE OF VENTILATION RATE FROM ANNUAL SIMULATION ASSESSMENTS

Simulations have been carried out using the computer program MOMIC and by using the Danish design reference year, DRY. Cumulative simulations have been successfully completed on an annual basis. The IEA-collector was used for these simulations as being representative of the traditional solar collector. The lEA-collector, which has a transparent area of $1.47 \, m^2$, should be operated at a 45° tilt and with a vertical separation of 1.4 m between the lower and higher positioned ventilation holes. The diameter of both ventilation holes is 7 mm. For the collector simulations, the optimized characteristic ventilation rate was calculated to be about 60 l/h for an insulation thickness of 50 mm. The ventilation rate is correlated with the transparent area of the collector cover because this is where condensation will form on the inside surface. The correlation will be adversely affected if the area acting as a moisture buffer differs significantly from the area of the collector cover. The separation of the cover and absorber is only of consequence if it becomes very small.

A larger collector cover area with the same vertical distance between the ventilation holes results in an increase in the optimized ventilation rate to values exceeding 60 l/h. Simulations have been carried out with and without a vapour barrier in front of the insulation. In addition, some simulations of collectors with a rear insulation thickness of 10 mm have been carried out. In the results presented in Figure 5.3-1, the lines either represent areal condensation exceeding $0.1 \, g/m^2$ or $1 \, g/m^2$ for the three collector configurations, i.e., traditional (T), with a vapor barrier (V), or with only 10 mm of insulation (I).

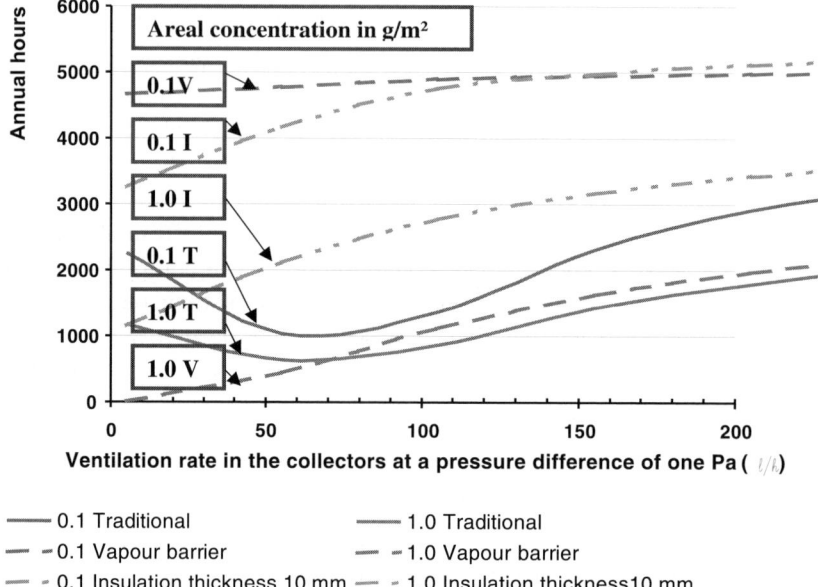

Figure 5.3-1. Results of simulations showing the annual hours of areal condensation concentrations of 0.1 and 1.0 g/m² on the cover as a function of flow rate. The solid lines represent the traditional (T) collector with 50 mm of insulation. Results for collectors with a vapor barrier (V) are shown with dashed lines; results for collectors with 10 mm of insulation (I) are shown by dash-double-dot line. Larger condensation concentrations are not shown because these will only appear at higher levels of ventilation rates.

In all cases, the annual hours of condensate concentration increases at ventilation rates exceeding 60 l/h. The simulations show that the annual hours of non-visible condensation (exceeding 0.1 g/m²) are decreased when using 10 mm of insulation material compared with using a vapor barrier. However, neither is as good as the traditional collector. At an areal concentration of 1 g/m², the collector with a vapor barrier is better than the other two types at flow rates of less than 60 l/h. In this situation, the vapor barrier eliminates the effect from the insulation by decreasing the amount of visible condensation at the lowest ventilation rates. From the data, the optimum ventilation rate for a traditional solar collector with a rear insulation thickness of 50 mm is about 60 l/h. Ventilation rates as low as 10 l/h can be used if the collector has a tight vapor barrier, thus improving the microclimate. At an areal condensation concentration of 0.1 g/m², the vapor barrier has a negligible influence as a function of ventilation rates. Changing the insulation thickness to 10 mm from the original 50 mm results in longer periods with an areal concentration of 0.1 g/m². The principal focus needs to be on water concentrations that result in forming a visible condensate. The annual hours that the areal condensate

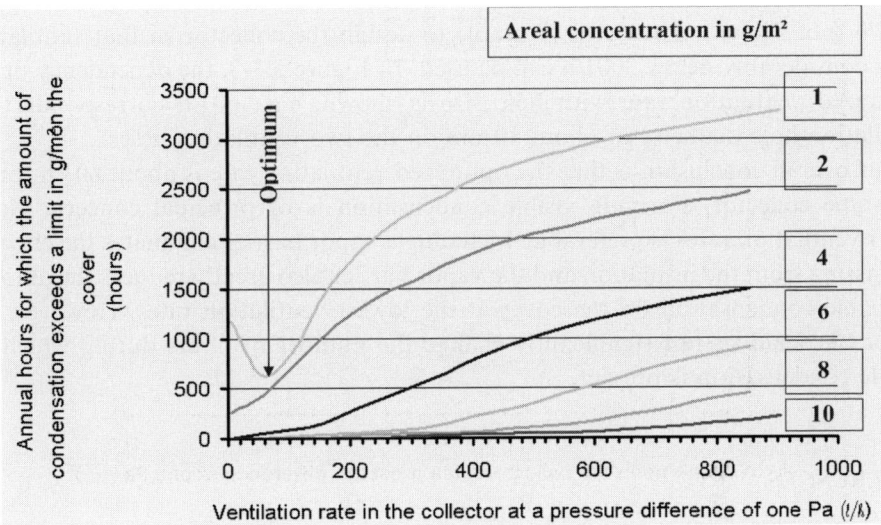

Figure 5.3-2. The annual hours that the areal concentrations exceed 1, 2, 4, 6, 8, 10 g/m² as a function of the flow rates. The arrow shows the optimum flow rate at 60 l/h and is based on simulations of a traditional collector with a transparent collector cover area of 1.47 m² and a rear insulation thickness of 50 mm. Higher ventilation rates result in a larger number of annual hours when condensation is on the cover.

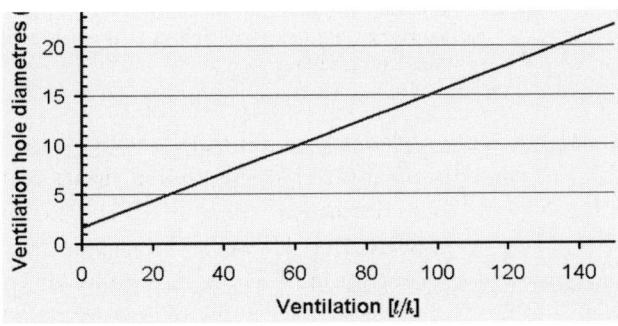

Figure 5.3-3. The size of the ventilation holes can be chosen for the predicted optimized ventilation rate characterized at a pressure difference of 1 Pa.

concentrations exceed 1, 2, 4, 6, 8, and 10 g/m² as a function of flow rate are shown in Fig. 5.3-2.

From those simulations of the annual hours of condensate the appearance of visible condensation on the cover with concentrations greater than 10 g/m² appear at ventilation rates of about 300 l/h using a tight vapor barrier will raise the rate to

450 l/h (not shown). Thus, it is desirable to design the collector so that ventilation rates considerably below 300 l/h can be used. In Figure 5.3-3, the dependency of the optimized ventilation rate with hole size is shown, e.g., 60 l/h corresponds to a ventilation hole diameter of about 10 mm on the two ventilation holes.

Our overall conclusion is that the optimized ventilation rate is about 60 l/h for an IEA-type collector. Forming visible condensation is of principal concern. Using lower ventilation rates is preferable. Including a vapor barrier eliminates the effect of outgassing from the insulation and the vapor barrier also greatly reduces the amount of visible condensation on the cover at the lowest ventilation rates. However, the vapor barrier does not significantly change the number of hours during which no visible condensation is present.

REFERENCES

EN 12975-2 (2001). Thermal Solar Systems and Components – Solar Collectors – Part 2: Test Methods, CEN publication date 2001-06-06, Europe, B-1050, Brussels, Belgium.
ISO 9806-2 (1995). Test methods for Solar Collectors – Part 2: Qualification Test Procedures Europe, CH-1211, Switzerland.
Holck O. (1999) MOMIC Computer Program, Modeling of Microclimate in Collectors, Report Number R-030, Thermal Insulation Laboratory, Technical University of Denmark, DK 2800 Lyngby, Denmark.

Chapter 5.4

Modeling of Microclimates

O. Holck and S. Svendsen

Department of Civil Engineering, Building 118,
Technical University of Denmark, DK-2800 Kgs. Lyngby, Denmark

Abstract: The purpose of this chapter is to present methods for simulations of the microclimate inside a solar thermal collector. To assess the behavior of airflow through the vents, CFD calculations with models for the water vapor on surfaces and the sorption of water vapor in the insulation were made at the Fraunhofer Institute for Solar Energy Systems. These results from CFD calculations support a reasonable test for air tightness of collectors. At the Building and Construction Research in the Netherlands, CFD results yielded dimensionless ratios of pressure differences. The dimensionless pressure coefficients (Cp) values together with climate data can be used to calculate the absolute pressure differences at a reference position for determining the flow through the collector. In a table, Cp values are presented for different wind directions. The structure of the balance model (MOMIC) based on an analytical heat and moisture balance is explained in detail. The model, which takes into account the condensation of water vapor on the inside of collector surfaces and the sorption of water vapor in the insulation material, has been validated by comparing result with measured data. Sensitivity tests of the balance model are presented by comparing results with varied parameters with measured data. The parameter sensitivity shows that it was essential to include the influence of the wind in the model and that insulation and vents size have an effect on the humidity in the air gap.

Keywords: Balance model, CFD, Pressure coefficients, Microclimate, Collectors, Condensation, Moisture, Wind

LIST OF ACRONYMS AND ABBREVIATIONS

Acronym	Definition
CFD	Computational Fluid Dynamics
CFX-F3D	CFD software package
DHW	Domestic Hot Water
DRY	Design Reference Year
IEA	International Energy Agency
ISE	Fraunhofer Institute for Solar Energy Systems

MOMIC	Compute program (Modeling of Microclimate in Collectors) based on an analytical heat and moisture balance in a collector
MSTC	Working group entitled "Materials in Solar Thermal Collectors"
PBL	Planetary Border Layer
TNO	Building and Construction Research in The Netherlands
TRY	Danish Test Reference Year

Symbol	Definition
A	amplitude of wind gusts
AH	absolute humidity
A_c	transparent surface area of the collector
a_c	solar absorptance of the cover
C	regression coefficient in airflow function
C_c	thermal heat capacity of the cover
C_p	thermal heat capacity of the absorber, the fluid and half of the rear insulation
C_p	dimensionless pressure coefficients
C_{pa}	specific heat capacity of air (J/kg°C)
C_s	Stefan–Boltzman constant (W/m^{2}K^4)
d	displacement length
ΔM_{ae}	mass of dry air removed by expansion of the air in the solar collector
ΔM_{wc}	increase of condensate on the inside of the cover
ΔM_{we}	mass of aqueous vapor removed by expansion of the air in the solar collector
M_{wv}	mass of aqueous vapor removed by ventilation
ΔP	pressure difference
Δt	time interval
e_g	thickness of the air gap (m)
e_i	thickness of the insulation
Φ_v	airflow
g	gravitational constant (m/s^2)
G	the solar irradiation on the solar collector (W/m^2)
Gr, Gr_g	grash of number
h	height of the air gap
h_{cb}	conduction through the rear insulation
h_{cg}	natural convection in the air gap
h_{co}	convection from the outside of the cover
h_{rg}	thermal radiation in the air gap

h_{ro}	thermal radiation from the outside of the cover
h_{vg}	ventilation of the air gap
K_a	Von Karman's constant
k_g	thermal conductivity of air (W/m°C)
k_i	the coefficient of thermal conduction of the rear insulation
K_o	thermal conductivity of the ambient air (W/m°C)
L	characteristic length (m)
M_{ag}	mass of dry air in the air gap
M_D	molecular weight of water
M_L	molecular weight of air
m_{vg}	total mass flow of aqueous vapor and dry air
M_{wi}	mass of aqueous vapor entering the solar collector
M_{wo}	mass of aqueous vapor leaving the solar collector
N	air exchange
v	kinematic viscosity (m^2/s^2)
n	exponent in airflow function
N_c	air exchange at a constant wind speed
N_g	air exchange for gusts
N_t	tightness coefficient $N_t = 10 \ (1000 \ V/C)^{1/n} \ 3600 \ A_c/V$
Nu_g	Nusselt number for the air gap
Nu_{lam}	Nusselt number for laminar flow
Nu_o	Nusselt number for convection
Nu_{turb}	Nusselt number for turbulent flow
p	total barometric pressure (Pa)
Pr	Prandtl number for air
p_w	partial pressure of aqueous vapor
p_{wa}	partial pressure of water in the ambient air
p_{wg}	partial pressure of water in the air inside the collector box
$p_{ws}(T_c)$	water vapor saturation pressure of the cover temperature (Pa)
$p_{ws}(T_{dewa})$	water vapor saturation pressure of the dew point temperature of the ambient air (Pa)
q_{ei}	evaporation heat loss from the inside of the cover
q_{eo}	evaporation heat loss from the outside of the cover
$\rho_{(Ta,pwa)}$	density of ambient air
RH	relative humidity
ρ_a	density of dry air
$\rho_{a((Tp+Tc)/2,pwg)}$	density of the air in inside the collector box
Ra_g	Raleigh number for air inside the collector box
r_c	heat of vaporization of water (J/kg)
Re	reynolds number for forced convection

Re'	Reynolds number for natural convection
Re_+	Reynolds number for forced and natural convection
ρ_o	density of dry air at $0°C$
S	slope of the collector (deg.)
T	time
T	temperture
$\tau\alpha$	transmittance–absorptance product of the solar collector
T_a, t_a, T_{amb}	ambient temperature
T_{abs}	absorber temperature
T_c, t_c, T_{cov}	temperature of the cover
T_g	temperature of the earth
T_{ge}	temperature inside the collector box at the end of a time interval
T_{gs}	temperature inside the collector box at the beginning of a time interval
T_i	temperature in the solar collector
T_p, t_p	temperature of the absorber
T_s	temperature of the "black body" of the sky
U_*	friction velocity
V	volume of air gap
$V(10)$	meteorological wind at an altitude of $10\,m$
$V(\Delta p)$	power function found in pressure tests
V_{lok}, v	local wind velocity
V_w, V_{col}	wind velocity at the collector cover (m/s)
V_z	wind velocity at level z
w	absolute humidity
w_g	absolute humidity of air inside the collector box
z	vertical co-ordinate
z_o	length of roughness
ε_c	emittance of the cover
ε_p	emittance of the absorber plate

5.4.1   INTRODUCTION TO COLLECTOR MICROCLIMATES

The use of solar thermal collectors is steadily increasing. To meet the customers needs for durable products, the quality of the collectors is important. To guarantee a service lifetime at a high performance, a dependable service lifetime prediction is needed for the collector. The principal component of every collector is the absorber

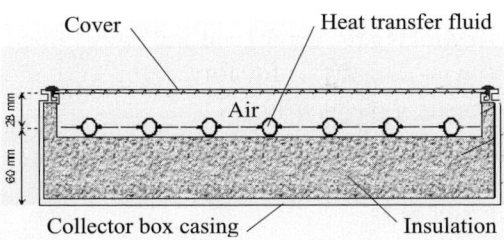

Figure 5.4-1. Cross-sectional view of a typical flate-plate collector. The air inside the collector box, its temperature, and its solar irradiance are the principal components of the microclimate for the absorber. We refer to the space between the inside surface of the cover and the support for the heat transfer fluid tubing as the air gap.

coating, and degradation of it is often the limiting factor for the service lifetime of the collector system because of corrosion. Corrosion results in a decrease in solar absorptivity and an increase in thermal emissivity, both of which result in an overall loss of performance. The microclimate inside the collector is important for determining the service lifetime of an absorber coating. Relative humidity, solar radiation, and the temperature inside the collector are the principal degradative stresses. A cross-sectional view of a common solar thermal collector is shown in Figure 5.4-1.

Because solar radiation is needed to raise the temperatures of the heat transfer fluid, the humidity inside the collector is the only controllable variable. During the night, the collector cover cools below the ambient temperature because of radiation to the cold sky. In climates where the night air becomes saturated with humidity, i.e., a relative humidity RH of 100%, condensed water will form on the outside and inside of the collector glazing. If too much condensation takes place on the inside of the glazing, it will drip onto the absorber surface. Water condensation on the inside of a collector cover is shown in Figure 5.4-2.

Test methods for assessing the microclimate in collectors have been developed and used as a part of the effort in the International Energy Agency (IEA) program entitled "Materials in Solar Thermal Collectors (MSTC)." Separate studies have also been carried out to assess the influence of wind, fluid dynamics, influence of insulation materials, air ventilation rates, and time of wetness. Airflow into and out of ventilation holes results in air and humidity transport into the collector. Airflow into and out of the collector box results from the temperature-differences during operation and the concomitant thermal expansion of the air inside the collector box. The ventilation rate, which is the measure for air exchange inside the box, is the volume of air that passes through the ventilation holes per unit time, e.g., in liters per hour (L/h). The ventilation rate depends on the geometry, operating conditions of the collector, and location and size of the

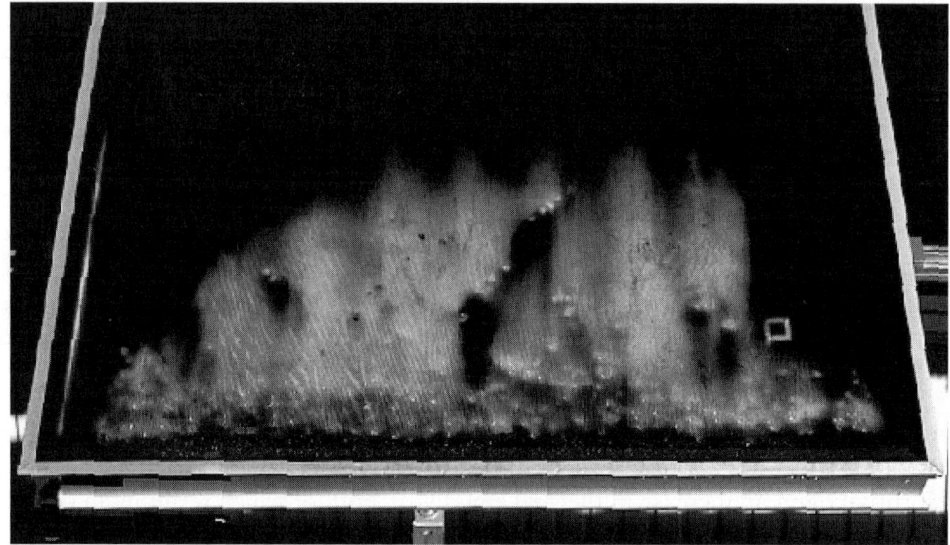

Figure 5.4-2. Condensation of water inside the collector cover.

ventilation holes. In the IEA MTSC program, measurements were carried out on identical collectors at laboratories and other test sites in Sweden, Denmark, The Netherlands, Switzerland, and Germany. The "identical" type of collector is referenced as an "IEA-collector". Collectors from industrial partners were also tested and modeled at the Fraunhofer Institute for Solar Energy Systems (ISE) in Freiburg, Germany. In addition, a compute model for calculating annualized data about the microclimate in the solar collector and the use of simulations with a Computational Fluid Dynamic (CFD) model are related activities established within the working group.

5.4.2 MODELING OF THE MICROCLIMATE IN THE SOLAR COLLECTOR

Simulations of the microclimate in collectors are useful to characterize quickly the microclimate and to understand the moisture content in ventilated collectors. A computer model that makes such calculations is useful for parametric sensitivity studies for accelerated testing and can be a valuable tool for the optimization of the collector and for the development of guidelines for solar collector design. From the modeling results, manufacturers can be advised if their solar collectors need more or less ventilation. Improved collector designs are essential for extending the long-term durability of solar collectors.

The amount of humidity inside the collectors is a major factor that can be optimized for establishing or retaining the most favorable microclimate condition for components inside the collector. For the collector design, the placement and size of ventilation holes, properties of the insulation materials, thickness of backside insulation, and dimension of the collector box are parameters that require optimization. The humidity in the air inside the collector box is determined from the humidity in the ambient air and the water that is sorbed in the insulation. The amount of humidity depends on the ventilation rate, and the tightness of the collector box, and the rate of desorption or sorption of moisture in the insulation. Control of the ventilation rate in the collector is not easy, even for the ventilation resulting from the chimney effect, which has a well-know theory. The effect from wind further complicated determining ventilation rates.

Using simulation with computational Fluid Dynamics (CFD) has been essential for rapidly assessing the behavior of airflow through the ventilation holes. Results from the CFD simulations were validated by experimentally measured air velocities in the vent holes, which are included in the computer program as a collector characteristic. CFD calculations have been made to clarify the pressure differences around the collector as well as the movement of air inside the collector box. The modeling results have supported the development of a reasonable test for the air-tightness of the collector and have been useful for modeling the microclimate inside the collector box. After the air-tightness of the collector is established, simulations of the microclimate can then be done most accurately. The years 1996 and 1997 were chosen to provide data for some parameters and to validate the use of the compute program. The special needs were to determine the influence of the wind, the rear-insulation thickness, and the effect of insulation on the humidity in the collector box.

5.4.2.1 *Computational Fluid Dynamics*

Simulations were carried out with a commercially available Computational Fluid Dynamics (CFD) software package CFX-F3D that was completed by importing models for the condensation of water-vapor on surfaces and for the sorption of water vapor in the rear-insulation of the collector. These phenomena strongly influence the microclimate in a collector. Stationary simulations showed the amount of loss during ventilation amounted to only about 1.0–1.6% of the total losses through the collector case. Both models for the condensation of water vapor on surfaces and for the sorption and desorption of water vapor on the insulation material reliably predicted those phenomena in realistic situations. The disadvantage of CFD simulations is that long computer times are needed. Therefore, a two-dimensional stationary simulation of the collector was used. The CPU time for a two-dimensional, transient simulation is about 1 h, but for three-dimensional simulations, the CPU time is about 80 h. By choosing the boundary conditions and using the collector geometry, absolute humidity (AH) and condensation can be predicted. The AH can be used for optimizing the size and position of the ventilation holes with a goal of obtaining a dry microclimate.

5.4.2.2   Steady state simulations

For a particular set of boundary conditions, e.g., the ambient and absorber temperatures, the ventilation rate of a collector can be determined by steady-state simulations. The calculated ventilation rates of the IEA-collector were compared with measurements for two specific sets of outdoor conditions, i.e., normal operating conditions such as on sunny winter days: $T_{amb} = 0°C$, $T_{abc} = 45°C$, $T_{cov} = 10°C$; and stagnation conditions: $T_{amb} = 5°C$, $T_{abs} = 120°C$, $T_{cov} = 43°C$. The calculated values differ the measured values by about 15%. Thus, we were encouraged to simulate ventilation rates of collectors that cannot be measured because of the number or position of their ventilation holes.

A vertically mounted collector enhances the ventilation rate compared with a horizontally mounted collector, because of the "chimney-effect." The simulations showed the losses from ventilation amounted to only 1.0 to 1.6% of the total heat loss of the collector at stagnation conditions; the percentage losses at normal operating conditions are even smaller. This indicated that the ventilation does not significantly influence the performance of the tested collectors.

5.4.2.3   Pressure Distribution around the Solar Collector

The ventilation rate is affected by thermal as well as wind effects. To support the development of a simulation model for the microclimate in a collector, measurements were carried out on a test collector mounted on a test roof at TNO. From these measurements, it was clear that the influence of the wind could be very large at high-wind speeds (Oversloot, 1997). Because making actual measurements on the mounted collector was difficult, CFD modeling was used to calculate the pressure differences around the collector box for different wind directions. The predicted pressure differences are expressed as dimensionless pressure coefficients (Cp value) relative to the wind pressure in the center of the collector cover. These values have been reported (Marcel, 1999).

In October 1998, a tightness test procedure was developed for solar collector boxed (see Chapter 5.5). The tightness parameters are necessary for calculating the ventilation rate in the collector. The calculated Cp values and tightness parameters were used to calculate the static part of the wind influence on a collector using data from a climate file. This approach can be initially used during the development phase of a collector to study the influence of the position of the ventilation openings and the resistance of the holes to compare and contrast different designs. The computer program MOMIC needs to be employed to obtain more detail on the ventilation rate and incorporating thermal effects for a complete microclimatic simulation model.

5.4.2.4 Tightness Test

For the tightness test, a reference collector was chosen for the microclimate measurements. This collector was tested in 1996 using the procedure described in Chapter 5.5. Extra openings were introduced for the test equipment, to supply air, and make the pressure difference measurement (see Chapter 5.5). This test normally results in a set of coefficients that describe the relation between pressure difference and ventilation rate, Φ_v, in the form of $\Phi_v = C \cdot \Delta P^n$. In this relation, the factor C is a coefficient related to the airflow at a 1 Pa pressure differential. The exponent n depends on the type of airflow through the openings. For large openings forming a low restriction, n will have a value close to 0.5. If the openings are formed by narrow slits or cracks, the value of n will increase to 1.

For the over-pressure test, the measured reference collector has a C of 268 L/h at 1 Pa and a value of $n = 0.60$. The test was also carried out for suction as well as for over-pressure. The differences are small, and are not taken into account by this calculation. In cases where large differences exist, the data needs to be combined because flow through the collector is a combination of inwardly and outwardly directed flow. This collector was prepared according to the instructions for the microclimate test setup, so the back plate of the collector was sealed along the edges to the frame of the collector box, thus resulting in a collector with low-ventilation characteristics.

5.4.2.5 The Pressure Coefficients for Roof Mounted Collector

The ventilation openings of the test reference collector are diagonally opposite and on the back plate of the collector. To calculate the net ventilation flow through the collector, the pressure difference between the collector openings is needed. The CFD result yielded the Cp values for different wind directions on the collector, which was divided into 9 sections over the back plate. These Cp values are dimensionless ratios of the pressure differences between a references position and a position of interest. The absolute pressure at a reference position can be calculated using normal climate data from a climate file. These are the wind speed at 10 m elevation and the air temperature. Together with the Cp values, these can be used for calculating the wind

Table 5.4-1. Pressure coefficients for the vent hoes at the lower left (lb) and upper right (rt) of the collector and reference wind speed at an elevation of 10 m made form CFD calculations. Zero is used for the wind direction from the north

Wind direction	Pressure coefficients (lb)	Pressure coefficients (rt)	Wind velocity (m/s) at an elevation of 10 m
0	−0.18	−0.09	5.90
45	0.03	0.10	9.07
90	−0.02	−0.04	6.43
135	0.20	−0.35	3.21
180	0.13	−0.17	6.43
225	0.10	0.03	9.07
270	0.18	−0.30	5.90
315	−0.09	−0.72	3.05

speed during the year to obtain an annual load curve. Normal climate data often take North as the zero direction, whereas in solar applications south is taken as the zero direction. Because the data in normal climate files are taking every hour, we only considered static effects; the dynamical effects based on the turbulent intensity numbers were neglected (see Table 5.4-1).

5.4.2.6 *Climatic Data*

For CFD calculations, the climate data must also contain the wind direction as a parameter. The often-used TRY files (Lund, 1985) for simulations in solar application do not contain this necessary parameter. The DRY files (Lund, 1995), that contain much more information, are an excellent source of data. Other sources can include measured data at specific locations, for example, such as those measured during a microclimate monitoring sequence. In the case for a wind speed at 10-m elevation, the reference height might differ from those generated by official metering stations and not be the undisturbed wind speed. Correction factors related to such locations are termed wind reduction factors. The ENV 1991-2-4, a European draft standard (CEM pREN 1991-1-4) contains detailed information about wind reduction factors for various locations in Europe.

Trial calculations were made using files generated from Meteonorm climate data. However, the data at several locations clearly contained artificially generated data by using an exponential decay for the wind speed and a very coarse approach to the wind direction. For Flushing, The Netherlands, the data did not have a proper relationship to the real wind speed and direction tables as published in official tables.

5.4.2.7 *Results Generated for Design Reference Year Copenhagen*

Using the wind speeds from the files for Design Reference Year Copenhagen, ventilation rates were calculated, as shown in Figure 5.4-3. The necessary data for wind speed, wind direction, and temperature were selected from the complete set. The calculations were formulated using a location-dependent wind reduction factor of 0.3, which can be taken as being normal for the type of surroundings found in a small village landscape, e.g., with trees and houses with less than two or three stories. The calculations reveal that natural wind speeds result in more than 100 h of ventilation rates exceeding 10 *l*/h (Oversloot, 1997). Compared with other locations, Copenhagen is a windy region and comparable with the coastal regions of Holland. The calculated results can be used for making comparisons among different collectors under constant conditions or for different climate sources. Measurements by TNO in The Netherlands confirm that high-ventilation rates can exist under certain windy conditions.

However, they measured more than 7000 occurrences with a ventilation rate of less than 5 *l*/h. This ventilation rate is larger than for thermal effects and for a larger number of hours. However, local circumstances can strongly influence the pressure differences and airflow around the collector. For example, Figure 5.4-4 shows the wind speed at the actual collector height. Clearly these are not extreme.

When compared with the ventilation rates measured, the CFD calculated ventilation rates are high in the microclimate experiments and, in the calculations using the model in MOMIC, the calculated ventilation rates are high. This probably is because most of the test sites either tests mounted the collector on racks instead of roofs or because the test sites are screened from the reported wind speed. For example, during the microclimatic test at TNO, the collector was on the part of the roof that is mostly screened from the wind.

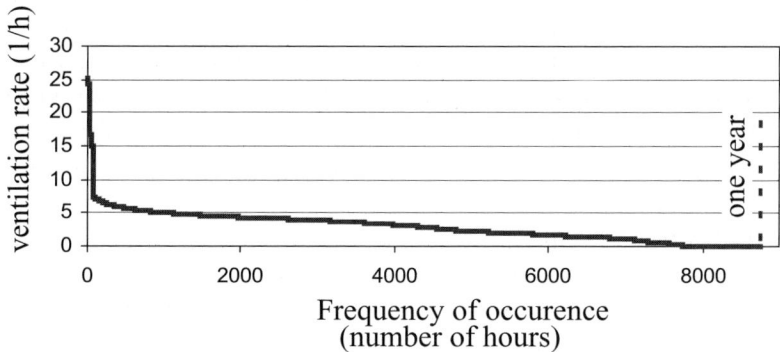

Figure 5.4-3. The number of hours at ventilation rates calculated from the wind speed over the collector for the Design Reference Year.

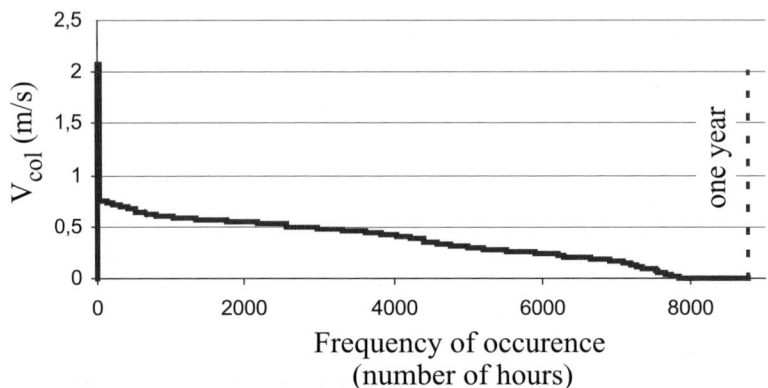

Figure 5.4-4. The number of hours of particular wind speeds, V_{col}, at the collector height and at the center position of the glazing.

However, a relationship exists between wind speed and the ventilation rate although it is less clear than calculated here (see also Section 5.4.5). The geometric influence of the surrounding close to the collector is large, which makes it difficult to generalize the results. The CFD method can still be used to study the effects and positions of the openings in the collector box on the ventilation rate, which cannot be studied in any other way.

5.4.3   MODEL OF HEAT AND MOISTURE BALANCE IN THE AIR IN THE COLLECTOR BOX (BALANCE MODEL)

To calculate the amount of condensate in the collector on an annual basis, a model of the microclimate in the collector is formulated, that is based on the heat balance, ventilation rate and moisture balance. From the moisture balance, the amount of water that condenses on and evaporates from the inside of the glass is obtained. The combined heat and moisture transfer balance in the insulation provides us with information about the buffering effect of moisture in this part of the collector.

5.4.3.1   *Heat Balance*
The heat balance is formulated for one square meter of a solar collector. In this approach, differences between the center and the corners of the collector are ignored. All of the thermal capacity of the collector is assumed to be in the cover and the absorber including the fluid and half of the back insulation. A cross-sectional view of the collector is shown in Figure 5.4-5.

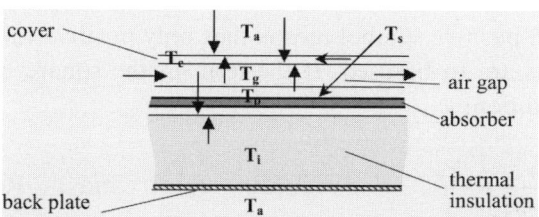

Figure 5.4-5. Cross-sectional view of a solar collector. The arrows show the transport of water and heat at in-service conditions without a heat transfer fluid flowing in the absorber.

To determine the temperatures, the following differential equations, which relate to Figure 5.4-5, need to be solved:

$$C_p \mathrm{d}T_p/\mathrm{d}t = G(\tau\alpha) - h_{cb}(T_p - T_a) - h_{cg}(T_p - T_c) - h_{rg}(T_p - T_c) \tag{1}$$

$$\begin{aligned} C_c \mathrm{d}T_c/\mathrm{d}t = {} & Ga_c + h_{cg}(T_p - T_c) + h_{rg}(T_p - T_c) \\ & - h_{vg}((T_p + T_c)/2 - T_a) - h_{co}(T_c - T_a) - h_{ro}(T_c - T_s) - q_{ei} - q_{eo} \end{aligned} \tag{2}$$

where the terms are defined in the list of symbols. In the differential equations, the heat transfer by convection and thermal radiation is the product of the heat transfer coefficient and the temperature difference to obtain a linearized form. The error resulting from linearization is small because the heat transfer coefficient is calculated for each time interval, and the time intervals are so short that the temperature changes are quite small. For the collector in operation, the boundary conditions for T_p are chosen to equal the measured value for T_p. In this case, one differential equation needs to be solved. The value of the solar transmittance–absorption product is used for normal incidence, but is corrected for the incidence angle by using incidence angle modifier coefficients. The heat resulting from conduction in the back insulation is calculated from $h_{cb} = k_i/e_i$.

The heat transfer coefficient for natural convection in the air gap is calculated from (Duffie and Beckman, 1991):

$$h_{cg} = \mathrm{Nu}_g k_g / e_g \tag{3}$$

where

$$\begin{aligned} \mathrm{Nu}_g = {} & 1 + 1.44 \left[1 - \frac{1708}{\mathrm{Ra}_g \cos s} \right]^+ \left(1 - \frac{(\sin 1.8 \cdot s)1.6 \cdot 1708}{\mathrm{Ra}_g \cos s} \right) \\ & + \left[\left(\frac{\mathrm{Ra}_g \cos s}{5830} \right)^{0.33} - 1 \right]^+ \end{aligned} \tag{4}$$

In equation (4), the positive symbol means that only positive values of the terms in the square brackets are to be used. If the term in the square bracket is negative, zero is used. In addition,

$$Gr_g = \frac{g\beta(T_p - T_c)e_g^3}{v^2}, \qquad \beta = 2/(T_p + T_c) \qquad \text{and} \qquad Ra_g = Gr_g Pr \qquad (5)$$

The heat transfer coefficient for thermal radiation in the air gap is calculated from:

$$h_{rg} = \frac{c_s(T_p^4 - T_c^4)}{(1/\varepsilon_p + 1/\varepsilon_c - 1)(T_p - T_c)} \qquad (6)$$

The heat transfer coefficient for thermal radiation from the cover to the sky and the ground is calculated from:

$$h_{ro} = 0.5(1 + \cos s)\varepsilon_c c_s(T_c^4 - T_s^4)/(T_c - T_s) + 0.5(1 - \cos s)\varepsilon_c c_s(T_c^4 - T_a^4)/(T_c - T_s) \qquad (7)$$

where $T_s = 0.0552 \cdot T_a^{1.5}$. The relation between the temperature of the air and the sky is taken from Duffie and Beckman (1991). and both temperatures are in Kelvin units. If measured data for the sky temperature are available, they should be used.

The heat transfer by convection from the cover to the ambient air is calculated from the following expressions (Krisher and Kast, 1978):

$$h_{co} = Nu_o k_o/L \qquad (8)$$

$$Nu_o = \sqrt{Nu_{lam}^2 + Nu_{turb}^2} \qquad (9)$$

$$Nu_{lam} = 0.664 Re_+^{0.5} Pr^{0.33} \qquad (10)$$

$$Nu_{turb} = 0.037 Re_+^{0.81} Pr /(1 + 2.443 Re_+^{-0.1}(Pr^{0.67} - 1)) \qquad (11)$$

$$Re_+ = \sqrt{(Re)^2 + (Re')^2} \qquad (12)$$

$$Re = V_w L/v \qquad (13)$$

$$Re' = 0.64 Gr^{0.5} \qquad (14)$$

$$Gr = Gr(L) = 2 \cdot (L)^3 g(T_c - T_a)/v^2(T_a + T_c) \qquad (15)$$

The heat transfer by evaporation/condensation at the outer surface of the cover is calculated from (Keller, 1984):

$$q_{eo} = h_{co} \frac{M_D r_c}{0.9 M_L C_{pa} p} \{P_{ws}(T_c) - p_{ws}(T_{dewa})\} \tag{16}$$

Equation (16) is only used for condensation. The expression is assumed to be valid for evaporation as well as for condensation. Of course, evaporation can only take place if the cover is wet. Because the heat of vaporization and the specific heat of air are nearly independent of temperature, typical values inserted in the equation (16) are:

$$T_c \geq 0 : q_{eo} = 0.017 \cdot h_{co} \{P_{ws}(T_c) - p_{ws}(T_{dewa})\} \tag{17}$$

Below the freezing point, the heat of vaporization is somewhat higher, so

$$T_c \leq 0 : q_{eo} = 0.019 \cdot h_{co} \{(p_{ws}(T_c) - p_{ws}(T_{dewa})\} \tag{18}$$

The water vapor saturation pressure is calculated using (ASHRAE Handbook, 1989):

$$p_{ws} = e^{(23.5771 - 4042.9/(T - 37.58))} \tag{19}$$

$$T_{dewa} = 37.58 - 4042.9/\{\ln(p_{ws}) - 23.5771\}; \tag{20}$$

The heat transfer by evaporation or condensation at the inside surface of the cover is calculated from the amount of transferred water that results from the moisture balance. The heat transfer coefficient resulting from ventilation is calculated from:

$$h_{vg} = m_{vg} C_{pa} \tag{21}$$

5.4.3.2 Ventilation Rate

Ventilation of the solar collector results from wind and thermally established pressure differences. The influence of wind can be divided into the influence of the mean wind speed and of wind fluctuations or gusts. The mean wind speed changes the pressure around the solar collector that results in ventilation of the solar collector through its cracks and apertures. A wind gust may move the cover ("pumping") that results in a pressure change inside the collector box, which is subsequently neutralized by airflow through cracks and apertures. The thermally established pressure difference results from the lifting action of warm air compared with inactive ambient air, thus moving air from below the collector at the temperature of the surroundings and ejecting heated air at the top. Furthermore, the thermal expansion or contraction from the heating and cooling of the solar collector will also result in ventilation.

The model used for calculating the ventilation rate can be divided into several terms related to the driving force that results in ventilation. Thus, ventilation can result from (1) thermal expansion of the air in the collector (breathing), (2) thermal buoyancy or the stack effect, (3) constant wind speed, and (4) pumping effects of wind gusts on the glazing. The ventilation rate of the air gap depends on the tightness of the collector casing. To measure the ventilation rate for a solar collector, a tightness test is used in which the pressure difference between the inside and outside of the collector box is measured for different values of airflow rates.

As stated before, we consider the steady and fluctuating components of the wind separately. For the steady or constant wind speed, we use the temporal average for 1 h. From meteorological data, we obtain the temporal mean wind speed at an elevation of 10 m. By using a logarithmic wind profile, we can predict the constant wind speed that interacts with a collector.

To obtain information about the fluctuating component of the wind when we only know the mean wind speed at the 10 m elevations, we use an energy spectrum for the geographical region in question from which we know the total kinetic energy contained between two frequencies. For a simplified model, we use the kinetic energy for the whole spectrum and the frequency, f, for which the spectrum has maximum density, and then model the fluctuating part of the wind with a harmonic oscillation at frequency f and the amplitude corresponding to the kinetic energy.

5.4.3.3   *Thermally Established Pressure Differences*

To move an air current for a given distance, e.g., at the elevation of the solar collector, there must be a driving pressure that is as large as the pressure loss over the distance. The driving pressure is the pressure difference, equivalent to the gravitational force that results from the difference in the density of air. When the inside pressure and the outside pressure are equal, no airflow will result. If there is a pressure difference, ΔP, then:

$$\Delta P = \Delta \rho g z = 273 \cdot \rho_0 \cdot g \cdot z \cdot (1/T_a - 1/T_i) \qquad (22)$$

The position of the neutral plane depends on the position of any leaks as well as the difference in resistance for suction and pressure. The ventilation rate is measured at variable pressure differences for the different solar collectors. If the pressure initially varies linearly through the collector box and leaks are evenly distributed, the neutral plane and the ventilation rate can be calculated from the static driving pressure. The airflow into and out of the collector box must be equal. Then, the integral of the ventilation rate from the neutral plane to the top of the collector box is equal to the integral of the ventilation rate from the neutral plane to the bottom of the collector box, i.e.,

$$\int_0^x V(\Delta P)_{\text{vacuum}} \mathrm{d}x = \int_0^{1-x} V(\Delta P)_{\text{overpressure}} \mathrm{d}x \tag{23}$$

The power functions determined by using pressure tests are designated, $V(\Delta P)$, so:

$$V(\Delta P) = B(\Delta P)^M \tag{24}$$

The driving pressure is obtained from equation (22), so the pressure variation expressed using the integration variable is $\Delta P(x) = \Delta P \cdot x$, where x is between 0 and 1. The integral becomes:

$$\int V(\Delta P)\mathrm{d}x = \left(\frac{\Delta P}{B}\right)^{1/M} \frac{M}{1+M}\left[x^{(M+1)/M}\right] = a[x^b] \tag{25}$$

where x is now found by the term:

$$a_{\text{vacuum}} x^{b_{\text{vacuum}}} = a_{\text{overpressure}}(1-x)^{b_{\text{overpressure}}} \tag{26}$$

and the volume that flows is:

$$V = a_{\text{overpressure}}(1-x)^{b_{\text{overpressure}}} \tag{27}$$

If the leakages are distributed equally around the top and the bottom of the collector box, the pressure difference for air entering the collector box is equal to the pressure difference for the aperture, therefore:

$$V = V(P/2)_{\text{overpressure}} \tag{28}$$

However, it is theoretically difficult to describe the distribution of leakages in solar collectors. During the tightness tests, the volume flow rate through the solar collector is measured at a series of pressure differences and a regression is expressed in terms of approximate functions of the form $\phi_v = C \cdot \Delta P^n$. Assuming that the exponent n is approximately ½, the pressure difference needed is only one fourth to one half the value obtained if the test is carried out using a collector with an aperture extending over the entire front surface. The ventilation rate is now determined from a given pressure difference, ΔP, the pressure from the functions require the pressure to be divided by 8 (see Chapter 5.5). On this basis, we are able to express the ventilation rate as a function of the pressure difference for each leak-tested solar collector $V = V(\Delta P/8)$. When the humidity in the air is considered, the differential pressure or the driving pressure is determined from:

$$\Delta P = \lfloor \rho\{T, P_{wa}\} - \rho\{(T_p + T_c)/2, P_{wg}\}\rfloor h \cdot g \tag{29}$$

The density of the air is calculated as given in Oversloot (1997) as: $\rho_a = (1/V)(1 + W)$ with $V = R_a T(1 + 1.6078\,W)/P$, and $W = 0.62198\,P_w/(P + P_w)$, where P_w is the partial pressure of water vapor. With data for T and P_w for the ambient air and the air in the collector box, we have defined the driving pressure for the ventilation rate. The mass flow of damp air is now calculated from:

$$m_{vg} = 0.5\lfloor \rho\{T, P_{wa}\} + \rho\{(T_p + T_c)/2, P_{wg}\}\rfloor V(\Delta P) \tag{30}$$

where a mean value is used for the air density of the air. Equations (22)–(29) form the mathematical model for the thermal driving pressure.

5.4.3.4   Wind

Wind is one factor that contributes to the ventilation of solar collectors. In this section, we will consider the behavior of the wind at the boundary layer closest to the ground. We will not consider the actual disturbance of the wind resulting from buildings and roof surfaces, but will only consider undisturbed wind profiles resulting from surface roughness of the ground and the wind speed at higher elevations.

Our model for the wind is simply a conversion of the meteorological mean wind speed into a constant wind speed and modeled gusts on the solar collector. Therefore, different wind types are omitted, e.g., inversion and convection. As a further simplification, we only consider one component of the wind direction, instead of a three-dimensional vectorial field. In our wind model, we treat separately the constant and fluctuating components. For a constant wind speed, we use the temporal average of the wind for 1 h. The mean wind speed increases with elevation above the ground and is described by a wind profile. The fluctuating component or turbulence has not been fully described theoretically, so we can treat the problem statistically and use empirical expressions derived from tests. During the day, the air is normally turbulent in the lowest 1 to 2 km of the atmosphere. In this layer, radical changes occur in the air movement, heat flow, and moisture content, and it is called the planetary border layer (PBL) or the friction layer (Panofsky and Dutton, 1983). At night, with a light breeze, the PBL is thinner, and may be less than 100 m.

On windy days and nights, especially when there is thick cloud cover, the PBL is quite turbulent, and its thickness is determined by the wind speed and the surface roughness. The depth of the PBL layer, which is the thickness of the turbulent air above the ground, is also called the "depth of the mixed layers". The part of the PBL, where the temperature increases with altitude, is called an inversion zone. Turbulence is suppressed in these zones. Neutral states, i.e., where the temperature increases with altitude, appear when there is a strong wind on cloudy days and nights. The friction speed in the neutral state is given by:

$$U_* = K_a V_z / \ln(z/z_0) \tag{31}$$

Under unstable conditions such as on sunny days with some wind, both mechanical turbulence and thermal convection are present. Under stable conditions on clear nights with a light breeze, only the lowest layer with air inversion is still turbulent. The part of the PBL closest to the ground is called the surface layer. There is no precise definition of the surface layer, which is the designation for that part of the PBL immediately over the ground level where vertical variations in the PBL can be ignored.

Normally, the lowest 10% of the PBL is used for the thickness of the surface layer. In the daytime with a strong wind, the layer may have a thickness of 100 m, and on a clear night with little wind, the PBL layer may be less than 10-m thick. A further simplification we apply to surface layer is that the direction of the wind does not change the height. Therefore, the mean wind speed can be described as a one-directional component alone and is only a function of the height z. The variation in wind speed with height is primarily controlled by three parameters: surface tension, vertical heat flux from the surface, and the roughness of the ground. The classical logarithmic wind profile is given by:

$$V = \frac{U_*}{K_a} \ln\left(\frac{z}{z_0}\right), \qquad z \geq z_0 \tag{32}$$

For mechanical turbulence, the logarithmic wind profile begins above the roughness elements, but below this height, the velocity falls more slowly. The logarithmic of V is extrapolated to a height where V is zero, and this height is defined as a new surface, i.e., $z = d + z_0$ above the visible ground surface, where d is the shear length.

The shear length is typically 80% of the height for the largest roughness element, e.g., houses. The wind profile is then given by:

$$V = \frac{U_*}{K_a} \ln \frac{z - d}{z_0} \tag{33}$$

U_* is determined from the meteorologically defined wind speed, i.e., at 10-m elevation, to be:

$$U_* = \frac{K_a}{\ln((10 - d)/z_0)} V(10) \tag{34}$$

In our model for ventilation resulting from wind, an empirical correlation between wind velocity and ventilation rate and between gusts amplitude and ventilation rate is used (equation (35)). For ventilation in a glass-covered collector, we define one coefficient to characterize the tightness of the collector box for each collector. We designate the tightness coefficeint as N_t. The tightness coefficient is determined from an air tightness test of the solar collector box. We have developed the empirical expressions for N_g and N_c from results of indoor experiments. N_g and N_c give the air ventilation per hour per square meter. For the thermal buoyancy, gusts, and constant wind speed we have a contribution to the ventilation rate expressed as:

$$\phi_v/A_c = C \cdot \Delta p^n/A_c + N_g + N_c \tag{35}$$

Equation (36) is obtained by including empirical expressions for N_g and N_c.

$$\phi_v/A_c = C \cdot \Delta p^n/A_c + (2.48 \cdot A - 1)/(N_t) + (0.33 \cdot V_{\text{loc}} + 0.18)/(N_t) \tag{36}$$

5.4.3.5 *Moisture Balance*

From the heat balance, we know the temperatures before and after each time interval. The moisture balance is based on the mean values of the temperatures. The mass of the water vapor in the air in the collector box after each time interval is equal to the mass of the water vapor in the air in the collector box before each time interval less the water vapor removed because of the stack effect, wind, expansion of the air in the collector box, and sorption by the insulation materials. The mass of the water in the collector box may be increased or decreased because of evaporation or condensation of water at the inside cover surface.

The mass of the water vapor removed by ventilation from the collector box because of the stack effect is calculated with the assumption that the mass flow of dry air into and out of the collector box must be identical.

$$M_{ai} + M_{wi} = m_{vg} \Delta t \tag{37}$$

where M_{ai} is the mass of dry air entering the collector in kg/m^2, M_{wi} is the mass of water vapor entering the collector in kg/m^2, and t is the time interval in seconds. Then,

$$W_i = M_{wi}/M_{ai} = 0.62198 \cdot p_{wi}/(p - p_{wi}) \qquad \text{and} \qquad p_{wi} = p_{ws}(T_d) \qquad (38)$$

where W is the humidity ratio of the air entering the collector, p_{wi} is the partial pressure in Pascal of water vapor in the air entering the collector, $p_{ws}(T_d)$ is the pressure of saturated water in Pascal at the dew point temperature, T_d, of the air entering the collector, and T_d is the dew point temperature (in °C) of the air entering the collector. We obtain:

$$M_{ai} = \frac{m_{vg}\Delta t}{1 + 0.62198(p_{wi}/p - p_{wi})}; \qquad M_{wi} = M_{ai}W_i;$$
$$M_{wo} = M_{ai}W_g; \qquad \text{and} \qquad \Delta M_{wv} = -M_{wi} + M_{wo} \qquad (39)$$

where W_g is the humidity of air in the collector, M_{wo} is the mass in kg/m^2 of water vapor leaving the collector by ventilation because of the stack effect, and ΔM_{wv} is the mass in kg/m^2 of water vapor removed by ventilation. The humidity ratios are assumed to be constant during each time interval. Their mean values are used.

The mass of water vapor removed by expansion of the air in the collector box resulting from an increase of temperature is calculated from

$$\Delta M_{we} = \Delta M_{ae}W_g; \qquad \Delta M_{ae} = V_g(\rho_a(T_{ge}) - \rho_a(T_{gs})); \qquad \text{and}$$
$$T_g = (T_p + T_c)/2 \qquad (40)$$

where ΔM_{we} is the mass in kg/m^2 of water vapor removed by expansion of the air in the collector because of the increase of temperature. ΔM_{ae} is the mass in kg/m^2 of dry air removed by expansion of the air in the collector because of the temperature increase. V_g is the volume of the gap, m^3/m^2, ρ_a is the density of dry air, kg/m^3, T_{ge} is the temperature in °C of the air in the collector box at the end of the time interval, T_{gs} is the temperature in °C of the air in the collector box at the beginning of the time interval, and $\rho_a = p/R_a T$.

The mass of water condensed on or evaporated from the inside of the cover is calculated as follows: If the partial pressure of water vapor in the air in the collector is higher than the pressure of saturated water at the temperature of the cover, condensation will take place and vice versa. The following relation between the humidity ratio and the partial pressure of water vapor is used.

$$p_{wg} = pW_g/(W_g + 0.62198) \qquad (41)$$

where p_{wg} is the partial pressure (Pa) of water vapor in the collector box air. If p_{wg} is greater than p_{wsc} at the cover temperature, condensation will take place until W_g has decreased to $W_{sg} = 0.62198 \cdot p_{wsc}/(p - p_{wsc})$. The increase of condensate is then calculated from $\Delta M_{wc} = (W_g - W_{sg})M_{ag}$, where W_{sg} is the humidity ratio of the air in the collector box at saturation, F_{wsc} is the pressure (Pa) of saturated water at the temperature of the cover, M_{ag} is the mass (kg/m²) of dry air in the collector box, and ΔM_{wc} is the increase of condensate (kg/m²) on the inside of the cover. The expression also can be used to calculate the amount of water evaporated from the inside of the cover.

5.4.3.6 Effect of Moisture and Ice on the Cover

The temperature at the cover must be adjusted because of the heat of solidification when the temperature is below the freezing point. The amount of condensation that can remain on the surface depends on the slope of the collector. On the outside collector surface, this amount is $m = 261.109 \cdot s^{-0.407}$, and on the inside collector surface the amount is $m = 225.828 - 34.247 \cdot \ln(s)$.

5.4.3.7 Moisture in the Insulation

A transient model was adopted to calculate the combined heat and moisture transfer in the insulation. The only thermal effect on the moisture flow results from the influence on vapor pressures. The construction is divided into three finite control volumes in layers with the same materials. The layers are indexed from the outer surface of the back plate, $i = 1$, to the surface at the absorber plate, $i = 3$. A vapor resistance (Z in m²s Pa/kg) is inserted at the absorber plate and at the back plate, as illustrated in Figure 5.4-6. The time intervals are the same as in the main program. The relation between vapor pressure and moisture content is given by the sorption curve. Therefore, the temperature must be known so that the vapor pressure can be calculated.

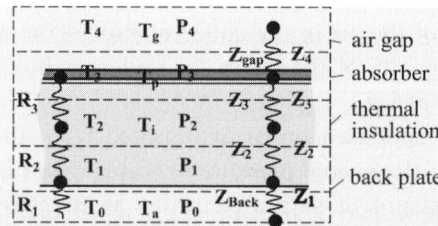

Figure 5.4-6. Model used for the calculation of the combined heat and moisture transfer in the insulation. The insulation is divided into the three finite control volumes in layers. R_x is the thermal resistance in the layer x, T_x is the temperature of the layer, p_x is the partial water vapor pressure in the layer, and z_x is the vapor resistance in the layer.

5.4.3.8   Computer Program

The computer model is used to elucidate the combined heat and moisture transport in the collector. Calculations with the model provide insight into the humidity and temperature for synthetic or realistic climate data (e.g., the humidity distribution during the year). This design tool makes it possible to calculate the effects of ventilation. An input module in the computer program named "MOMIC", an abbreviation of modeling of Microclimate in Collectors, has been developed to accept meteorological data from data files. Three types of data files can be input, two for synthetic climate data in a form used by the Danish Test Reference Year, TRY, and the Design Reference Year, DRY, and one for climate data, designed to receive weather data from the measurements on the reference collector, IEA MSTC. For synthetic climate data, the variation of the operating conditions is limited to the maximum value of the absorber plate temperature. When using real data, the measured absorber temperature is used as a fixed value and variations from the operating conditions can be substituted. The back plate temperature can also be input as a fixed value, if needed, e.g., if the collector is built for integration. To make parametric sensitivity studies, a parameter can be chosen from a table. Several calculations are then made with this parameter varied over a suitable range. The program is written in the Object Pascal language Delhi 3 for Windows 95.

5.4.4   VALIDATION OF THE MICROCLIMATE MODEL

Long-term collector tests with measurements of the microclimate in the IEA-collector have been used to validate the microclimate model. At the test site the collector is operated in a system that does not allow the absorber temperature to be greater than above 60°C. These measurements were carried out in 1996 and 1997. In the autumn of 1997 the test facility was used for measurements during various operating conditions and dry-out situations. These cases were also included in the work with the validation. Time periods were chosen that permitted comparing the measurements of the microclimate with the simulations in detail. Wetness of the collector occurs when the microclimate condition results in accumulation of water on the collector. By using a high-ventilation rate or ventilating for a sufficient time, the absolute humidity in the collector box will be the same as the ambient absolute humidity. At this time, an increase of the ventilation rate will only result in a faster air exchange, which will yield a negligible influence on the temperature and no change in the humidity.

Water accumulates in the collector at night when the temperature of the cover is lower than the ambient dew point temperature. Hygroscopic materials are moisture sources or sinks, depending on the moisture balance with the surrounding air. The insulation can accumulate moisture during the day when, after a sunny period, the insulation was dry and air flowing through the collector box is dried by moisture diffusion into the insulation. In a typical North European winter with high-relative humidity at night, the inside of the cover may be sufficiently below the dew point that moisture condenses on the surface. If the condensate is not reevaporated by ventilation or solar heat gain, water drops onto the absorber plate and may result in corrosion.

In the winter, changes in relative humidity are small, e.g., it remains in the range of 85 to 100%. In the summer, the relative humidity varies between 60 and 100% and ambient absolute humidity ranges from 2 to 14 g/kg. Ventilation affects the microclimate in the collector box when water can accumulate or at high temperatures when the net water will be removed. If the cover temperature is below the dew point, the amount of condensation will increase and the condensate may be visible.

When measurements were begun in August 1997, the temperature typically ranged from 15°C at night to 34°C during the day. The ambient humidity was typically 95 to 100% at night. Several tests were performed during that time with constant forced ventilation at different ventilation rates. The data in the computer program were modified by choosing various ventilation rates for a specified time. Good agreement was found when the measured values of the relative humidity inside the collector were compared with calculations using the computer program. During the day, the air temperature in the collector box was high and the measured relative humidity was not reliable because the instrument range was exceeded. Computer simulations were also carried out under conditions for removing accumulated moisture in the collector box, i.e., a "dry-out" test. Simulating the dry-out test is complicated because it is necessary to indicate the amount and physical state of the moisture in the collector at a specified time. Saturated air at any temperature limits the amount of moisture in the air and the water content in the insulation is limited by its sorption capacity, if we are not considering the wetness of the insulation to be in the over-hygroscopic range. The remaining moisture condenses on the inner surface of the cover. From the measurements, some stratification of the humidity apparently occurs in the collector box.

5.4.4.1 Comparing the Microclimate Model with Measurements

The correlation between microclimate and ambient climate is described by the ventilation rate or rate of airflow through the collector box. The driving force for ventilating the collector results from a difference in air pressure between the surroundings and inside the collector box. Thermal buoyancy, wind, or wind gusts might also result in a pressure difference.

In the participating laboratories, the air tightness of identical reference collects was determined by applying a defined pressure difference. In the air tightness test, measurements are made during the airflow through the collector box for different pressure differences between the inside of the collector and the ambient air. The results from the air tightness tests agreed well with other tests of the collectors. However, using these tests for the assessment of the ventilation rate is questionable because the pressure range of interest for this airflow is zero to only a few Pascal. We then carried out CFD calculations to investigate the behavior of the fluid dynamics in the collector in detail. The ventilation rate was determined by using steady-state simulations for two given sets of boundary conditions, i.e., one set for normal operating conditions and the other at stagnation conditions. The theory of chimney effects was used to relate the pressure difference (Δp) to the temperature (T) of the collector.

From the CFD simulations, a linear relationship is found between Δp and T. By incorporating this relationship in the computer program, better agreement is obtained with the measured microclimate data. For example, the temperatures calculated for the collector cover and inside the collector box range between the measured temperature at the bottom and the top of the collector. In addition the calculated humidity corresponds closely to the measured humidity. Figure 5.4-7 shows the computed humidity and the measured humidity in the collector box and the ambient humidity.

The data in Figure 5.4-7 were obtained between December 9 and 15, 1996. Because the period selected in December 1996 was humid, we conducted an additional test

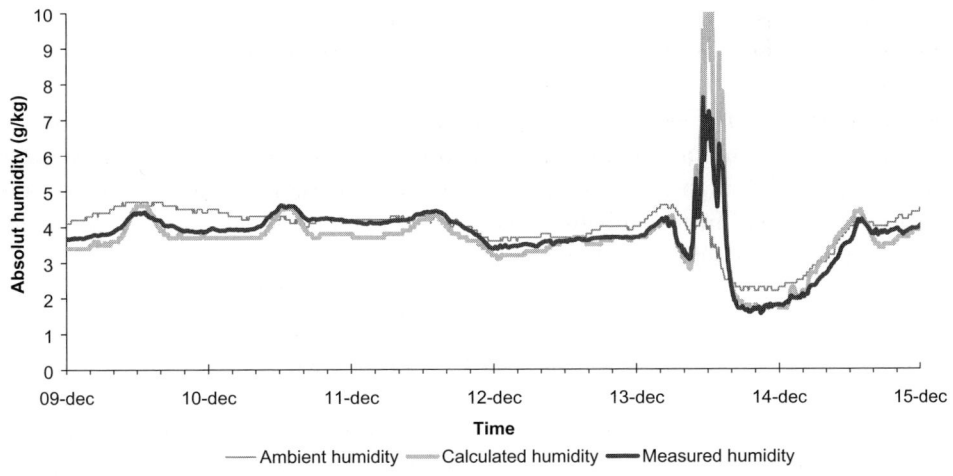

Figure 5.4-7. Ambient, computed, and measured humidities in the collector box and wind speed for 6 days in December 1996.

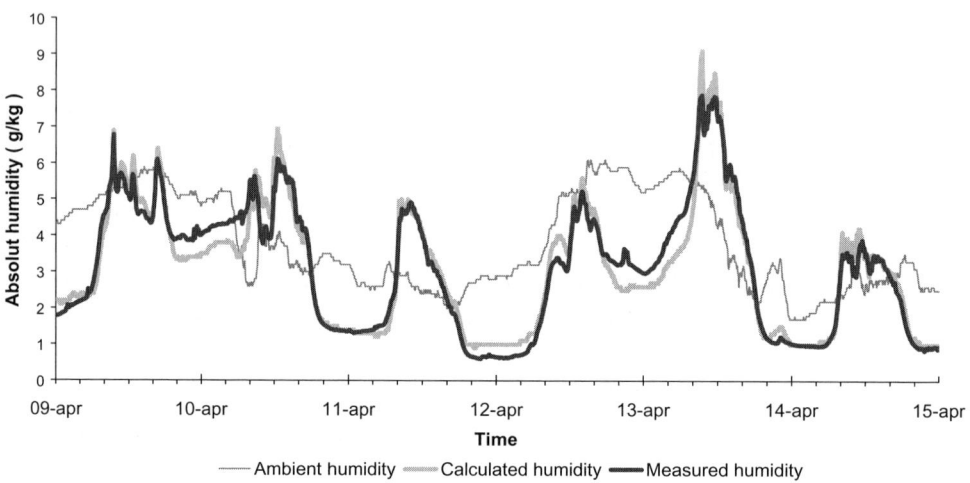

Figure 5.4-8. Ambient, computed, and measured humidities in the collector box and wind speed for 6 days in April 1997.

during April 1997. The data collected are shown in Figure 5.4-8 for April 9 through April 15, 1997.

As is seen, the ambient humidity ranged between 20 and 96%. We can also see that the calculated humidity corresponds closely to the measured humidity. We present similar figures in Section 5.4.5.

5.4.5 SENSITIVITY TEST OF THE COMPUTER PROGRAM

5.4.5.1 *Influence of Wind on the Microclimate*

The influence of wind on the microclimate inside the collector box was investigated during periods with high wind speed. From statistical analyses of the measured data, no correlation could be found between wind components and the humidity inside the collector. For modeling purpose, however, it is important to consider the influence of the wind in detail. A model provides us with an empirical dependence between wind speed and ventilation rate. By comparing the simulated and measured data and by scaling up the influence of wind dependence in the model, we were able to improve the agreement in the data. The result of the calculations was remarkably good, so we tentatively concluded that the statistical analyses of the measured data were not correct. To support over conclusion, data taking during several other periods of time were investigated. We demonstrated that both the ventilation rate for the reference collector and the humidity inside the collector depend on the wind speed in a

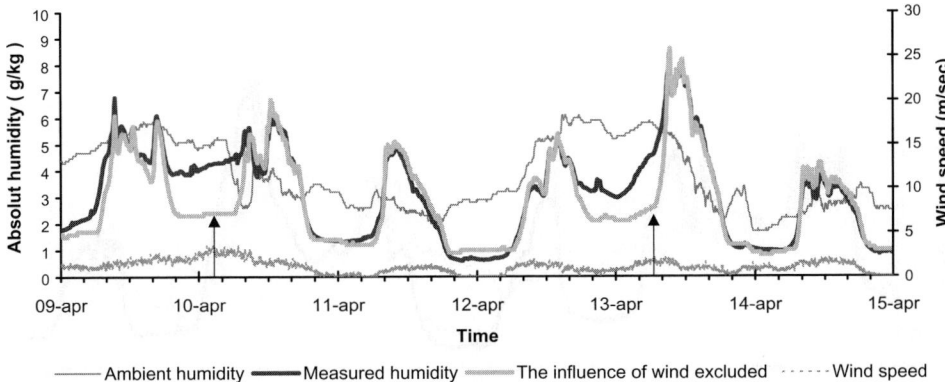

Figure 5.4-9. Ambient, computed, and measured humidities in the collector box and wind speed for 6 days in April 1997. Ignoring the influence of wind results in a disparity between the lines representing the measured and computed humidity (see the arrow at 3 p.m. on April 10th).

systematic way. If the influence of wind is ignored, as was done for the plots in Figure 5.4-9, we find a great degree of disparity between the measured and calculated data for windy nights.

5.4.5.2 *Influence of Insulation Materials on the Microclimate*

From measured data on a "customer" collector, we obtained good correlation between the absorber temperature and the absolute humidity in the air in the collector box between the absorber and the cover. The collector studied was airtight and, in addition, the ambient humidity was almost constant, so the humidity in the collector box could only originate from the mineral wool used in insulation (see Chapter 3.4). Insulation materials can decrease the humidity in a collector box during cold periods, but may also release water that can form dew on the collector cover when the collector is heated. As stated before, damage to the collector may result from dew and the visual impression of the collector is certainly compromised. In Figure 5.4-10, the data show that a calculated higher humidity resulted from reducing the insulation thickness to half of the original thickness of the IEA-collector.

5.4.5.3 *The influence of Changing the Size of the Ventilation Holes*

Calculations were made to investigate the influence of changing the size of the ventilation holes. These are shown in Figure 5.4-11 for hole sizes double the original

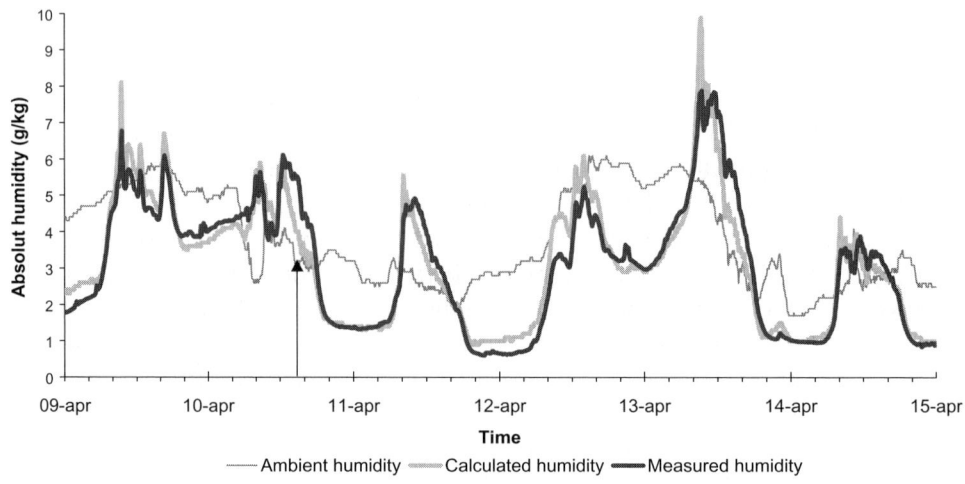

Figure 5.4-10. Ambient, computed, and measured humidities in the collector box and wind speed for 6 days in April 1997. Incorporation of the rear insulation thickness of half of the original thickness indicates a level of humidity closer to the ambient humidity in the afternoon (see the arrow at 3 p.m. on April 10th).

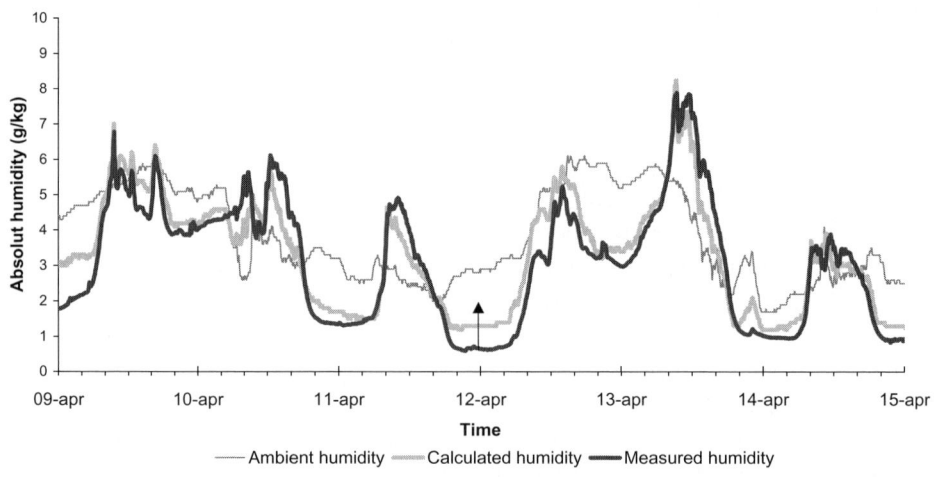

Figure 5.4-11. Ambient, computed, and measured humidities in the collector box and wind speed for 6 days in April 1997. Incorporation of a hole size double the original hole size indicates a level of humidity closer to the ambient humidity during windy nights (see the arrow at midnight on April 12).

hole size. The effect of doubling the diameter of the vents is similar to the effect shown in Figure 5.4-10 for insulation that is only half the original thickness. For ventilation holes with a size half the original size of the IEA collector, lower humidities are obtained.

5.4.6 CONCLUSIONS

The choices of insulation materials and ventilation rate affect the durability of solar collectors. The ventilation rate and the insulation materials used in collectors affect the relative humidity and the formation of condensation inside the collector, both of which influence the durability of the collector. Calculations using a microclimate model for the collector show that the humidity inside the collector depends on the wind speed, and that the insulation materials can lower the relative humidity during cold periods and later increase the humidity when the collector is warm. The principal output from the simulations are that the duration and amount of condensation inside the collector and the duration of humidity for specific values in the collector gap vary with air permeability and insulation thickness. The results have to be interpreted with some care because of the uncertainty about the weather conditions and the low-pressure range for the ventilation rate. Optimization of the rear insulation thickness is related to the sorption properties of the insulation materials. Consequently, different insulation materials will yield different results for the optimal insulation thickness. Most collectors are not operated at optimal humidity. Many collectors have problems with condensation on the inside cover, and this can adversely influence the long-term durability of the absorber. Many manufacturers are not pleased with the poor visual impression potential customers have for collectors covered with condensate of varying amount and distributions. Simulations obtained using computer programs were useful for improving the design of collectors. We did not perform enough work to recommend all the choices for designing an ideal collector, but some choices and their effects on the operation of solar collectors are given in Chapter 5.3.

REFERENCES

ASHRAE Handbook (1989) *Fundamentals*, SI edn, pp. 6.1–6.17, American Society of Heating, Refrigerating and Air Conditioning Engineers, Inc., Atlanta, GA, USA.

CEN prEN 1991-1-4 (Under Development), Actions on Structures – Part 1–4: Actions on Structures – Wind Actions, Publication ENV 1991-2-4:1995, Europe, B-1050, Brussels, Belgium.

Duffie, J.A. & Beckman, W.A. (1991) *Solar Engineering of Thermal Processes*, 2nd edition, Wiley Interscience, New York, NY, USA

Keller, J. (1984) Characterization of the Thermal Performance of Uncovered Solar Collectors by Parameters Including the Dependence on Wind Velocity, Humidity and Infrared Sky Radiation as well as on Solar Irradiance, Swiss Federal Institute for Reactor Research CH-5303 Wuerenlingen, Switzerland.

Krisher, O. & Kast, W. (1978) *Die wissenschaftlichen Grundlagen der Trocknungstechnik*, Springer, 14197 Berlin, Germany.

Lund, H. (1985) The Test Reference Year TRY, Commission of the European Communities, Report no. EUR 9765.

Lund, H. (1995) The Design Reference Year User Manual, International Energy Agency Solar Heating and Cooling Programme, Report No. IEA-SHCP-9E-1, Technical University of Denmark, Dept. of Civil Engineering, DK 2800 Kgs. Lyngby, Denmark.

Marcel, L. (1999) Numerical Investigation into the Pressure Distribution of a Solar Collector on a Pitched Roof, Report TNO –Bouw(BBI-DEGO), Building and Construction Research, the Netherlands, Postbus 6050, NL-2600 JA Delft.

Oversloot, H. (1997) Influence of Wind on the Ventilation of Collectors, Progress Report International Energy Agency Solar Heating and Cooling Programme – Working Group: Materials in Solar Thermal Collectors, TNO Building and Construction Research, The Netherlands, Postbus 6050, NL-2600 JA Delft.

Panofsky, H.A. & Dutton, J.A. (1983) *Atmospheric Turbulence*, John Wiley and Sons, New York, NY, USA.

Chapter 5.5

Ventilation Rate Testing Procedure

O. Holck

Department of Civil Engineering, Building 118,
Technical University of Denmark, DK-2800 Kgs. Lyngby, Denmark

Abstract: The purpose of this chapter is to present a standard test procedure of a tightness test for ventilated solar thermal collectors. This is needed for calculation of the ventilation rate in collectors, which is one of the major effects that influence the microclimate in collectors and thereby influence the loads for degradation of components in solar collectors. The correlation between microclimate and ambient climate is described by the ventilation rate, which means the airflow rate into the collector. The driving forces for ventilation of the collector are primarily an air pressure difference between the surroundings and the air gap. The procedure was tested using a round robin format. Four laboratories participated in the tests. The purpose of the round robin testing was to test the procedures in the standard test method for measuring the air tightness of collector boxes. In the standard test method, the procedure requires setting a ventilation rate through the collector and measuring the pressure difference between the inside and outside of the collector box. In general the agreement among the measurements was good.

Keywords: Test procedure, Air tightness, Ventilation, Collector box, Microclimate, Durability

LIST OF ABBREVIATIONS

a	coefficient obtained from C in $a = 4/C^{1/n}$
C	regression coefficient in air flow function
n	exponent in air flow function
ΔP	pressure difference
Φ_v	airflow

5.5.1 INTRODUCTION TO THE VENTILATION RATE TEST PROCEDURE

The collector box is tested for air tightness to assess the ventilation of the collector. The collector box is the space enclosed by the frame, the cover, and the back plate of the collector and protects the components inside the box from the outdoor weather. The ventilation rate is defined as the air-exchange per unit time and is frequently reported for a pressure difference of 1 Pa. A high-ventilation rate may result from unintended openings, i.e. leaks, and superfluous vents (see Chapter 5.2).

The test procedure for determining the air tightness, a controllable air source capable of over-pressure and under-pressure, an airflow meter, and equipment to measure the pressure difference are used. The basic test procedure is to connect the air source to the collector box and measure the pressure differences between the surroundings and the collector box for a range of airflow values. The procedure results in a series of two measurements, i.e., one with over-pressure inside the box and the other with an under-pressure inside the box. The procedure is then repeated with the vents of the collector closed by using tape or some other suitable means. The complete test results in a series of four measurements, i.e., two with the box at over-pressure and two with under-pressure in the box and with open and covered vents. The results from the air tightness test are needed to classify the microclimate in the collector. The results are normally expressed in terms of functions of the form $\Phi_v = C \cdot \Delta P^n$ where Φ_v is the ventilation rate, ΔP is the pressure difference, C is a coefficient related to the air flow at a 1 Pa pressure difference, and the exponents n depends on the type of air flow through the openings. Van der Linden used $n = 1/2$ (Van der Linden *et al.*, 1990).

5.5.2 VENTILATION RATE TESTING PROCEDURE

5.5.2.1 *Preparation of the Collector*
A plastic hose is connected directly from the air source to the collector. Additional leaks from the air space in the collector box to other parts of the collector must not be introduced. The hose is closed around the hole with a silicone sealant. The inner diameter of the hose is approximately 6 mm. A pressure sensor is mounted in a similar way.

5.5.2.2 *Testing Apparatus*
The testing apparatus consists of a controllable air source that is capable of supplying a known value of positive or negative pressure to the collector for different flow rates to simulate collector leakage. An airflow meter with an accuracy of 2% is

installed to measure the "leakage" flow rate, and a micromanometer with an accuracy of 2%, which is capable of measuring a gas pressure of 1 Pa, is installed to measure the pressure difference inside and outside the collector. The air source, airflow meter, and micromanometer are connected to the collector. The tightness test is carried out in two steps. In the first case, air is forced into the box to produce over-pressure inside the box, and in the other case, air is expelled from the collector to produce under-pressure inside the box.

5.5.2.3   *Test Procedure and Instrumentation*
A controllable air source is needed that is capable of drawing air out of the collector at different flow rates. An airflow meter is installed to measure the rate of airflow. The air source and airflow meter are connected to the collector. Leaks between the flow meter and the collector are not allowed. First, the collector is tested with over-pressure in the collector box at six different rates of airflow. In practice, it is sometimes easier to establish a pressure difference and then measure the flow rate, depending on the equipment being tested. However, six different airflow rates should be used, in general. The two lowest airflow rates are optional if the range is not practicable with the apparatus, but then at least 5 airflow rates should be used with a sufficient spread in the range of interest and practicability. Four readings should be taken of the pressures for each stabilized airflow rate. The collector is then tested in the same way, but by establishing an under-pressure in the collector box instead of an over-pressure. The values of airflow rates needed are about 50, 100, 200, 400, 600, and 1000 L/h. A lower range of airflow rates is preferable if it is practicable with the apparatus. The same procedure for testing with the collector box at under-pressure and over-pressure is repeated with the vents closed.

5.5.2.4   *Test Conditions for Temperature and Wind Speed*
The collectors are tested for tightness at room temperature ($\sim$18°C) with no wind. The ranges of measured pressures and flow rates are listed in Table 5.5-1.

5.5.2.5   *Test Report*
The test results are listed in four table sections with clearly identified measurement conditions, i.e., vents open with over-pressure or under-pressure, and vents closed

Table 5.5-1. Ranges of pressure and flow-rate range for testing

The pressure range is:	The flow-rate range is from:
$\pm$0.5 to $\pm$15 Pa	60 l/h to 1200 l/h

Table 5.5-2. Typical documentation of air tightness test operating with over-pressure and under-pressure of the air inside the collector

Collector Test								
Four readings of the pressure difference ΔP (Pa)				Four readings of the airflow rate Φ (L/h)				
1	2	3	4	1	2	3	4
Holes open								
0.64				127				
0.73				138				
Over-pressure								
1				170				
1.5				209				
4				393				
6.6				613				
Holes open								
0.3				85				
0.8				161				
Under-pressure								
1.2				206				
3.6				403				
6.1				600				
13.5				1023				
Holes closed								
2.1				65				
3.2				85				
Over-pressure								
4.1				100				
8.4				153				
9.5				164				
12.7				197				
Holes closed								
1.9				56				
3.2				82				
Under-pressure								
4.2				99				
7.6				147				
12.3				203				

with over-pressure or under-pressure. The units for the pressure differences, ΔP, are given in Pascal and the flow rates Φ_v are given in liters per hour, as illustrated in Table 5.5-2.

5.5.2.6　*Transfering the Results from Test to In-service Conditions*
In the test, air flows into the collector box through the additional aperture of the measuring apparatus and can then flow out through all the vents and other unintended leaks. However, at actual in-service conditions, air enters the box through some of the holes and exits through others. If the leaks are divided equally around the top and bottom of the solar collector, the pressure difference across the

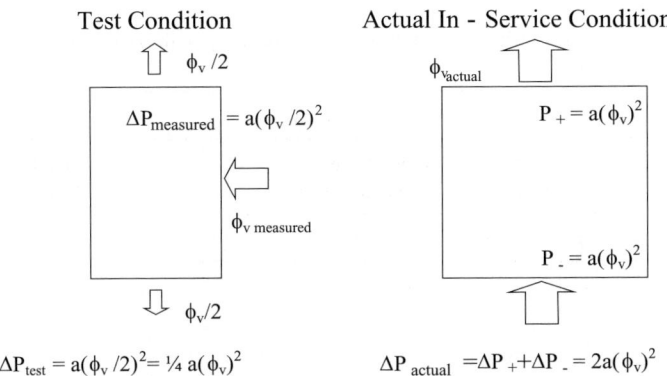

Figure 5.5-1. Illustration of two-experimental arrangements in which one is for the test condition and the other is at an actual in-service condition. In both cases, the total flow rate is the same through the collector.

total entrance aperture at the bottom is equal to that across the total aperture at the top of the collector. The air stream has to overcome both the pressure differences on the way into and on the way out of the collector. This means that the following relations for the pressure difference can be formulated for the actual in-service and test conditions for the same total flow rate through the collector. In equations (1) and (2), we assume that in the pressure ranges for the test, the exponent n is approximately 1/2, i.e.,

$$\Delta P_{test} = a(\phi_v)^2/4 \tag{1}$$

$$\Delta P_{actual} = 2a(\phi_v)^2 \tag{2}$$

The coefficient, a, is obtained from C in $a = 4/C^{1/n}$. The actual and test conditions are illustrated in Figure 5.5-1.

This means that the pressure differential under actual in-service conditions is 8 times as large as the testing pressure differential for the same airflow rate, i.e., $\Delta P_{actual} = 8 \cdot \Delta P_{test}$. The ventilation rate is subsequently found for a given pressure difference ΔP by inserting the actual pressure divided by 8 in the functions given by the test, i.e., $\Phi_v = C \cdot (\Delta P/8)^n$.

5.5.3 VALIDATION OF THE APPLICABILITY OF THE PROCEDURE

The procedure was tested using a round robin format. Four laboratories participated in the tests, i.e., the Fraunhofer Institute for Solar Energy Systems (ISE) in

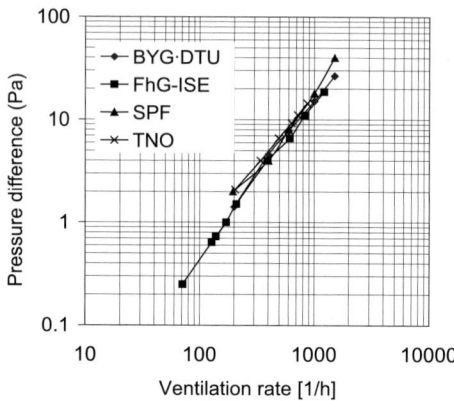

Figure 5.5-2. Dependence of ventilation rates on pressure differentials as measured at four different laboratories, with over-pressure in the collector box and the vents open.

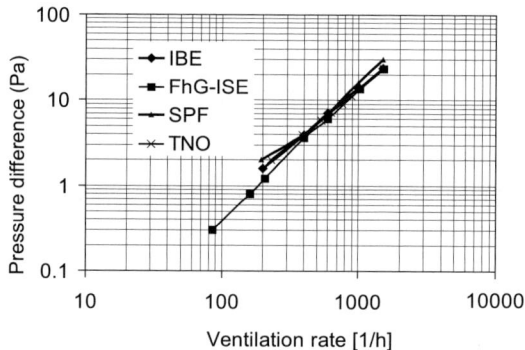

Figure 5.5-3. Dependence of ventilation rates on pressure differentials as measured at four different laboratories, with under-pressure in the collector box and the vents open.

Germany, Department of Civil Engineering, Technical University of Denmark (IBE) in Demark, Institut für Solartechnik HSR Hochschule, Rapperswil (SPF) in Switzerland, and Building and Construction Research (TNO) in the Netherlands.

The collector to be tested was selected by ISE in Germany. The collector box was selected because it consisted of a single full metal casing without any seams. The main sources of leakage were expected to be at the edge around the glazing and the inlet and outlet connections. The box itself has two 4-mm vents at the bottom and two at the top of the collector. The collector was carefully packed and shipped in a solid crate so that any measured differences could not be attributed to damage during transit. The results of the series of measurements by the participating laboratories were collected. To avoid a confusing range of closely packed lines, the

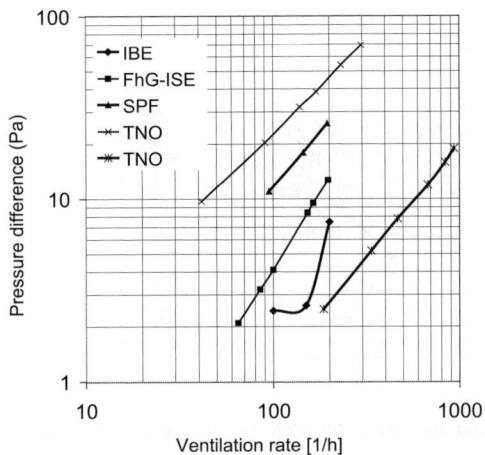

Figure 5.5-4. Dependence of ventilation rates on pressure differentials as measured at four different laboratories, with over-pressure in the collector box and the vents closed.

graphs are presented separately for each of the four types of test, and are shown in Figures 5.5-1–5.5-5. For the data taken with the collector vents open as shown in Figures 5.5-2 and 5.5-3, the agreement among the measurements above 5 Pa is good.

For pressure differences below 5 Pa, there are differences between ISE and IBE and SPF and TNO. For the under-pressure data, the ISE measurements by ISE below 5 Pa are more clearly different from the other three laboratories. In this pressure region, it is difficult to measure the pressure differences accurately. In the ISE data a different pressure-measuring instrument appears to have been used above 500 L/h because of the change in slope in the plot. The results from SPF are fit best with a curve rather than a straight line.

In general, the small disagreements are attributed to the calibrations of the pressure-monitoring instruments. The instruments may have been used beyond their optimal range because of the low-pressure differentials measured. However, for the purpose of the round robin test, we conclude that the general agreement among the measurements is good. When the vents are closed, the data in Figures 5.5-4 and 5.5-5 are in marked contrast with that in Figures 5.5-2 and 5.5-3.

The difference between over-pressure and under-pressure data for each laboratory is negligible compared with the difference among the laboratories. During the measurements at TNO, it became clear that the openings around the inlet and outlet were not completely closed, so two sets of curve exist. The data from TNO are the two plots in Figures 5.5-4 and 5.5-5, which are the farthest apart. Because plastic covers surround the tubes, the best results are obtained by using an adhesive plastic medium like plasticine in combination with soft, stretchable tape. The measurements

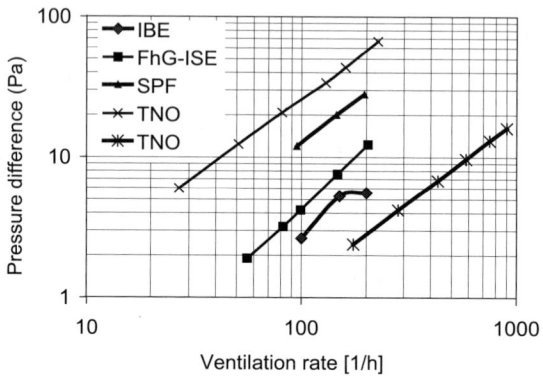

Figure 5.5-5. Dependence of ventilation rates as measured at four different laboratories, with over-pressure in the collector box and vents closed. Ventilation characteristics as measured by the various laboratories, with under-pressure and vents closed.

Table 5.5-3. Values of C in $\phi_v = C \cdot \Delta P^n$ from measurements at four laboratories

		ISE Lab 1	IBE Lab 2	SPF Lab 3	TNO Lab 4	TNO (repeated)
C	over-pressure, vents open	$170 \pm 4.6\%$	$149 \pm 6.1\%$	$141 \pm 13.3\%$	$119 \pm 1.3\%$	
	Under-pressure, vents open	$184 \pm 3.3\%$	$143 \pm 1.3\%$	$132 \pm 11.3\%$	$132 \pm 0.4\%$	
	Over-pressure, vents closed	$42 \pm 0.8\%$	$79 \pm 30.3\%$	$13 \pm 0.9\%$	$4 \pm 2.0\%$	$89 \pm 1.9\%$
	Under-pressure, vents closed	$37 \pm 1.3\%$	$46 \pm 19.2\%$	$12 \pm 0.9\%$	$5 \pm 2.1\%$	$82 \pm 0.5\%$

at IBE are questionable in the mid-range of pressure differentials. The errors at the mid-pressure differentials are also indicated in the coefficients in Tables 5.5-3 and 5.5-4. The uncertainties in C and n are based on standard error analysis. The larger uncertainties, e.g., at laboratories 1 and 3, resulted from the measurement problems already discussed. With one point excluded from the already small set of data from IBE, the regression does not result in vaild coefficients, as can be seen from the case with the exponent $n = 0.47$. The errors probably resulted from moving or loosening of the covering materials because of the pressure differences used during the test.

The purpose of the round robin testing was to test the procedures in the standard test method for measuring the air tightness of collector boxes. In the standard test method, the procedure requires setting a flow rate or ventilation rate through the collector and measuring the pressure difference between the inside and outside of the collector box. No indication of significant differences resulted for data with with

Table 5.5-4. Values of n in $\phi_v = C \cdot \Delta P^n$ from measurements at four laboratories

		ISE Lab 1	IBE Lab 2	SPF Lab 3	TNO Lab 4	TNO (repeated)
n	Over-pressure, vents open	$0.66 \pm 4.6\%$	$0.69 \pm 6.1\%$	$0.66 \pm 13.3\%$	$0.75 \pm 1.3\%$	
	Under-pressure, vents open	$0.66 \pm 3.3\%$	$0.74 \pm 1.3\%$	$0.73 \pm 11.3\%$	$0.79 \pm 0.4\%$	
	Over-pressure, vents closed	$0.61 \pm 0.8\%$	$0.47 \pm 30.3\%$	$0.84 \pm 0.9\%$	$1.00 \pm 2.0\%$	$0.81 \pm 1.4\%$
	Under-pressure, vents closed	$0.69 \pm 1.3\%$	$0.78 \pm 19.2\%$	$0.85 \pm 0.9\%$	$0.89 \pm 2.1\%$	$0.86 \pm 0.4\%$

open vents at four different laboratories. However, when the tests were made with closed vents widely differing results were obtained among the four laboratories. The differences are attributed to leaks around the vents and closing the vents can still be difficult if sharp edges or complex forms are involved. In such cases, an adhesive plastic medium that can be removed afterwards is a great help. The adhesive can be further secured by using a strechable, soft tape to keep the adhesive in the intended position. Finally, calibration of pressure-monitoring instruments in the low-pressure range needs special care prior to measurements.

REFERENCES

Van der Linden J., de Geus A.C. & Kowalczyk W. (1990)
  Microclimate in Solar Collectors (Draft Final Progress Report), Report of Task X. Solar Materials Research and Development, International Energy Agency, Solar Heating and Cooling Programme, TNO-report TPD-HWM-RPT-90-052, Delft, 2628 CK, The Netherlands.

Part 6

Case Studies on Polymeric Glazings

Part 6 consists of two chapters, which are concerned with case studies about using polymeric glazings for reflector applications. In the first chapter, the results of preliminary and ongoing screening tests for a large number of candidate polymeric glazing materials are presented. In the second chapter, a summary is given for data obtained by outdoor exposure and indoor testing of different polyvinyl chloride (PVC) and polycarbonate (PC) materials. It is shown how the methodologies developed in Chapter 4.6 can be applied to predict accurately the optical performance of the PVC and PC materials during in-service use.

Chapter 6.1
Screening Tests of Candidate Polymeric Glazings

G. Jorgensen

National Renewable Energy Laboratory, 1617 Cole Boulevard,
Golden, CO 80401-3393, USA

Abstract: The purpose of this chapter is to discuss a number of candidate polymeric glazings that have been identified, and to present results from preliminary and ongoing durability screening tests of these materials. The economic viability of solar collector systems for domestic hot water (DHW) generation is strongly linked to the cost of such systems. Installation and hardware costs must be reduced by 50% to allow significant market penetration. An attractive approach for reducing costs is to replace glass and metal parts with less expensive, lighter weight polymeric components. Weight reduction will decrease the cost of shipping, handling, and installation. The use of polymeric materials will also allow the benefits and cost savings associated with well-established manufacturing processes, along with savings associated with improved fastening, reduced part count, and overall assembly refinements. A key challenge is to maintain adequate system performance and assure requisite durability for extended lifetimes. Nevertheless, this concept has begun to be recognized and accepted throughout the world.

Keywords: Polymeric glazings, Outdoor exposure testing, Accelerated exposure testing, Risk analysis, Screening tests, Service lifetime prediction, Durability

LIST OF ABBREVIATIONS AND ACRONYMS

AM	air-mass
E-CTFE	ethylene-chlorotrifluoroethylene
ETFE	ethylene-tetrafluoroethylene
PC	polycarbonate
PE	polyethylene
PEN	polyethylene naphthalate
PET	polyethylene terephthalate
PFA	perfluoroalkoxy fluorocarbon
PS	polystyrene

PVC	polyvinyl chloride
PVDF	polyvinylidene fluoride
RH	relative humidity (%)
t	time
T	temperature
UV	ultraviolet light (or radiation)
WOM	an exposure chamber capable of controlling light intensity, T, and RH levels
λ	wavelength (nm)
$\tau_{2\pi}$	hemispherical transmittance

6.1.1. INTRODUCTION

Polymeric glazings offer significant potential for cost savings both as direct substitutes for glass cover plates in traditional collector systems and as an integral part of all-polymeric systems. Cost savings result from lower base material costs and lower costs associated with shipping, handling, and installation, which is because of their light weight and they are not fragile. The cost goal for low-cost collector system glazings is $10/m^2 in 2002 dollars. Glazings should have high transmittance across the solar spectrum and must be able to survive 10–20 y exposures to service conditions including operating at temperatures of 55–90°C and in solar ultraviolet (UV) light. They must retain mechanical integrity e.g., impact resistance and flexural rigidity, under these harsh environmental stresses. A review of candidate glazing materials that existed in 1984, along with relevant initial (unweathered) physical properties is available (Kutscher *et al.*, 1984). In the current efforts, the emphasis is to identify new or improved candidate glazings and to evaluate the optical and mechanical durability during exposure to actual and simulated in-service conditions.

6.1.2. DURABILITY EXPOSURE TESTING

Samples of candidate polymeric glazings are being subjected to in-service outdoor and accelerated laboratory exposure conditions. Outdoor testing is carried out at three sites in the United States (Golden, CO; Phoenix, AZ; and Miami, FL). Accelerated exposure testing is also carried out at 60°C and 80% relative humidity (RH) using an Atlas Ci5000 Weather-Ometer® (WOM). A xenon-arc lamp, which is filtered to provide a close match to a terrestrial air mass 1.5 spectrum, is used in the WOM. The measured irradiance for the light source is shown in Figure 6.1-1, and is

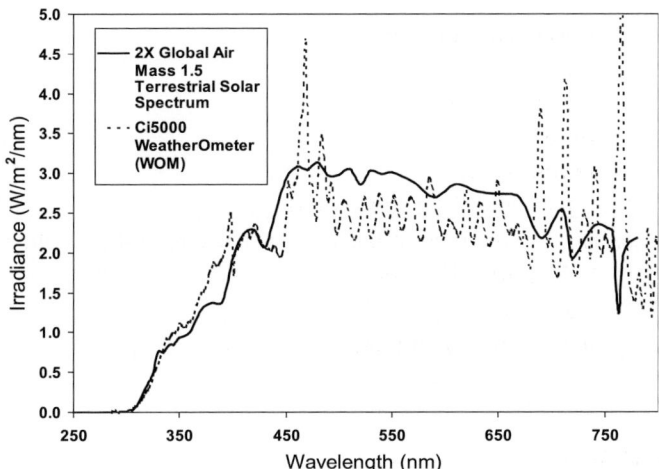

Figure 6.1-1. Spectral irradiance of the light source in a WOM accelerated exposure chamber and that for a 2X global air mass 1.5 terrestrial solar spectrum.

compared with a 2X global terrestrial standard spectrum (ISO Standard 9845-1, 1992; ASTM G159-98, 2003).

Glazings for solar collector applications must exhibit high hemispherical transmittance. Thus, the measure of performance that was chosen to quantify degradation was hemispherical transmittance ($\tau_{2\pi}$) as a function of exposure time, t, weighted by a terrestrial air-mass (AM) 1.5 global solar spectrum (λ from 300 to 2500 nm) (ISO Standard 9845-1, 1992; ASTM G159-98, 2003). All samples are optically characterized prior to exposure ($t = 0$) and then periodically as a function of weathering. For samples exposed outdoors, half of the exposed sample surface is measured as received (uncleaned) and then after cleaning; the other half of the exposed sample surface is measured as received, i.e., it is never intentionally cleaned, to provide an indication of the effect of cumulative soiling. Optical measurements were made according to standard test procedures (ASTM E903-96, 2003).

6.1.3. CANDIDATE SAMPLES

Depending on the collector design, polymeric glazings with a number of different forms can be considered. These include thin films, rigid sheets, or multi-channel constructions. A number of candidate materials have been identified by reviewing the literature and through discussions with experts within the polymer and solar manufacturing industries. In Table 6.1-1 the polymer materials, which are being subjected to both outdoor and accelerated exposure testing, are listed. The material,

Table 6.1-1. Candidate polymeric glazing materials

Material	Product	Description	Thickness (mm)	Form	Initial solar-weighted hemispherical transmittance $\tau_{2\pi}(\Delta\lambda = 300 - 2500, t=0)$
PET[1]	Mylar D	Non UV-stable	0.18	Film	86.7
PET	Melinex 442/400	Non UV-stable	0.10	Film	86.2
PET	Melinex D 387	UV-stabilized	0.03	Film	85.4
PET	Melinex D 389	UV-stabilized	0.03	Film	85.4
PEN[2]	Kaladex	Biaxially oriented and heat set	0.10	Film	84.6
ETFE[3]	Tefzel 150 ZMC	Heat-stabilized	0.04	Film	93.8
ETFE	Tefzel 250 ZMC	Heat-stabilized	0.06	Film	94.0
ETFE	Duralar CS50	Non-oriented	0.05	Film	93.7
ETFE	Duralar E	Mono-axially oriented	0.05	Film	93.9
E-CTFE[4]	Halar Clear NP		0.05	Film	92.9
PFA[5]	Teflon PM	Heat-stabilized	0.05	Film	95.8
PFA	Teflon PH	Heat-shrinkable	0.05	Film	95.7
PVDF[6]	Kynar		0.03	Film	93.9
Acrylic	Korad Klear	UV absorbers	0.05	Film	89.3
Polycarbonate	Lexan HP92WDB	UV/mar coating	0.18	Film	89.2
Polycarbonate	Lexan HP92WDB	UV/mar coating	0.51	Film	86.6
Polycarbonate	Lexan Thermoclear	UV coating	5.99	Twin-wall	74.8
Polycarbonate	Lexan Thermoclear	UV coating	7.92	Twin-wall	77.0
Polycarbonate	Lexan Thermoclear	UV coating	10.03	Twin-wall	76.4
Polycarbonate	Lexan XL 10	UV/mar coating	3.00	Sheet	79.6
Polycarbonate	APEC 9351	Heat stabilized	3.18	Sheet	83.0
Polycarbonate	APEC 9353	UV & Heat stabilized	3.35	Sheet	79.9
Polyetherimide	Ultem 1000		0.10	Film	83.5
Polyetherimide	Ultem 1000		0.18	Film	78.7
Polyethylene		UV coating	0.13	Film	86.8
Polystyrene				Sheet	85.1
PVC[7]	DuraGlas		1.02	Sheet	82.9

[1]PET = PolyEthylene Terephthalate; [2]PEN = PolyEthylene Naphthalate; [3]ETFE = Ethylene-TetraFluoroEthylene; [4]E-CTFE = Ethylene-ChloroTriFluoroEthylene; [5]PFA = PerFluoroAlkoxy fluorocarbon; [6]PVDF = PolyVinyliDene Fluoride; [7]PVC = PolyVinyl Chloride.

product name and description, nominal thickness, form, and initial solar-weighted hemispherical transmittance are provided.

Polyethylene terephthalate (PET, a polyester) films have high optical clarity, low cost, and exhibit good mechanical properties. However, the permitted continuous use temperature is only ~90°C and its resistance to UV-induced photothermal degradation is generally poor. In recent years, UV-stabilized products have been developed by incorporating UV absorbers into the bulk films. One concern with this approach is that such additives cannot prevent deleterious reactions from occurring at the surface. The viability of such materials for use as solar glazings has not been demonstrated; consequently, durability exposure testing of a variety of unstabilized and stabilized PET films are being conducted.

Polyethylene naphthalate (PEN), which has improved thermal and mechanical properties and improved UV stability, has been developed as a more expensive alternative to PET. However, the resistance to loss in optical transmittance to UV exposure is questionable and must be demonstrated.

Many fluoropolymers exhibit excellent UV stability and resistance to thermal degradation. Unfortunately, fluoropolymers are relatively expensive and, therefore, can only be considered for use in thin film form. Commercially available fluoropolymers such as Tefzel, Duralar, Halar, Teflon, and Kynar (Table 6.1-1) exhibit very high spectral transmittance and many have sufficient tear resistance to be considered as collector glazings. Samples of those shown in Table 6.1-1 are being subjected to exposure testing to quantify their durability at in-service conditions.

As with fluoropolymers, acrylic polymers are also inherently resistant to UV-induced degradation. However, acrylic polymers generally cannot withstand the operating temperatures experienced by solar collector glazings. In addition, acrylic polymers are fairly brittle and therefore susceptible to hail impact damage, even in sheet form. Korad acrylic film may be too brittle to be used alone as a glazing but it has excellent UV-screening properties. As such, Korad is being considered as a candidate UV-screening layer in laminate constructions. In laminate constructions, a synergetic relationship can be achieved between the UV-screening layer and the substrate to which it is bonded. The UV-screening layer precludes UV light from reaching the substrate glazing that otherwise might be unusable because of its lack of resistance to UV exposure. The substrate glazing provides structural integrity and a thermal barrier for the UV-screen. The UV-screen can also provide protection for other elements in the light path that might otherwise be non-UV weatherable, e.g., a polymeric absorber, thereby making an all-polymeric construction a possibility as illustrated in Figure 6.1-2.

Two commercially available UV-screening films have been evaluated, i.e., Korad and Tedlar, which is a fluoropolymer. Both of these UV-screening polymers are available with a pressure-sensitive adhesive for ease of lamination to a substrate material. The effectiveness of these UV screening films was evaluated by laminating them to glass substrates and exposing them to accelerated weathering in the Ci5000 WOM. Any change in transmittance with exposure resulted from a loss of functionality in the UV-screen and adhesive combination only, i.e., the transmittance

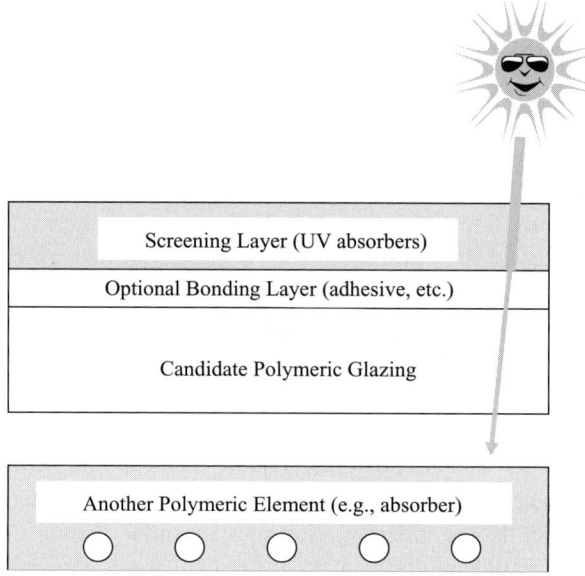

Figure 6.1-2. UV-screening polymeric glazing.

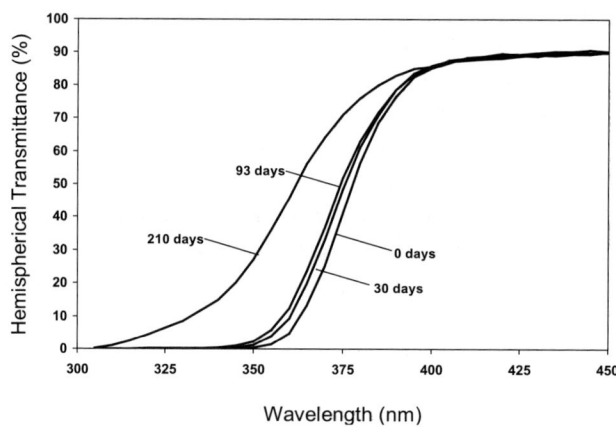

Figure 6.1-3. Hemispherical transmittance of a Tedlar UV-screening polymer laminated to glass as a function of Ci5000 exposure times.

of the glass does not change. The spectral hemispherical transmittances of the Tedlar/glass and Korad/glass laminates are shown in Figures 6.1-3 and 6.1-4, respectively. Several important features are evident. First, the unweathered cut-on wavelength of Tedlar is lower (~378 nm) than Korad (~390 nm); in addition, the cut-on profile of Tedlar is considerably less sharp (378 ± 16 nm) than Korad

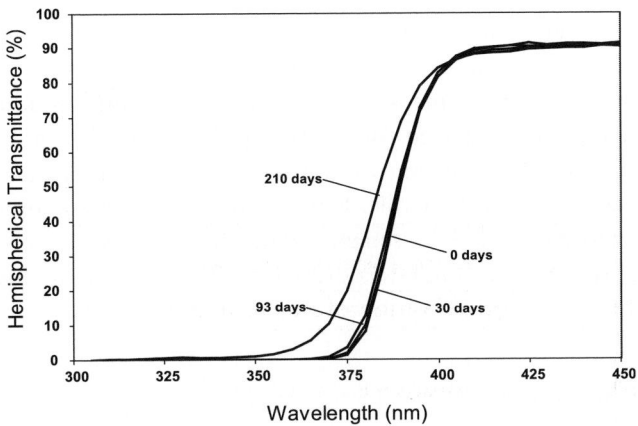

Figure 6.1-4. Hemispherical transmittance of a UV-screening polymer laminated to glass as a function of Ci5000 exposure times.

(390 ± 10 nm). This means that considerably more UV light is transmitted by Tedlar than by Korad; consequently, Korad provides a more effective UV-screen. Furthermore, upon weathering, Tedlar loses its functionality much more rapidly than Korad. Thus, Korad was chosen as the preferred UV-screening polymeric film and has been used in conjunction with a number of polymeric glazing laminates.

Polycarbonate (PC) has high optical clarity and excellent impact strength. However, it will yellow and become brittle when exposed to UV radiation. Recently, stabilized versions of polycarbonate have been developed. For example, Bayer has two products designated APEC 5391 and APEC 5393. The first is a thermally stabilized formulation, which is claimed to permit continuous use up to 180°C, and the second is both temperature and UV stabilized. General Electric has incorporated an integral UV-screening coating, which is also mar-resistant, into a number of their Lexan products such as HP92WDB (film), XL10 (sheet), and Thermoclear (twinwall multichannel). These are also being tested outdoors and in a WOM.

Polyetherimide is a high performance thermoplastic material with high temperature stability and low moisture absorption. It is thermoformable and heat-sealable to a wide range of other polymers. The weatherability of this material, with and without UV-screening layers, is largely uncertain and requires further investigation.

Finally, several commodity polymeric materials are being considered based upon their very low cost. Generally, these materials do not weather well outdoors, but we are investigating whether combining them with UV-screening layers will provide adequate protection to make these inexpensive candidates viable for solar applications. A polyethylene (PE) film with an integrated UV-screening coating, which is intended for solar greenhouse applications, has been identified as a candidate polymeric glazing. This material, with and without an additional Korad UV-screening layer, is being tested for durability. DuroGlas is a polyvinyl chloride (PVC) product for which significant outdoor weathering data, i.e., 10 years at two locations in Switzerland, are available. PVC is known to degrade upon exposure to UV so the effectiveness of UV screens with this material is also of interest. Polystyrene (PS) is used as a standard dosimeter within the weathering industry and is being tested to provide a calibration of our testing protocols.

6.1.4. PRELIMINARY TEST RESULTS

Most of the materials listed in Table 6.1-1 have not been tested long enough to establish useful results. The 9351 and 9353 APEC PC sheets have been exposed, both with and without Korad UV-screens, for extended lengths of time in the Ci5000 and at outdoor sites. In Figures 6.1-5 through 6.1-7, the measurement values are indicated by the different symbols used; these symbols are connected by lines to allow easy association of common data and to help identify trends. As shown in Figure 6.1-5, the unscreened 9351 PC material degrades very rapidly; the unscreened 9353 also degrades but at a reduced rate. This may result because the addition of UV stabilizers cannot prevent the occurrence of harmful surface reactions. Lamination of Korad onto 9351 PC provides good protection for about 75 days, but a precipitous loss in $\tau_{2\pi}$, occurs thereafter. When a UV-screen is used with the 9353

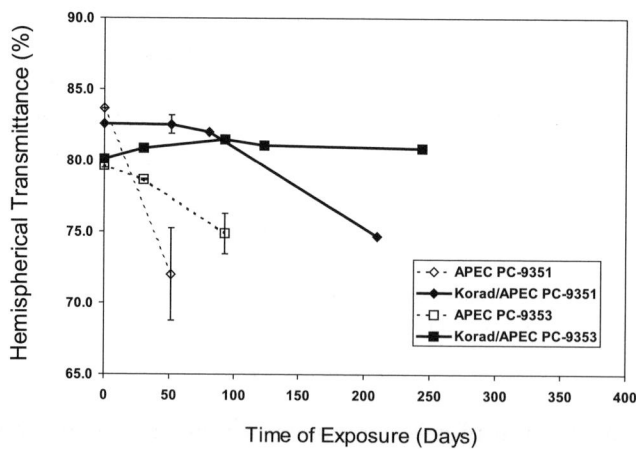

Figure 6.1-5. Solar-weighted hemispherical transmittance for APEC PC-9351 and PC-9353 glazings with and without a Korad UV-screening polymer as a function of Ci5000 exposure. The lines are only to guide the eye.

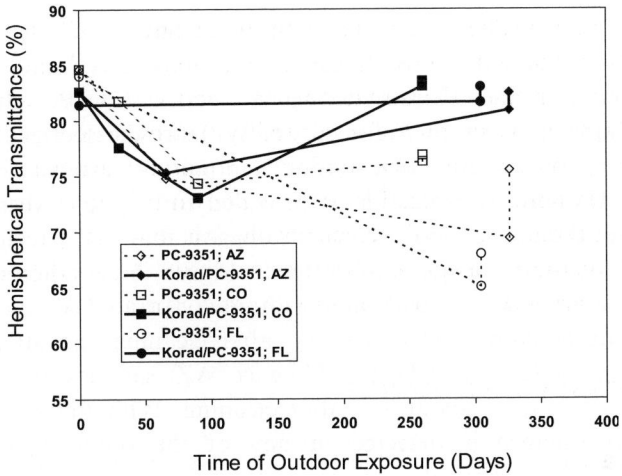

Figure 6.1-6. Solar-weighted hemispherical transmittance for an APEC PC-9351 glazing with and without a Korad UV-screening polymer as a function of outdoor exposure. The lines are only to guide the eye.

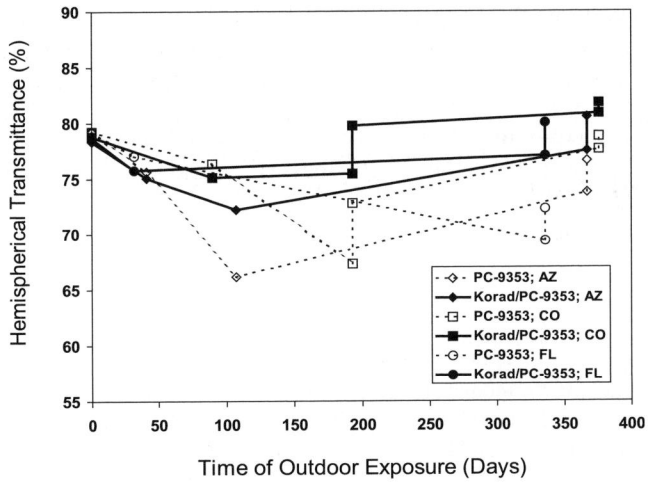

Figure 6.1-7. Solar-weighted hemispherical transmittance for an APEC PC-9353 glazing with and without a Korad UV-screening polymer as a function of outdoor exposure. The lines are only to guide the eye.

PC, excellent protection is afforded and no loss in $\tau_{2\pi}$ occurs for up to 250 days of exposure.

For samples exposed outdoors, measurements of the hemispherical transmittance, $\tau_{2\pi}$, were made on the samples as received after field exposure, and then after cleaning. This results in two data points at the same exposure time. The lower value was obtained before cleaning and the higher value after cleaning. The effectiveness of the Korad UV-screen with 9351 PC during outdoor weathering is shown clearly in Figure 6.1-6. Without the Korad, significant loss in transmittance resulted after only about 3 months of exposure at all outdoor sites; with the UV-screen, protection is provided for up to a year. However, as discussed above, the accelerated test results suggest that UV-screened 9351 is not adequate for long-term solar applications. Figure 6.1-7 shows the outdoor exposure results for 9353 PC with and without a Korad UV screen. Without a UV screen, samples show little loss in $\tau_{2\pi}$ after cleaning and after more than a year of exposure in Golden, CO and Phoenix, AZ; samples exposed in Miami, FL show a loss in $\tau_{2\pi}$ of about 7% after cleaning. With the Korad UV-screen, no loss in performance is detected at any of the outdoor sites, and is in agreement with the accelerated exposure test results.

6.1.5. CONCLUSIONS

The time-dependent changes in hemispherical transmittance of several candidate polymer glazing constructions have been assessed for exposure at outdoor test sites and in laboratory-controlled environmental conditions. Preliminary results show that the candidate laminates studied are promising and merit further consideration and evaluation. In particular, the effectiveness of Korad as a UV-screening layer has been demonstrated. The polymers studied provide attractive options for polymeric glazings and for all-polymeric solar collector systems.

REFERENCES

ASTM G159-98 (2003). Standard Tables for References Solar Spectral Irradiance at Air Mass 1.5: Direct Normal and Hemispherical for a 37° Tilted Surface. *Annual Book of ASTM Standards 2003*, **14.04**, American Society for Testing and Materials, West Conshohocken, PA.

ASTM E903-96 (2003) Standard Test Method for Solar Absorptance, Reflectance, and Transmittance of Materials Using Integrating Spheres. *Annual Book of ASTM Standards 2003*, **12.02**, American Society for Testing and Materials, West Conshohocken, PA.

ISO Standard 9845-1:1992, Solar energy – Reference solar spectral irradiance at the ground at different receiving conditions – Part 1: Direct normal and hemispherical solar irradiance for air mass 1.5. International Organization for Standardization, Geneva, Switzerland.

Kutscher, C., *et al.* (1984). *Low-Cost Collectors/Systems Development Progress Report.* SERI/RR-253-1750, National Renewable Energy Laboratory, Golden, CO.

Chapter 6.2

Case Study on Polymeric Glazings

G. Jorgensen[1], S. Brunold[2], B. Carlsson[3], M. Heck[4], and K. Möller[3]

[1]*National Renewable Energy Laboratory, Golden, CO 80401, USA*
[2]*SPF, Institut für Solartechnik Prüfung Forschung,*
HSR Hochschule für Technik Rapperswil, CH-8640 Rapperswil, Switzerland
[3]*Swedish National Testing and Research Institute,*
P.O. Box 857, S-501 15 Borås, Sweden
[4]*Fraunhofer Institute for Solar Energy Systems, D-7800 Freiburg, Germany*

Abstract: Results of preliminary and ongoing screening tests for a large number of candidate polymeric glazing materials were presented in Chapter 6.1. In this chapter, we focus on two specific glazings and demonstrate how the methodologies developed in Chapter 4.6 can be applied to predict accurately the optical performance of these materials during in-service use. In Section 6.2.1, a summary is given for data obtained by outdoor exposure and indoor testing of different polyvinyl chloride (PVC) and polycarbonate (PC) materials. In Section 6.2.2, an initial risk analysis is given for the two materials. In Section 6.2.3, screening tests and analyses for service lifetime prediction are discussed. A methodology that provides a way to derive correlations between degradation experienced by materials exposed to controlled accelerated laboratory exposure conditions and materials exposed to in-service conditions is given in Sections 6.2.4. In Section 6.2.5, a validation is presented for the durability methodology based upon durability test results for PVC and PC.

Keywords: Polymeric glazings, Outdoor exposure testing, Accelerated exposure testing, Risk analysis, Screening tests, Service lifetime prediction, Durability

LIST OF ABBREVIATIONS AND ACRONYMS

a	acceleration factor
AM	air-mass
D	test index (subscript)
DSC	differential scanning calorimetry
E	activation energy (kJ/mole)
EFF	effective mean value (subscript)

EDX	energy dispersive X-ray analysis
FTIR	Fourier transform infrared
G_D	diffuse solar horizontal irradiation (kWh/m^2)
G_H	global solar horizontal irradiation (kWh/m^2)
I	photoactive irradiance (W/m^2)
I_{UV}	ultraviolet irradiance (W/m^2)
IEA	International Energy Agency
k	Boltzmann constant
NIR	near-infrared light
p	material-dependent constant (exponent)
PC	polycarbonate
PO	photooxidation
PVC	polyvinyl chloride
P_D	probability of detection
P_O	probability of occurrence
R	risk
RH	relative humidity (%)
s	index for service (subscript)
S	severity
SEM	scanning electron microscopy
t	time
T	temperature
T_a	ambient temperature ($^\circ$C)
TOF-SIMS	time-of-flight secondary ion mass spectroscopy
UV	ultraviolet light (or radiation)
UV-A	the region of the electromagnetic spectrum between 315 nm to 400 nm
UV-B	the region of the electromagnetic spectrum between 280 nm to 315 nm
VIS	visible light
WIRE	World-wide Information system for Renewable Energy
WOM	an exposure chamber capable of controlling light intensity, T, and RH levels
XPS	X-ray photoelectron spectroscopy
τ	transmittance
τ_{sol}	solar transmittance
$\tau_{(400-600)}$	transmittance between 400 nm and 600 nm

6.2.1 SUMMARY OF DATA

From 1993–2002, numerous samples of PVC and PC materials were exposed to accelerated life testing in laboratories and to the environment at test sites located in Europe and in the USA by the laboratories now participating in IEA Task 27 (see the Preface). In addition, different types of indoor tests have been performed by these laboratories. An overview and summary of these efforts has been reported (Jorgensen *et al.*, 2003). The hemispherical transmittance of the polymeric specimens was measured before, after, and at periodic intervals during the testing.

All spectral data have been converted to the same wavelength grid. Because presenting plots of all the spectra is beyond the scope of this document, the files are downloadable from the World-wide Information system for Renewable Energy (WIRE), which is an internet site at http://wire0.ises.org, an initiative of the International Solar Energy Society (ISES). Additionally, the resulting integral values are summarized at WIRE in an ASCII table (tab delimited). Integral values are the solar transmittance ("τ_{sol}") and the mean value of the transmittance in the interval from 400 nm to 600 nm ("$\tau_{(400-600)}$").

In general, polymeric materials exhibit their greatest sensitivity to stress exposure-induced loss in optical performance from 400 to 600 nm, as shown in Figure 6.2-1. Consequently, $\tau_{(400-600)}$ is plotted versus testing time in the subsequent transmittance figures in this chapter.

Each spectrum has a name from which the material, the treatment, and the responsible laboratory can be determined as described in Table 6.2-1.

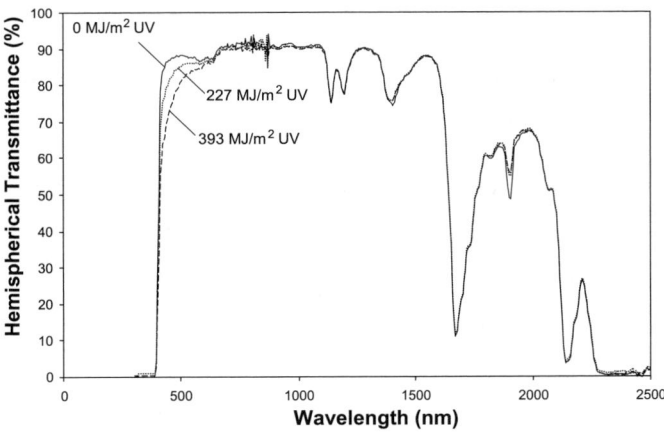

Figure 6.2-1. Change in hemispherical transmittance as a function of cumulative UV dose for UV-stabilized polycarbonate.

Table 6.2-1. Identification of the transmittance spectra

Material AAAA	Type	Trade name	Miscellaneous	*file name = AAAABBCCDDEEEEFG.asc* Manufacturer/Distributor
PCA1	PC	APEC 9351	$d = 4$ mm	Bayer
PCA3	PC	APEC 9353	$d = 4$ mm	Bayer
PCL1	PC	Lexan	$d = 4$ mm	Wachendorf
PCL2	PC	Lexan	$d = 0.375$ mm; AR coating outside	Wachendorf
PCL3	PC	Lexan	$d = 0.375$ mm; AR coating inside	Wachendorf
PCL4	PC	Lexan	$d = 4$ mm ; 'new type'	Wachendorf
PCM1	PC	Macrolon	$d = 4$ mm	Notz (Röhm)
PCM2	PC	Macrolon	$d = 3$ mm; AR coating outside	Röhm
PCM3	PC	Macrolon	$d = 3$ mm; AR coating inside	Röhm
PVCD	PVC	Duroglas	$d = 1$ mm	Gurit-Worbla

Treatment (outdoor exposure)

BB	Location of exposure	CC	UV protection	*file name = AAAABBCCDDEEEEFG.asc* DD	cleaning
RA	Rapperswil (CH)	—	direct exposure	bc	before cleaning
DA	Davos (CH)	Fi	with UV blocking filter – Korad acrylic film in US tests – standard glass in Swiss tests	Ac	after cleaning
FR	Freiburg (D)				
CO	Golden, Colorado (USA)				
AZ	Phoenix, Arizona (USA)				
FL	Miami, Florida (USA)			Nc	never cleaned

*file name = AAAА**BB**CC**DD**EEEEFG.asc*

	CC	DD	
	Temperature	rel. humidity	
	XX	Xx	
	Temperature (°C)	relative humidity (%)	

Treatment (indoor testing)
BB — *Irradiation*
UV — with UV irradiation
DK — Dark test (no light)

Alternatively used (in case of a Weather-Ometer® test):
BBCCDD — *Weather-Ometer®*
WOM___ — Weather-Ometer® test with irradiation
WOMFi_ — Weather-Ometer® test with UV filtered radiation
SPART_ — Weather-Ometer® – SPART 14 test (see text)

*file name = AAAА**BB**CC**DD**EEEEFG.asc*

Duration of treatment
EEEE — *Time*
xxxx — xxxx days; instead of a 'decimal point' the 'underline' is used, i.e 21.5 – > 21_5
xxxh — xxx hours
xxxm — xxx months
xxxy — xxx years

Sample number
G — *replicate sample number*
x — used to distinguish between samples of the same material undergoing the same test

*file name = AAAА**BB**CC**DD**EEEE**FG**.asc*

Laboratory
F — *Laboratory*
C — SPF-HSR, Oberseestr. 10, CH-8640 Rapperswil, Switzerland
G — FhG-ISE, Oltmannsstr. 5, D-79100 Freiburg, Germany
U — NREL, 1617 Cole Blvd./MS-3321, USA Golden, Colorado 80401, USA
S — SP, Swedish National Testing and Research Institute, S-50115 Borås, Sweden

6.2.1.1   Sample Selection and Testing

A number of candidate materials have been identified by reviewing the literature and through discussions with experts in the polymer and solar manufacturing industries. Polycarbonate has high optical clarity and excellent impact strength. However, it will yellow during UV exposure and become brittle. Recently, stabilized versions of PC have been developed. For example, Bayer has two products designated APEC 5391 and APEC 5393. The first is a thermally stabilized formulation, which is offered for a maximum continuous use temperature of up to 180°C, and the second is stabilized for UV exposure and elevated temperatures. General Electric has incorporated an integral UV-screening coating, which is also mar-resistant, into a number of their Lexan products.

Samples of candidate polymeric glazing materials were subjected to in-service outdoor and accelerated laboratory exposure conditions. Outdoor testing was carried out in Switzerland (C) at the Institut für Solartechnik (SPF), Germany (G) at the ISE in Freiburg, and at three sites in the United States (U), namely, Golden, CO; Phoenix, AZ; and Miami, FL. A precise and detailed knowledge of the specific environmental stress conditions experienced by weathered samples is needed to allow understanding of site-specific performance losses and to permit service lifetime prediction of candidate glazings. Consequently, each operational exposure site is fully equipped with the appropriate meteorological and radiometric instrumentation and data-logging capability.

6.2.1.2   Outdoor Exposure Testing

The materials tested are for the intended use in solar thermal flat-plate collectors. Thus, the samples for outdoor exposure were fixed onto mini-collector boxes, as illustrated in Figure 6.2-2. To simulate the elevated temperature collector covers are exposed to, the "mini-collectors" are made of solar selective coated stainless steel. A thermocouple is affixed to the glazing material to monitor sample temperature, and a reflective light shield hood is used to prevent direct heating of the thermocouple.

The samples prepared in this way were exposed to the ambient climate at locations in Europe and in USA at an inclination angle between 45° and 60° facing south. The spectral transmittance of all samples was measured prior to exposure. After some time, some of the samples were remeasured and exposed again without any cleaning. Other samples were measured before and after cleaning and then exposed again.

The values obtained for $\tau_{(400-600)}$ are plotted in Figure 6.2-3 (a through z and aa) for APEC 9351, APEC 9353, several different thicknesses of Lexan, Macrolon, and Duroglas after exposures at the different test sites for periods of up to 10 y. For the

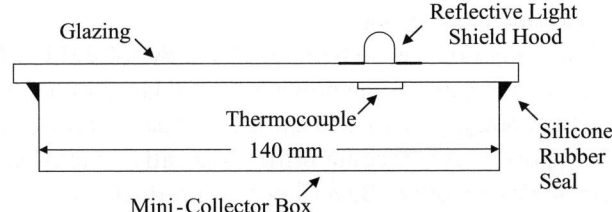

Figure 6.2-2. "Mini-collectors" used for outdoor exposure of transparent cover materials.

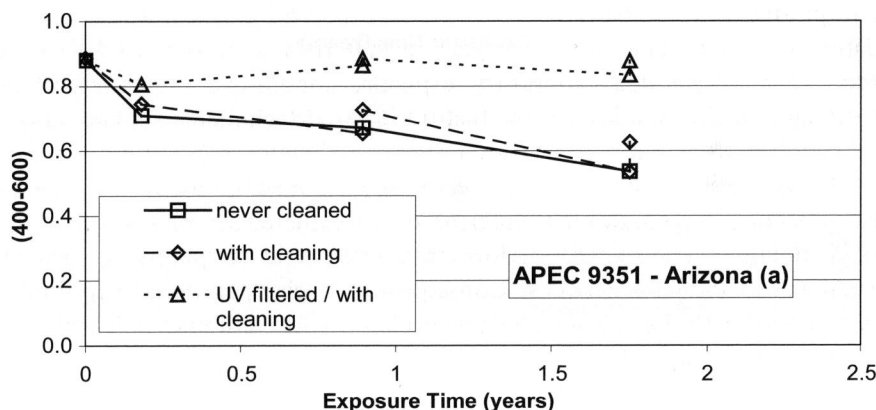

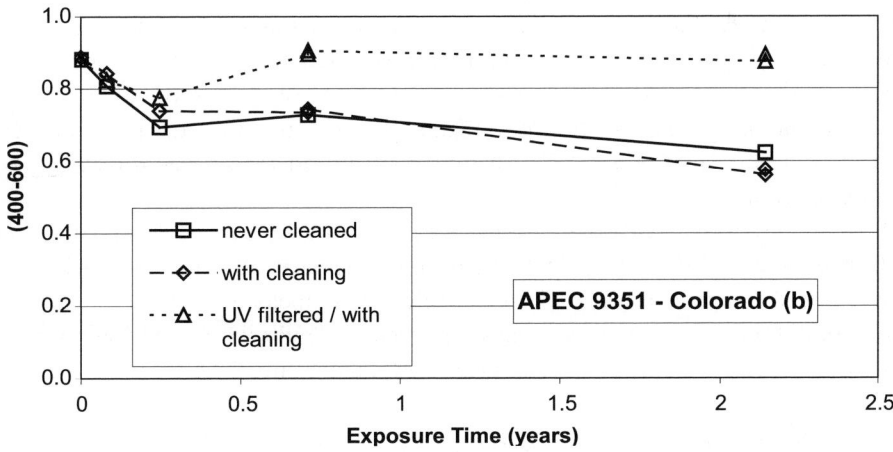

Figure 6.2-3. (a through aa). Changes in $\tau_{(400-600)}$ with exposure time APEC 9351, APEC 9353, several different thickness of Lexan and Macrolon, and Duroglas at different test sites.

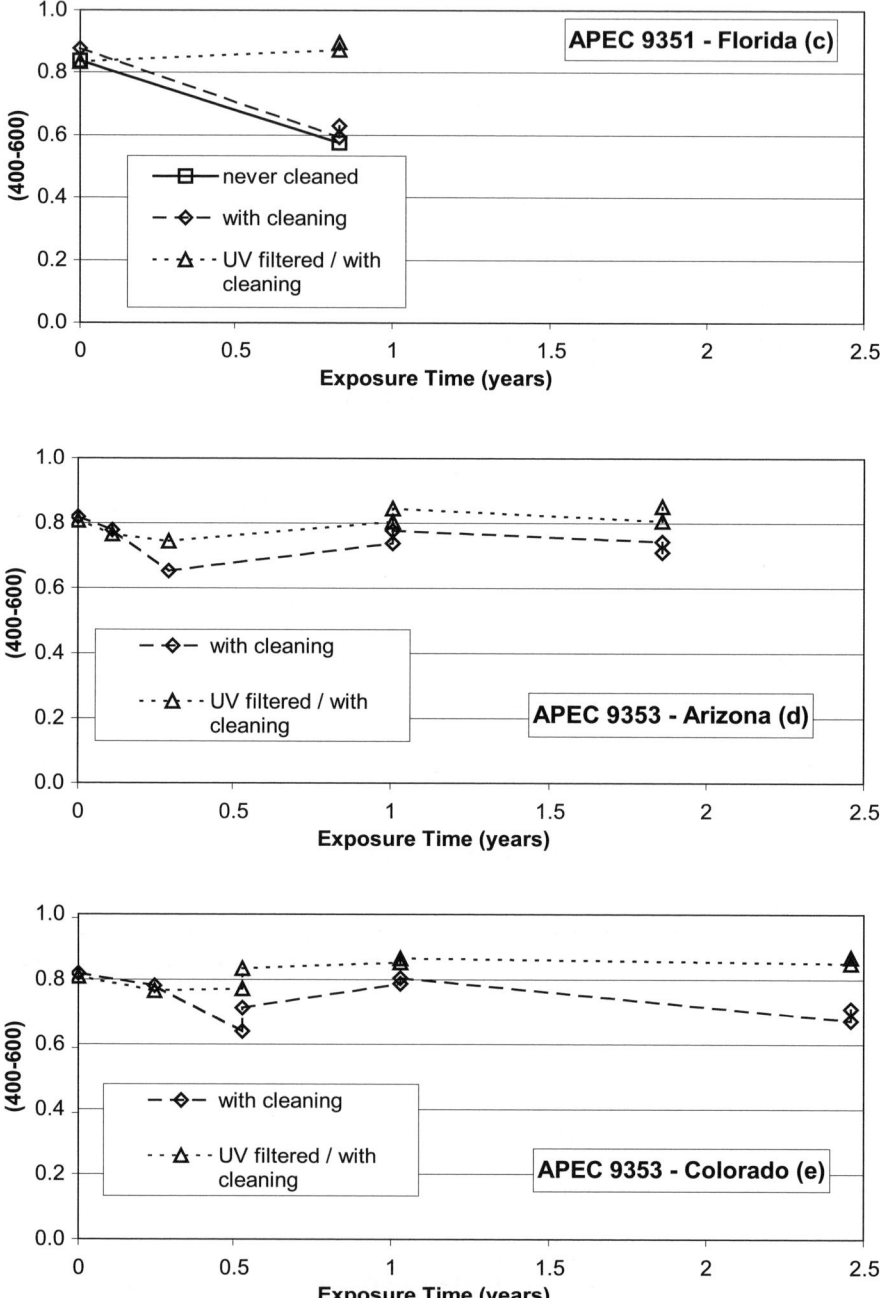

Figure 6.2-3. Continued.

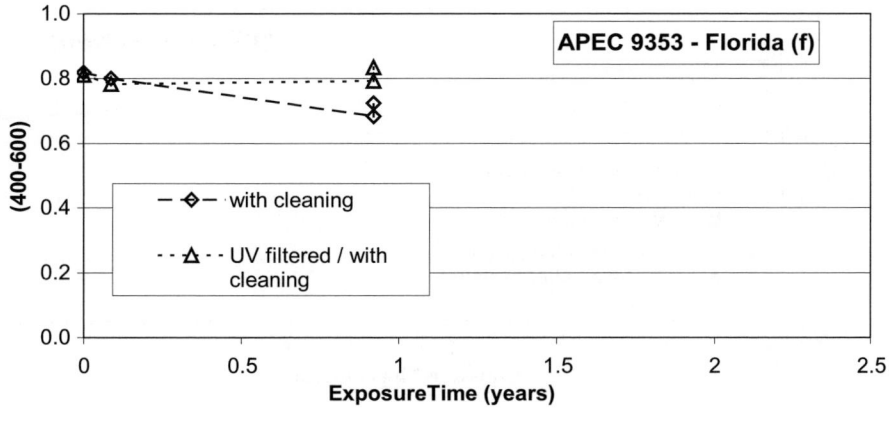

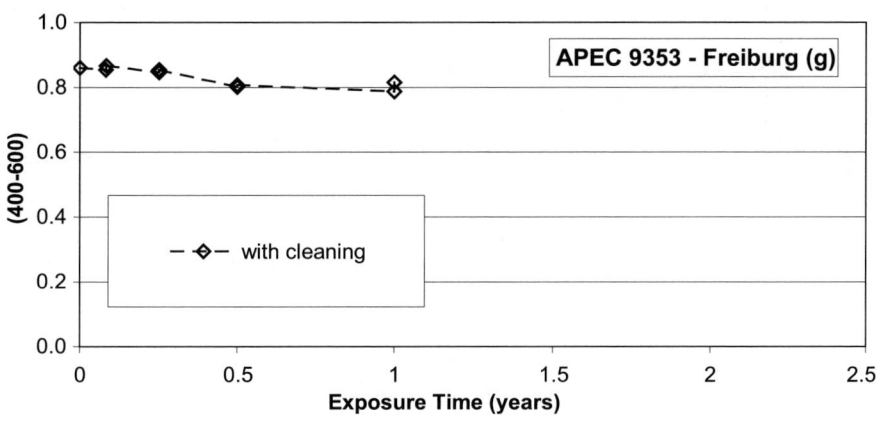

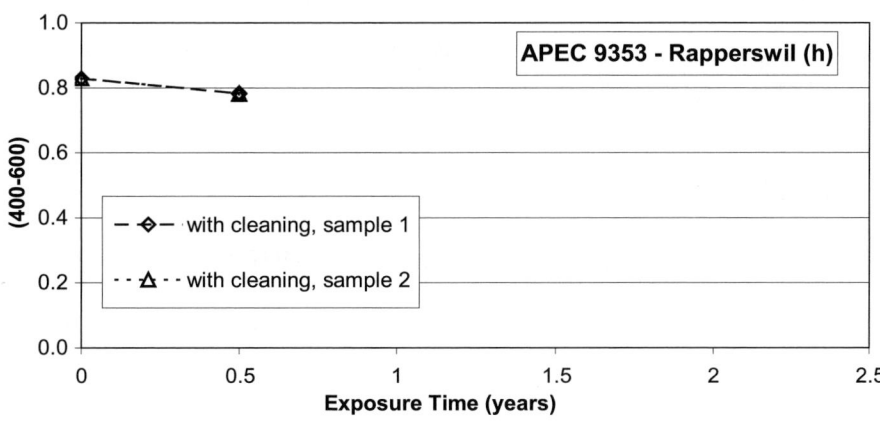

Figure 6.2-3. Continued.

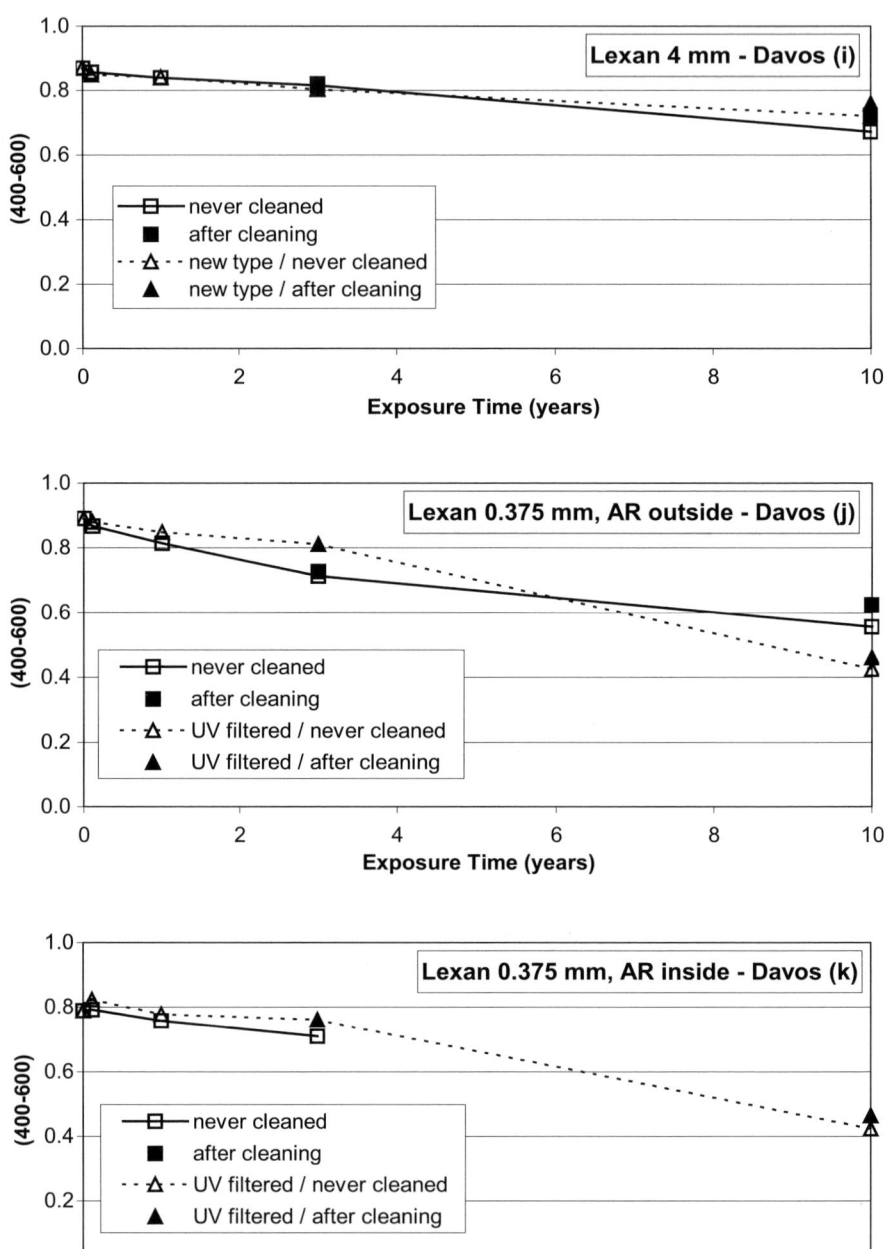

Figure 6.2-3. Continued.

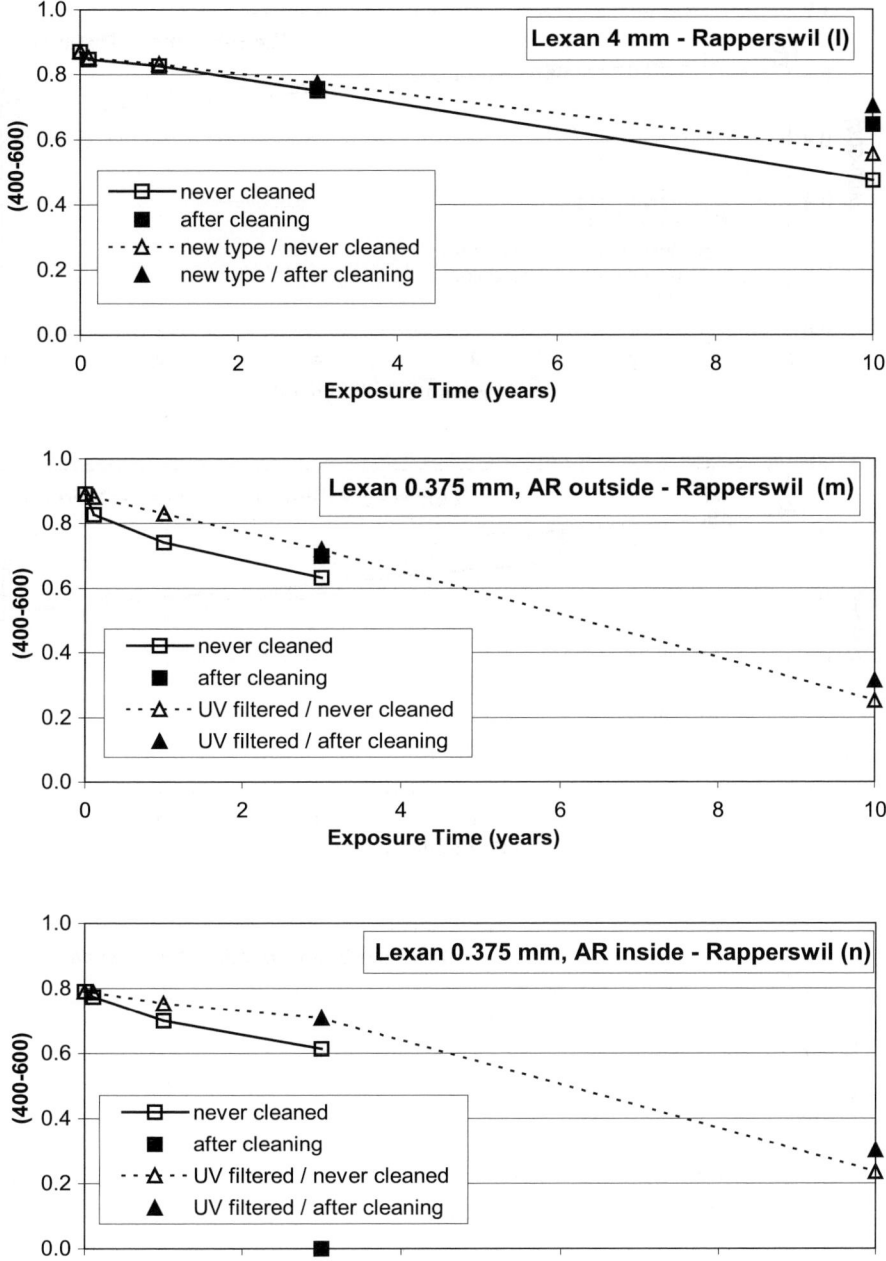

Figure 6.2-3. Continued.

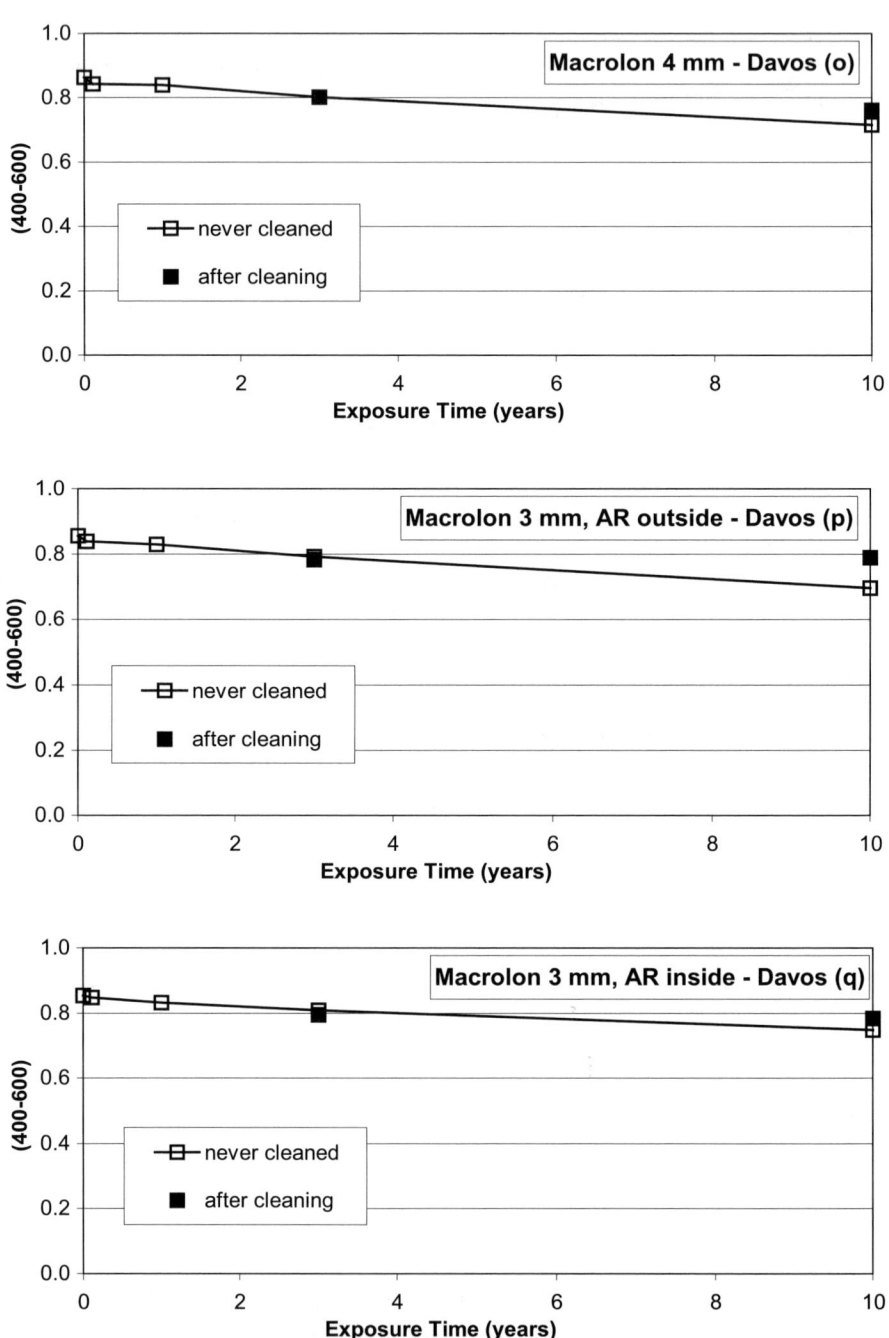

Figure 6.2-3. Continued.

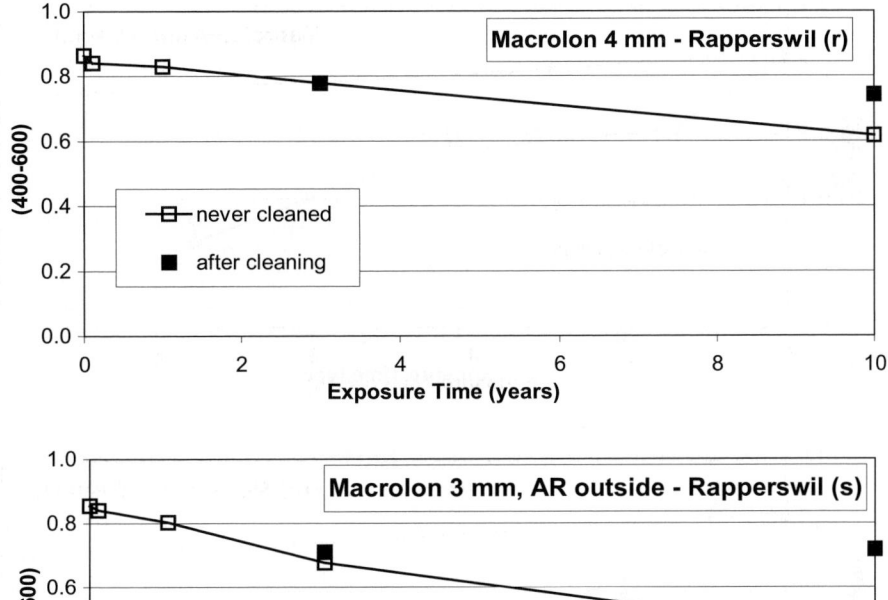

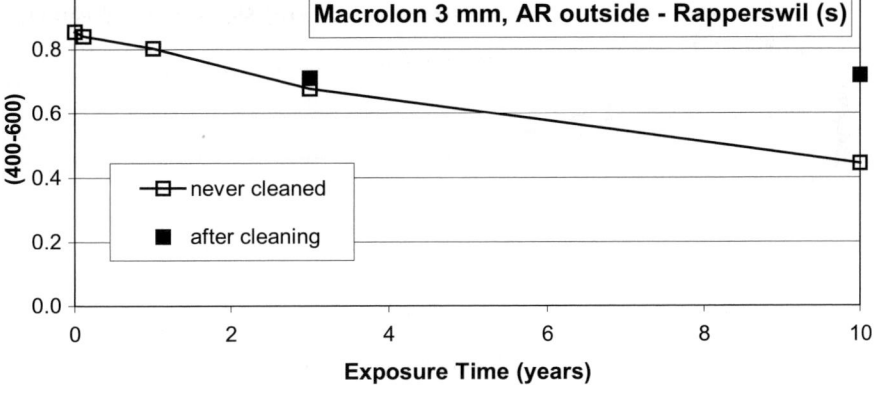

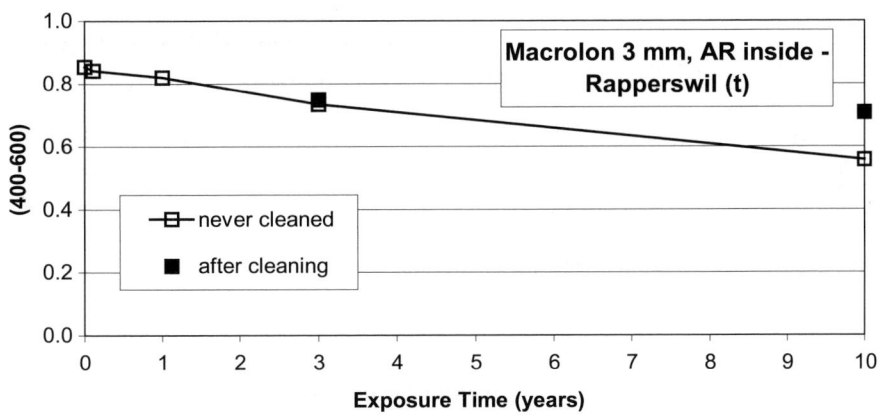

Figure 6.2-3. Continued.

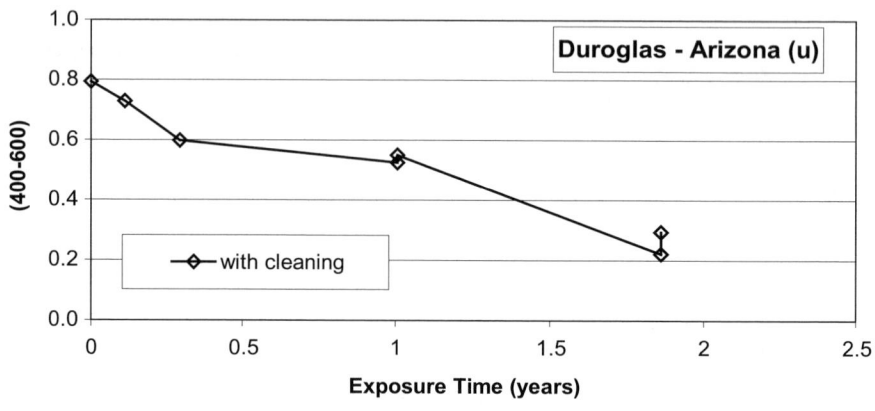

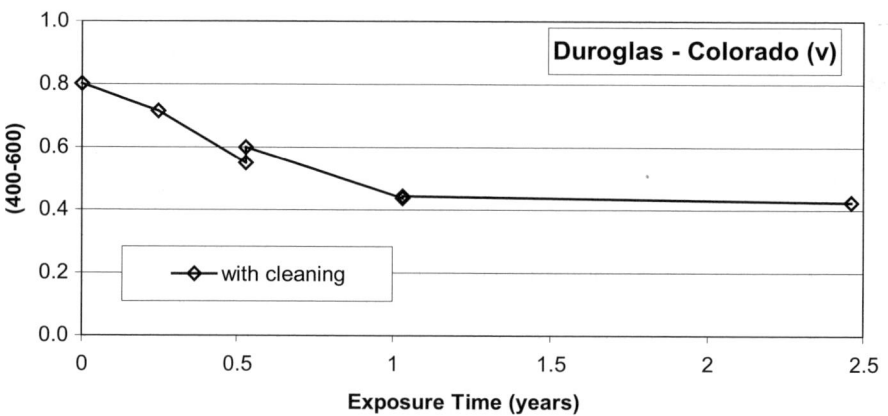

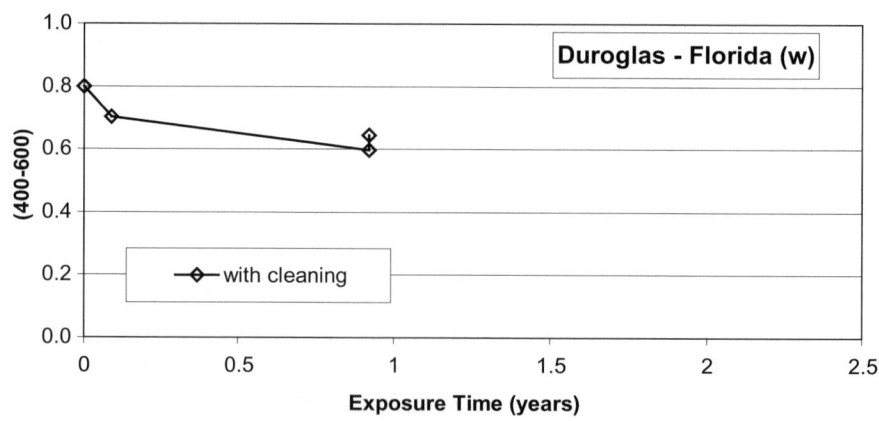

Figure 6.2-3. Continued.

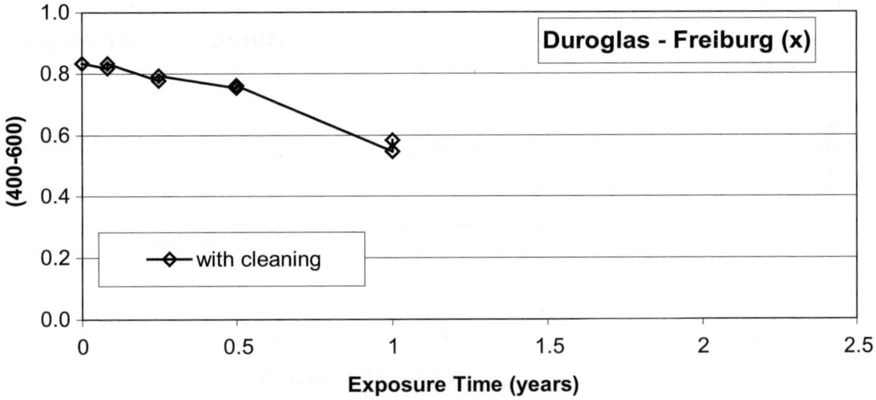

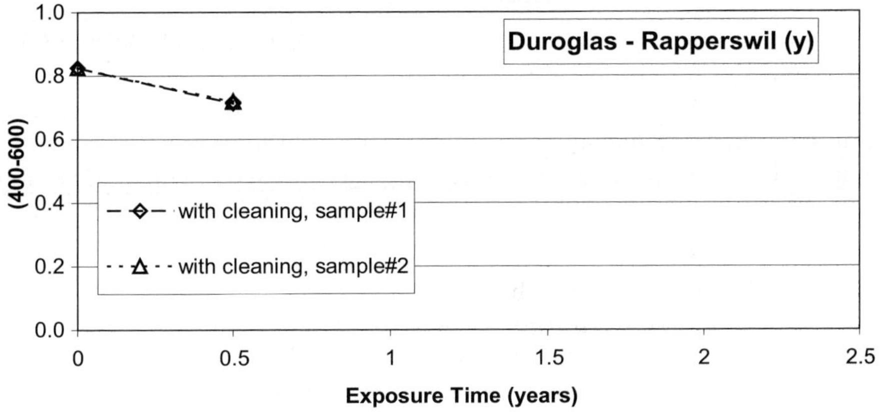

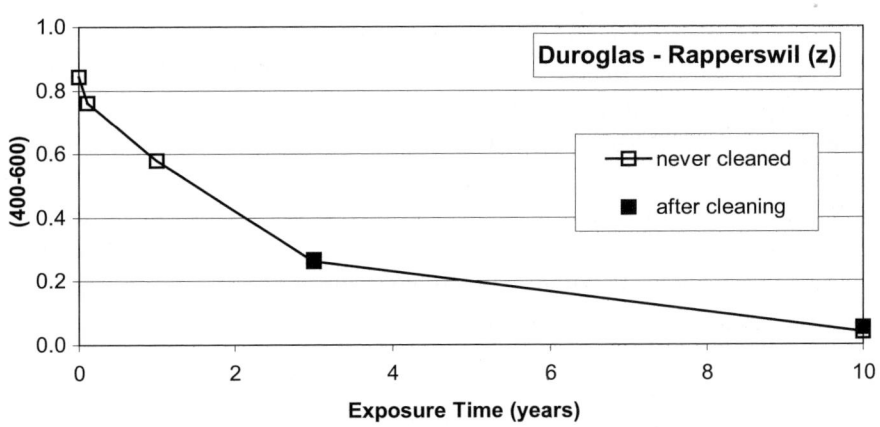

Figure 6.2-3. Continued.

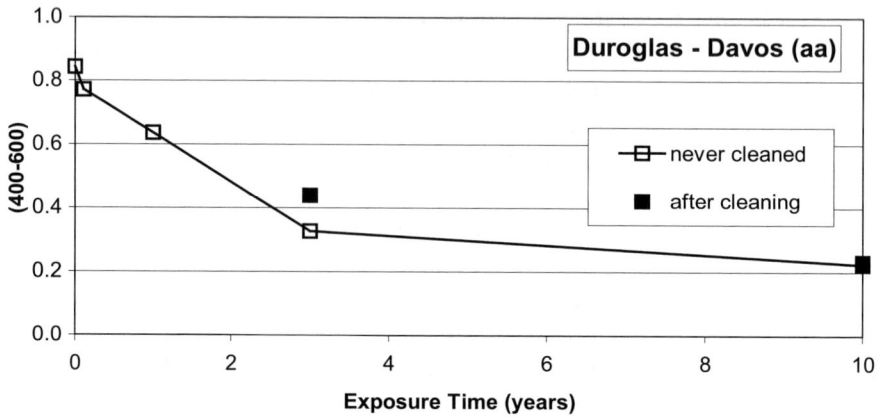

Figure 6.2-3. Continued.

ordinates in these plots, the factor $\tau_{(400-600)}$ has a maximum value of 1.0 at 100% transmittance and a minimum value of 0.0 at 0% transmittance.

Figures 6.2-3 a through c show the difference between APEC 9351, a heat-stabilized but non-UV stabilized PC, exposed outdoors in the United States with and without an affixed UV-screening layer (Korad acrylic film). The unscreened APEC 9351 degrades rapidly, whereas the screening layer significantly improves the outdoor durability of this product. The same results occur for APEC 9353, a heat-stabilized and UV-stabilized PC, exposed outdoors in the United States (Figures 6.2-3 d through f), although the performance of the unscreened material is better than the unscreened APEC 9351. Figures 6.2-3 g and h show a similar loss in $\tau_{(400-600)}$ for samples of APEC 9353, without a Korad UV screening layer, exposed outdoors in Europe. Figures 6.2-3 i through n display the ten-year performance of various thicknesses of Lexan PC at two locations in Switzerland. Samples exposed at Rapperswil experience greater degradation than those exposed at Davos. Results for Macrolon PC also exposed for ten years at these two locations are shown in Figures 6.2-3 o through t. Weathering at Rapperswil is generally also more deleterious than Davos for Macrolon PC. Duroglas, a PVC glazing material, degrades very rapidly when exposed to UV light. Figures 6.2-3 u through aa present outdoor test results for Duroglas exposed in the United States (u through w) and Europe (x through aa). Samples of Duroglas exhibited roughly a 25% loss in $\tau_{(400-600)}$ after only one year exposure at all outdoor sites.

The monthly mean meteorological data of the test sites presented in Table 6.2-2 are also taken from Meteonorm (Meteonorm, 1995).

6.2.1.3 Accelerated Laboratory Exposure Testing

Accelerated indoor testing was carried out with different types of test equipment available at the participating laboratories. Several test protocols were performed using corresponding types of exposure chambers. In the first type of test, samples were exposed only to elevated temperature in standard laboratory ovens, without any irradiation, at 40°C, 60°C, 65°C, 70°C, 75°C, and 80°C. In the second type, UV exposure was combined with various combinations of elevated temperature and a defined level of relative humidity (RH), i.e., 60°C/80% RH, 80°C/40% RH, and 50°C/95% RH. These tests were performed in climatic cabinets with an unfiltered metal halide (HMI) lamp as a light source. The intensity of the irradiation compared to an air mass (AM) 1.5 solar spectrum is about 3 times as much UV-A and 7 times as much UV-B. In the third type of exposure test, an Atlas Ci5000 Weather-Ometer® (WOM) was operated at 60°C and 60% RH, and an irradiation level close to twice an AM 1.5 solar spectrum throughout the UV and visible portion of the spectrum. Tests were performed on samples with and without an additional UV filter (Korad acrylic film) laminated to the surface of the sample. In the final test protocol, an Atlas XR35 WOM – SPART 14 test was used. The SPART 14 test procedure was originally developed for clear coats in automotive paint systems. The test is a weatherability test that includes acidic rain spraying. In test method SPART 14, which is a modification of SAE J1960 (SAE J1960), the Xenon arc light source is filtered through borosilicate filters and has an irradiance level of 0.5 W/m^2 at 340 nm. The test cycle is comprised of: (a) 40 min of light only; (b) 20 min of light with water sprayed on the front surface of the sample; (c) 60 min of light only; and (d) 60 min of no light with water sprayed on the back surface of the sample. Every fourteenth cycle, the water used to spray the front of the samples is acidic, with a pH of 3.2. The black standard temperature and relative humidity during light periods are 70°C and 75%, respectively. The air temperature during light periods is about 47°C because the temperature inside the Weather-Ometer is controlled by the black standard temperature, so the air temperature can vary slightly. The temperature and relative humidity during the dark periods are 38°C and 95%, respectively. The temperature of the tested glazing materials was estimated to reach 50–55°C during the "light only" exposure.

The test specimens were exposed for up to 4000 h (~ 24 weeks). An exposure time of 1000 h (~ 6 weeks) in the accelerated Weather-Ometer® test is estimated to correspond to about 1.3 years of outdoor testing in Miami, Florida for automotive paints. Thus, 4000 h of SPART 14 testing corresponds to about 5 years outdoors in Florida. However, one can assume that the temperature of an automotive coating will be at least 10°C higher than for transparent low light absorbing glazing materials. Consequently, the acceleration factor for the glazing can be estimated to

Table 6.2-2. Monthly Mean Meteorological Data of the Test Sites

	Jan	Feb	Mar	Apr	May	Jun	Jul	Aug	Sep	Oct	Nov	Dec	Year
Location: **RA** – Rapperswil, Switzerland													
Altitude: 417 m													
Longitude: − 8.82°													
Latitude: 47.20°													
Solar horizontal irradiation, global G_H, diffuse G_D; ambient temperature T_a													
G_H (kWh/m²)	26.9	44.5	90.0	114.0	148.0	152.0	168.0	142.0	99.2	59.9	28.5	20.4	1093.0
G_D (kWh/m²)	19.3	30.3	51.2	64.2	93.6	88.1	92.1	82.1	55.3	38.9	19.4	15.0	649.4
T_a (°C)	1.3	1.4	5.0	7.9	12.3	15.1	18.9	18.4	15.3	10.8	5.0	2.8	9.5
Location: **DA** – Davos, Switzerland													
Altitude: 1556 m													
Longitude: 9.83°													
Latitude: 46.80°													
Solar horizontal irradiation, global G_H, diffuse G_D; ambient temperature T_a													
G_H (kWh/m²)	52.0	74.6	125.0	150.0	173.0	171.0	182.0	155.0	119.0	86.0	51.7	41.4	1381.0
G_D (kWh/m²)	21.6	29.2	43.2	62.3	95.7	85.5	99.4	79.4	49.4	37.1	24.8	20.4	648.0
T_a (°C)	−6.4	−5.6	−1.5	1.6	6.1	9.0	12.9	12.2	9.2	5.2	−1.3	−4.7	3.1
Location: **FR** – Freiburg, Germany													
Altitude: 308 m													
Longitude: − 7.51°													
Latitude: 48.00°													
Solar horizontal irradiation, global G_H, diffuse G_D; ambient temperature T_a													
G_H (kWh/m²)	27.5	44.4	81.1	111.6	146.6	162.0	171.1	148.8	102.2	63.2	34.6	21.6	1114.7
G_D (kWh/m²)	19.7	27.7	51.7	68.8	87.8	93.0	93.0	77.1	54.9	36.5	21.7	13.8	645.7
T_a (°C)	1.3	2.7	6.0	9.9	14.3	17.1	19.3	18.5	15.2	10.2	5.2	2.5	10.2

Location: **AZ** – Phoenix, Arizona, USA
Altitude: 340 m
Longitude: 112.12°
Latitude: 33.41°
Solar horizontal irradiation, global G_H, diffuse G_D; ambient temperature T_a

G_H (kWh/m²)	101.9	122.3	174.0	216.6	248.3	244.7	229.0	213.4	179.3	149.6	107.3	91.5	2078.0
G_D (kWh/m²)	33.7	35.9	48.0	48.5	54.9	59.9	71.2	70.1	55.3	43.4	35.3	30.3	586.3
T_a (°C)	11.9	14.2	16.8	21.0	25.9	31.1	34.1	33.0	29.7	23.5	16.6	12.2	22.5

Location: **FL** – Miami, Florida, USA
Altitude: 5 m
Longitude: 80.04°
Latitude: 25.59°
Solar horizontal irradiation, global G_H, diffuse G_D; ambient temperature T_a

G_H (kWh/m²)	108.5	117.7	158.7	179.9	184.8	168.2	180.9	173.3	146.7	134.3	108.8	101.3	1763.0
G_D (kWh/m²)	48.7	54.9	71.2	79.5	92.4	84.4	86.4	87.5	76.8	73.4	50.9	39.1	845.3
T_a (°C)	19.5	20.2	22.0	23.9	25.9	27.4	28.1	28.2	27.7	25.7	23.0	20.5	24.3

Location: **CO** – Golden, Colorado, USA (the meteorological data are for Denver, CO, approximately 20 km from Golden)
Altitude: 1585 m
Longitude: 104.54°
Latitude: 39.50°
Solar horizontal irradiation, global G_H, diffuse G_D; ambient temperature T_a

G_H (kWh/m²)	73.8	91.5	136.3	166.5	192.1	205.4	207.7	185.4	149.3	117.8	78.7	66.4	1670.9
G_D (kWh/m²)	30.0	35.1	50.3	68.9	89.3	86.1	86.2	78.9	57.8	44.8	30.9	26.7	685.1
T_a (°C)	-1.2	0.9	4.0	9.1	14.1	19.5	23.2	22.0	16.9	10.9	4.0	-0.5	10.3

be a factor of 2 higher. Accordingly, 1000 h of artificial weathering corresponds to 2.5 y outdoors and 4000 h corresponds to 10 y outdoors.

Parallel testing with the relevant stress factors of UV, temperature, RH, and acid spray at different levels was intended to allow the sensitivity of materials degradation to these factors to be quantified, and allow damage function models to be evaluated. These in turn can be used to compare the time-dependent performance of these materials with measured results from in-service outdoor exposure.

Highly accelerated exposure testing of selected samples was also performed at NREL using a unique UV concentrator (Jorgensen *et al.*, 1999, 2002). It consists of an array of faceted mirrors that tracks the sun in two axes and redirects sunlight back to a sample exposure chamber attached by three structural support tubes. The concentrator is designed to provide up to 100X concentration having a uniform flux at high UV intensity and low visible (VIS) and near-infrared (NIR) intensity. The concentrated UV flux is achieved by coating the mirror facets with a custom-designed 37-layer film that consists of alternating high and low refractive index materials and yields the desired high UV reflectance and low VIS/NIR reflectance.

The values obtained for $\tau_{(400-600)}$ are plotted in Figure 6.2-4 (a through t) for APEC 9351, APEC 9353, and Duroglas after different types of exposure for up to 600 days. For the ordinates in these plots, the factor $\tau_{(400-600)}$ has a maximum value of 1.0 at 100% transmittance and a minimum value of 0.0 at 0% transmittance.

Figures 6.2-4 a and b show the difference between APEC 9351 exposed in NREL's Atlas Ci5000 WOM with and without an affixed Korad UV-screening layer. The unscreened APEC 9351 (a) degrades more rapidly than the screened sample (b), although the screened material does start to degrade after about 75 days of

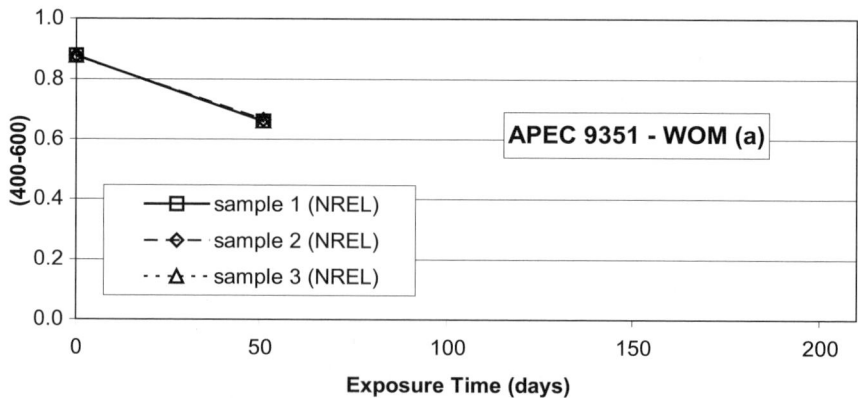

Figure 6.2-4. (a through t) Changes in $\tau_{(400-600)}$ with exposure time of APEC 9351, APEC 9353, and Duroglas in various accelerated weathering chambers.

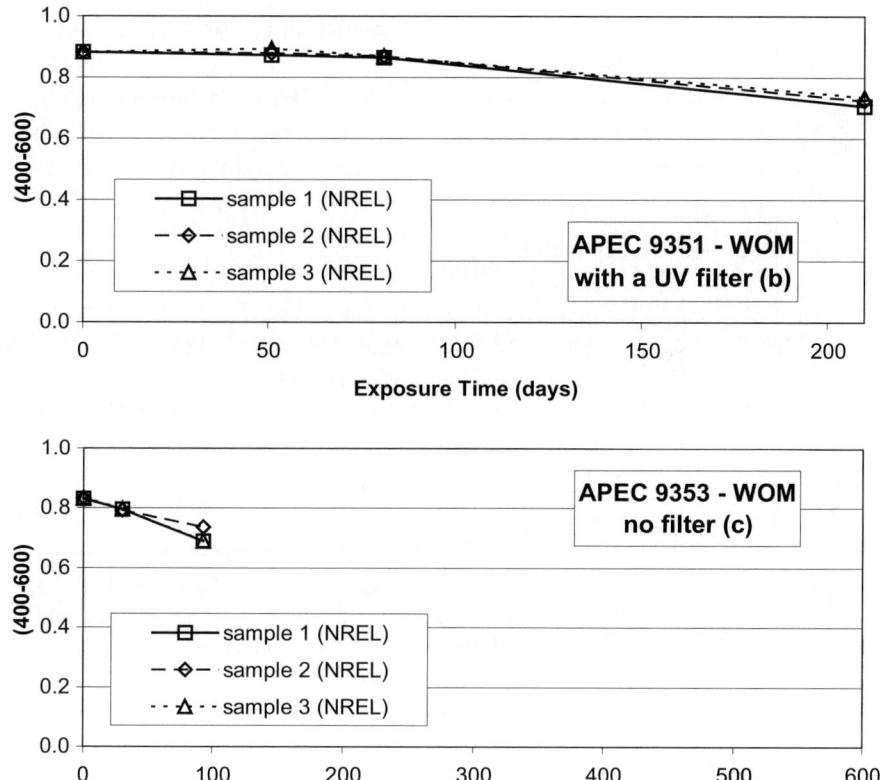

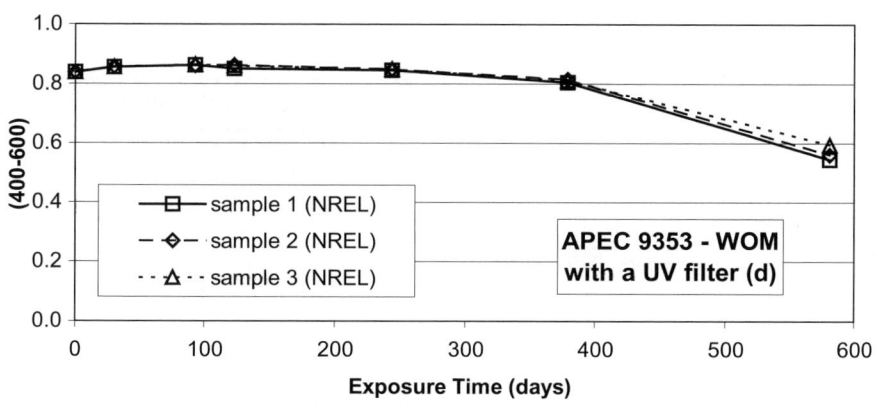

Figure 6.2-4. Continued.

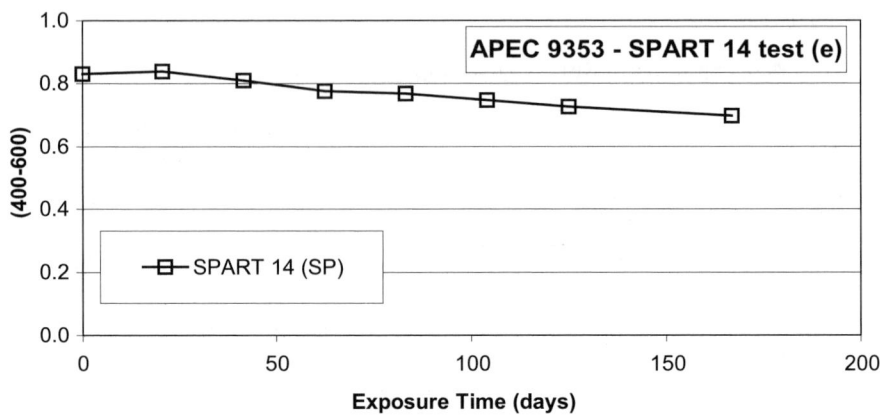

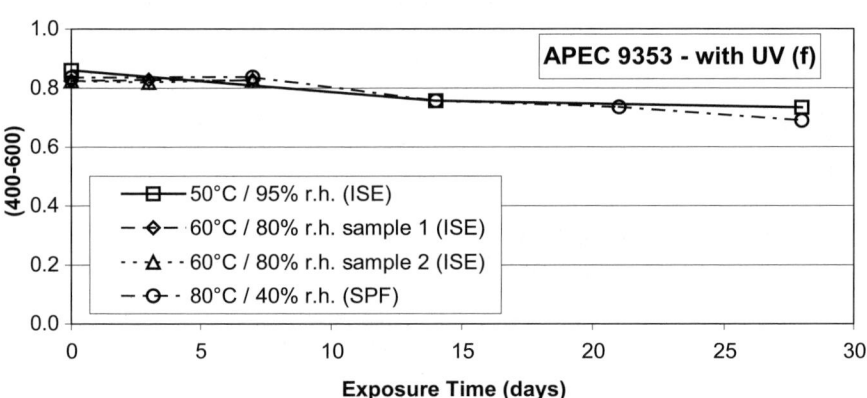

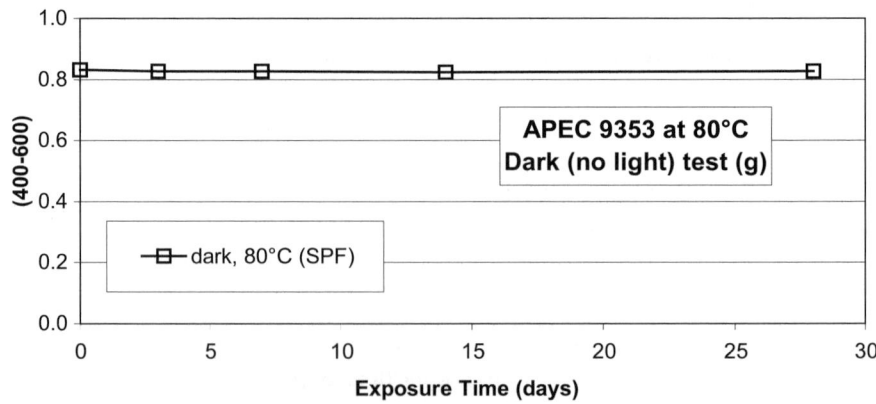

Figure 6.2-4. Continued.

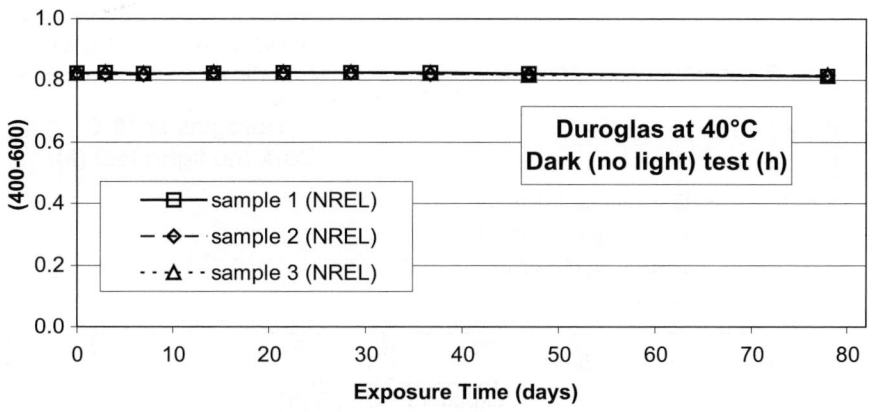

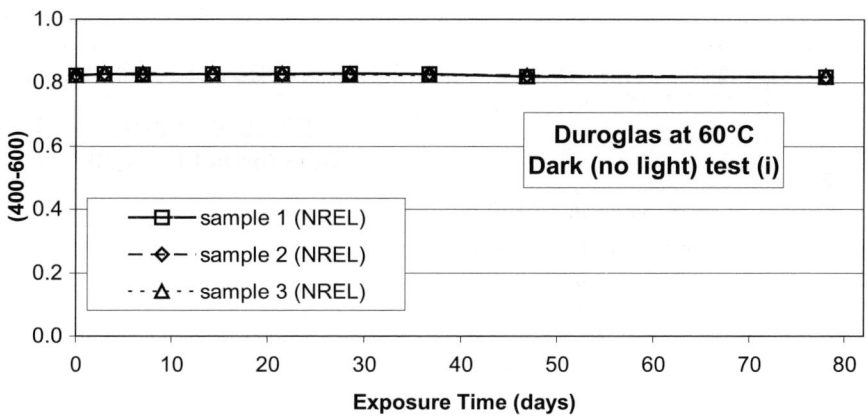

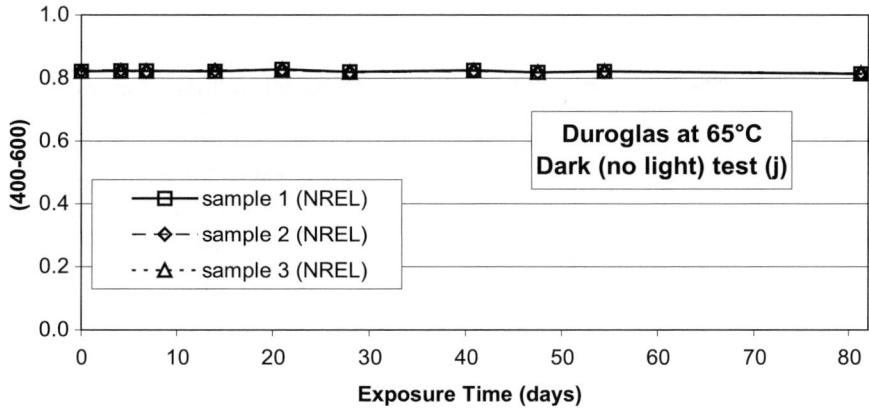

Figure 6.2-4. Continued.

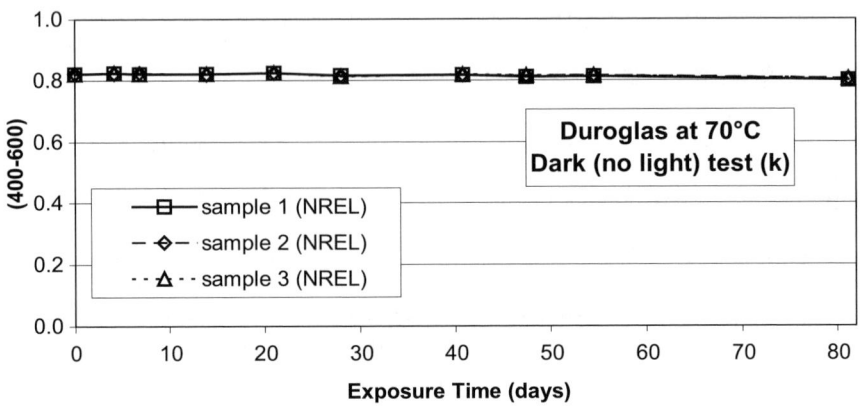

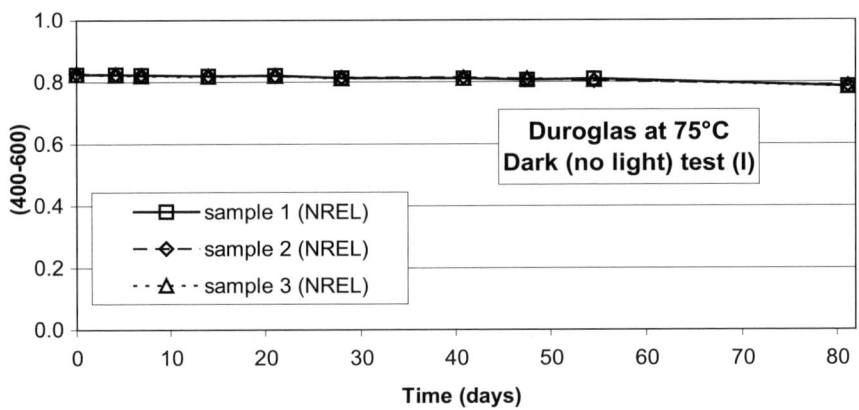

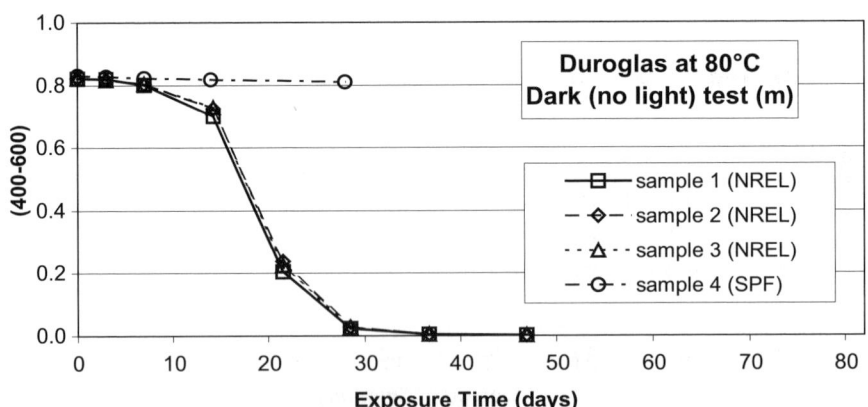

Figure 6.2-4. Continued.

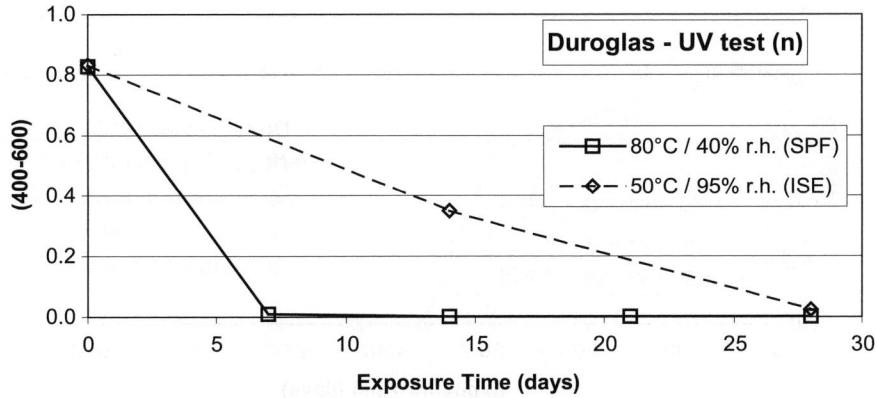

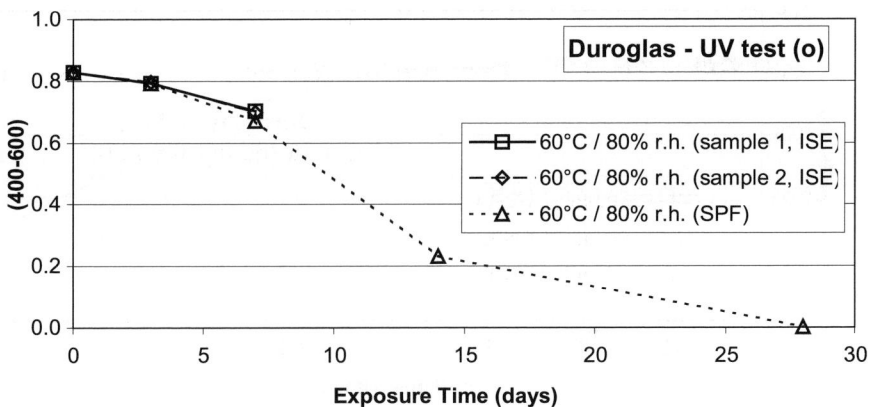

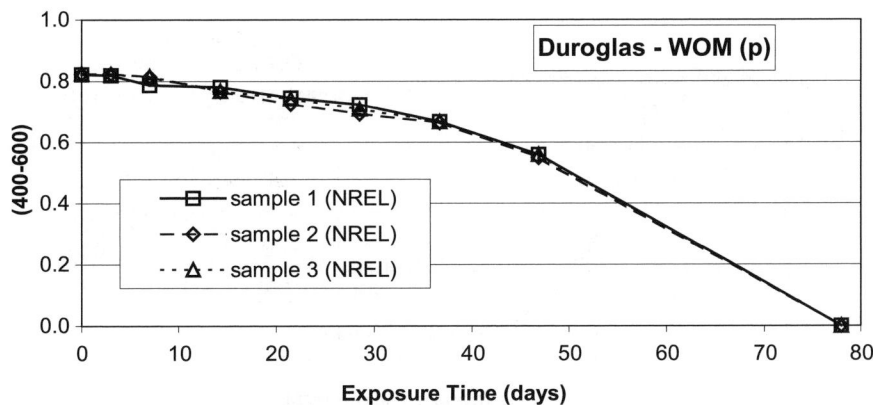

Figure 6.2-4. Continued.

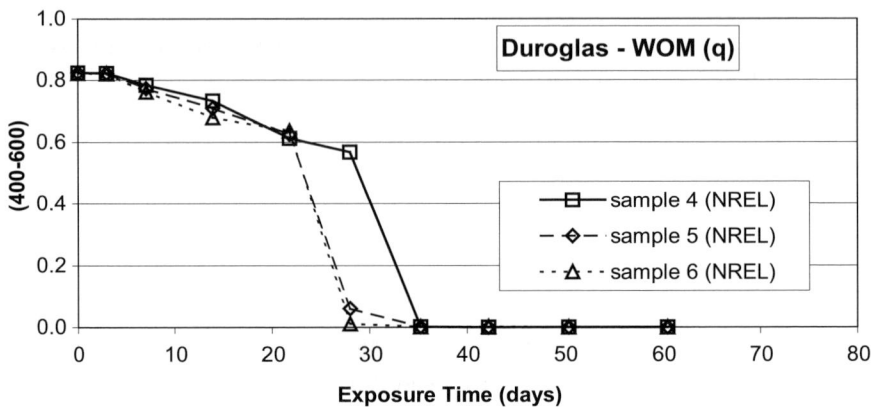

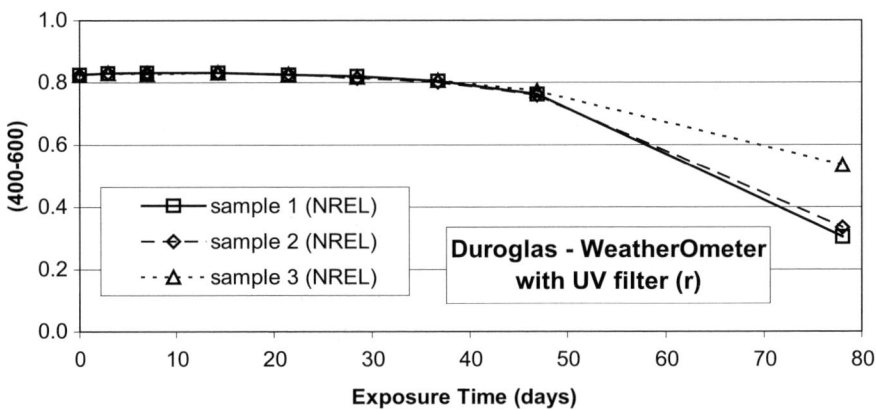

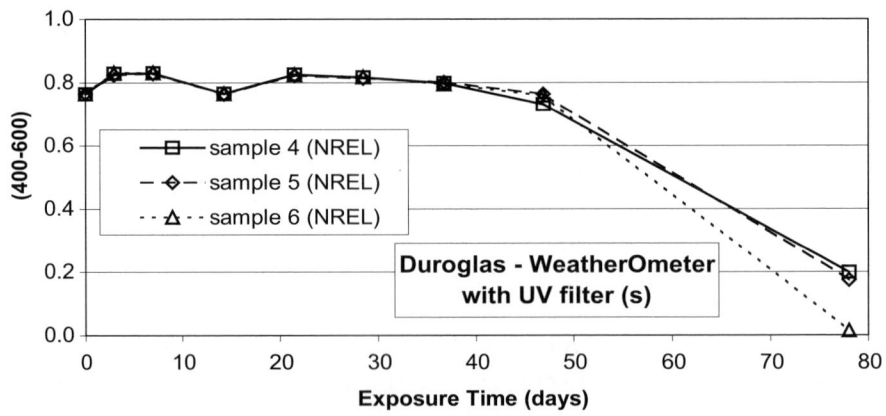

Figure 6.2-4. Continued.

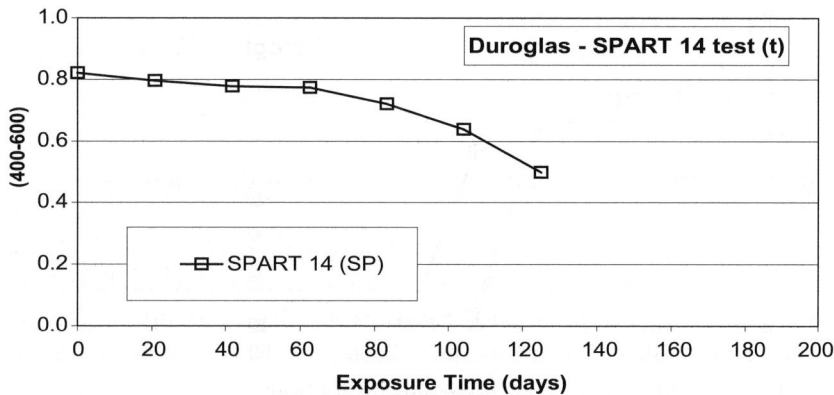

Figure 6.2-4. Continued.

accelerated exposure. Similar results occur for APEC 9353 (Figures 6.2-4 c and d). The measured loss in performance of the unscreened UV-stabilized material (c) takes roughly twice as long to occur as for the unscreened non UV-stabilized APEC 9351 (a). The screened APEC 9353 (d) demonstrated improved UV resistance but does start to degrade after about a year of accelerated exposure. Results for unscreened APEC 9353 exposed in the SPART 14 chamber (Figure 6.2-4 e) are in good agreement with Ci5000 data (c). However, exposure of APEC 9353 in the unfiltered metal halide chambers (Figure 6.2-4 f) is much more severe than in the Ci5000 and SPART 14. With the unfiltered metal halide light source, a 15–20% loss in $\tau_{(400-600)}$ occurs after only 25 days, whereas it took roughly 100 days for an equivalent loss to occur in the Ci5000 and SPART 14. The effect of temperature only on APEC 9353 was tested by exposure at 80°C with no light (Figure 6.2-4 g); no loss in $\tau_{(400-600)}$ was measured.

Accelerated exposure test results for Duroglas PVC are provided in Figures 6.2-4 h through t. The effect of temperature only (no light) on Duroglas was tested by exposure at 40, 60, 65, 70, 75, and 80°C; results are presented in Figures 6.2-4 h through m, respectively. Thermal degradation was found only at a temperature above the glass transition temperature of PVC (m). Exposure of Duroglas in the unfiltered metal halide chambers at a variety of temperature and RH conditions produced rapid degradation (Figure 6.2-4 n and o). Results for Ci5000 WOM exposures were less severe (Figure 6.2-4 p and q), although precipitous degradation did occur in fairly short time periods. The addition of a Korad UV screen does not significantly improve the optical durability of Duroglas (Figure 6.2-4 r and s); apparently light at wavelengths greater than 385 nm that are transmitted by Korad can damage the PVC material. Results for Duroglas exposed to the SPART 14 chamber conditions are shown in Figure 6.2-4 t. Degradation is considerably less than for the unfiltered metal halide and Ci5000 chamber exposures. In Section 6.2.5 we show how the accelerated test data can be used to derive correlations for Duroglas and APEC 9353 with outdoor exposure test results presented in Figure 6.2-3.

6.2.2 INITIAL RISK ANALYSIS OF POTENTIAL FAILURE MODES

The results of the IEA Working Group Materials in Solar Thermal Collectors on PVC and UV-stabilized PC cover plate materials are now used to illustrate how the general methodology can be applied for assessing the durability of polymeric-type materials. The details of this process have been previously reported (Carlsson *et al.*, 2003). The first step in the general methodology is to analyze the potential failure modes. In this work, we proceeded by (a) specifying, from an end-user point of view, the expected function of the component and its materials, its performance and its service life requirement, and the intended in-use environments; (b) identifying the important functional properties by defining the performance of the component and its materials, relevant test methods and requirements for qualification of the component with respect to performance; (c) identifying the potential failure modes and degradation mechanisms, relevant durability or life tests, and requirements for qualification of the component and its materials with respect to durability; and (d) estimating the risks associated with different failure modes. After considering the polymeric glazing materials studied in the IEA Solar Materials Working Group, the results of the initial risk analysis are summarized in Tables 6.2-3 through 6.

Table 6.2-3. Specification of end-user requirements for the glazing materials studied in the IEA MSTC case study

Function and general requirements	General requirements for performance and service life	In-use conditions and severity of environmental stress
Efficiently transmit solar radiation to the absorber Suppress heat losses by creating an insulating air gap Project the absorber against the outdoor environment	Loss in material performance should not result in reduction of the solar system performance (solar fraction) by more than 5%, in a relative sense, during the service life of the material Material service life should exceed 25 years	Above the solar absorber, which means exposure to temperatures that may reach $\sim 200^\circ$C (stagnation) Exposed to condensation of water Exposed to outdoor environments, which means UV-radiation, humidity, rain, snow, hail, wind loads, air pollutants, high and low temperatures, etc. Exposed to mechanical loads like tensile stress because of its own weight, wind loads, thermal expansion, and impact of various objects, e.g., birds, stones.

Table 6.2-4. Specification of functional properties and requirements for the glazing materials studied in the IEA MSTC case study

Critical functional property	Test methods for the functional property	Requirement for functional capability or performance
Solar transmittance, τ_{sol}		$-\Delta\tau \leq 0.05$
Tensile strength:	ISO 179	
Stress at break		50 MPa
Elongation at break		2%
Impact strength	ISO R 180, ISO/R 179-1961	10 kJ/m^2
Shape performance under thermal load-heat deflection temperature		> 150°C
Critical extension		2%
Maximum continuous service temperature		> 150°C
Brittleness temperature		−40°C

6.2.3 SCREENING TESTS AND ANALYSES FOR SERVICE LIFE PREDICTION

The analysis for the combined stress and failure or damage mode forms the basis for identifying suitable durability tests; as a rule, durability tests for accelerated life testing have to be customized to the materials being tested and the environmental stresses they are expected to experience during service. Screening testing (see Chapter 4.4), which is the first step in accelerated life testing, is conducted for the purpose of qualitatively assessing the importance of the different degradation mechanisms and degradation factors identified in the initial risk analysis of potential life-limiting processes. The main purpose of screening testing is to identify which stress factors contribute to deterioration in performance of the studied component or part(s) of a component. From the result of the tests, the most suitable range for the different degradation factors may also be determined for the purpose of life testing. When selecting the most suitable test methods for screening testing, it is important to select those with test conditions representing the most critical combination of degradation factors. We show in Tables 6.2-8 and 9 for PC and PVC, respectively, how the program of screening testing may be presented and is based on the recommendations made in the general methodology of the IEA Task 27 Project B1 (Carlsson, 2002).

Using artificially aged samples from the screening testing, changes in the key functional properties or the selected degradation indicators are analyzed with respect to the associated changes in the materials. The analyses were made to identify the predominant degradation mechanisms of the materials. The techniques that can be

Table 6.2-5. Risk assessment for each potential failure mode or process using FMEA

Failure/Mode Degradation process	Severity (S)		Probability of Occurrence (P_O)		Probability of Detection (P_D)		Risk (R) ($R = S \cdot P_O \cdot P_D$)
	S	Rating #	P_O	Rating #	P_D	Rating #	
For each failure mode/degradation process listed, estimate S, P_O, and P_D	No effect on product	1	Unlikely that failure will occur	1	Failure that is always noted; probability of detection > 99.99%	1	Compute R from estimates for S, P_O, and P_D
	Minor effect on product	2–3	Very low probability for failure to occur	2–3	Normal probability of detection 99.7%	2–4	
	Risk of failure in product function	4–6	Low probability for failure	4–5	Certain probability of detection > 95%	5–7	
	Certain failure in product functioning	7–9	Moderate probability for failure to occur	6–7	Low probability of detection > 90%	8–9	
	Failure that may affect personal safety	10	High probability of failure to occur	8–9	Failures will not be found; cannot be tested	10	
			Very high probability for failure to occur	10			

Table 6.2-6. FMEA initial risk analysis for the PC glazing studied in the IEA MSTC case study

Failure/Damage mode/Degradation process	Degradation indicator	Critical degradation factors in the environmental stresses	Estimated risk of the failure or damage mode from FMEA			
			S	P_O	P_D	R
Unacceptable loss in optical performance	$PC = -\Delta\tau$					
High T oxidation	UV-VIS-NIR τ	High T*	7	5	7	245
Photooxidation	UV-VIS-NIR τ	UV-radiation, high T*	7	8	7	392
Hydrolysis combined with photooxidation	UV-VIS-NIR τ	Humidity, condensation, air pollutants	7	3	7	147

*In many high temperature applications, durability is limited because of the physical and/or chemical loss of stabilizers, e.g., antioxidants, UV-absorbers, HALS, etc.

Table 6.2-7. Risk assessment and service reliability for PVC glazing

Failure/Damage mode/Degradation process	Degradation indicator	Critical degradation factors in the environmental stresses	Estimated risk of the failure or damage mode from FMEA			
			S	P_O	P_D	R
Unacceptable loss in optical performance	$PC = -\Delta\tau$					
High T degradation – Dehydrochlorination – Formation of conjugated double bonds – Polyenes	UV-VIS-NIR τ	High T*	7	8	7	392
Photooxidation	UV-VIS-NIR τ	UV-radiation high T*	7	8	7	392
Effects of acid rain – Acid catalyzation of degradation – dehydrochlorination	UV-VIS-NIR τ	Humidity, condensation, air pollutants	7	2	7	98
Physical aging	Changes in the glass transition temperature, DSC	Long-term relaxation to equilibrium by amorphous materials	7	2	8	112

*In many high-temperature applications, durability is limited because of the physical and/or chemical loss of stabilizers, e.g., antioxidants, UV-absorbers, HALS, etc.

used to study the degradation of the PC and PVC glazing materials are shown in Tables 6.2-10 and 11.

In the work undertaken by the IEA Working Group on Solar Materials, the conclusions concerning degradation of PC and PVC were based almost entirely on UV-VIS-NIR transmittance spectroscopy. It was, for example, assumed that hydrolysis plays a minor role in the degradation of PC in solar collector applications.

Table 6.2-8. Program for screening testing of PC in the IEA MSTC polymeric glazing case study

Possible degradation mechanism	Critical periods of high environmental stress	Suitable accelerated test methods and range of degradation factors
High *T*, thermal oxidation*	Stagnation conditions of solar collector at high levels of solar irradiation (no heat outflow from the collector)	Constant-load, high T exposure tests in the range of 90–150°C
Photooxidation*	At high levels of solar irradiation	WOM test – ISO 4892 UV, T, and RH
Hydrolysis combined with photooxidation	Under humidity conditions involving condensation of water on the glazing	WOM SPART 14 test – an acid rain modification of SAE J1960 (SAE J1960)

*During outdoor exposure it is impossible to distinguish between high temperature and photoinduced oxidation.

Table 6.2-9. Program for screening testing of PVC in the IEA MSTC polymeric glazing case study

Possible degradation mechanism	Critical periods of high environmental stress	Suitable accelerated test methods and range of degradation factors
High T (photo) degradation* – Dehydrochlorination – Formation of conjugated double bonds – Polyenes	During stagnation conditions of solar collector at high levels of solar irradiation (heat outflow from the collector)	Constant load high T exposure tests up to a maximum of 80°C**
(Photo)oxidation***	At high levels of solar irradiation	WOM tests – ISO 4892 UV, T, and RH
Effects of acid rain – Acid catalyzation of degradation – Dehydrochlorination	During humid conditions involving condensation of water on the glazing	WOM SPART 14 test – an acid rain modification of SAE J1960 (SAE J1960)

*The dehydrochlorination degradation is a nonoxidative process leading to yellowing. In the presence of oxygen the formed polyenes may oxidize leading to bleaching.
**Concerning PVC a temperature of 80°C might be too high. A more reasonable maximum temperature is 70°C.
***During outdoor exposure high temperature and photo-induced dehydrochlorination and oxidation occur at the same time.

The two most dominating degradation factors were thus assumed to be high temperature and UV radiation. Techniques besides UV-VIS-NIR spectroscopy are included in Table 6.2-10 and 11 to give a more complete list of possible analytical methods available and possible results of these methods.

To be able to predict expected service life of the component and its materials from the results of accelerated aging tests, the degradation factors during service

Table 6.2-10. Techniques for analysis of materials changes upon durability testing of PC

Degradation mechanism	Techniques for analysis of materials changes	Expected results if degradation is by the assumed mechanism
High T oxidation	UV-VIS-NIR τ spectroscopy	Reduction of τ_{sol} – yellowing
	FTIR-spectroscopy	Chemical changes, e.g., formation of OH^-
	DSC (OIT)	Changes in stabilization
	SEM	Small changes in surface topography
	TOF-SIMS, XPS	Chemical changes – oxidation
	Mechanical testing*	Changes on mechanical properties
Photooxidation	UV-VIS-NIR τ spectroscopy	Reduction of τ_{sol}
	FTIR-spectroscopy	Chemical changes, e.g., formation of OH^-
	DSC (OIT)	Changes in stabilization
	SEM	Small changes in surface topography
	TOF-SIMS, XPS	Chemical changes – oxidation
	Mechanical testing*	Changes in mechanical properties
Hydrolysis combined with photooxidation	UV-VIS-NIR τ spectroscopy	Reduction of τ_{sol}
	FTIR-spectroscopy	Chemical changes, e.g., formation of OH^-
	DSC (OIT)	Changes in stabilization
	SEM	Small changes in surface topography
	TOF-SIMS, XPS	Chemical changes
	Mechanical testing*	Changes in mechanical properties

*tensile and impact strength

conditions need to be assessed from measured data. The techniques used in the IEA Solar Material Working Group case study are given in Table 6.2-12.

6.2.4   SERVICE LIFE PREDICTION FROM RESULTS OF ACCELERATED AGING

6.2.4.1   Mathematical Modeling

Possible mechanisms of degradation of the PC glazing were assumed to be (a) photooxidation (b) thermal oxidation, and (c) combined photooxidation and hydrolysis. From the screening tests, it was concluded that only photooxidation contributes significantly to the service life of the glazing. A suitable time-transformation function is shown in Table 6.2-13.

Concerning the PVC glazing, degradation mechanisms that could reduce the service life were assumed to be (a) dehydrochlorinization, (b) photooxidation, and (c) physical aging. For (a), the mechanism is a chain reaction type because hydrogen chloride formed from the dehydrogenation reaction acts also as a catalyst for this reaction. The reduction is consequently difficult to model mathematically in a simple way and thus, it is also difficult to express the rate of degradation in terms of a time-transformation function. During screening testing, significant degradation was

Table 6.2-11. Techniques for analysis of materials changes upon durability testing of PVC

Degradation mechanism	Techniques for analysis of materials changes	Expected results if degradation by the assumed mechanism occurs
High T oxidation – Dehydrochlorinaton	UV-VIS-NIR τ spectroscopy	Reduction of τ_{sol} – yellowing
	FTIR-spectroscopy	Chemical changes, e.g., formation of OH^-, $C=O$
	Dehydrochlorination Measurements-ISO 182	Residual stability (pHTS)
	DSC	Physical aging
	SEM	Small changes in surface topography
	TOF-SIMS, XPS	Chemical changes – oxidation
	Mechanical testing	Changes in mechanical properties
Photooxidation	UV-VIS-NIR τ spectroscopy	Reduction of τ_{sol} – yellowing
	FTIR-spectroscopy	Chemical changes, e.g., formation of OH^-, $C=O$
	Dehydrochlorination Measurements-ISO 182	Residual stability (pHTS)
	DSC	Physical aging
	SEM	Small changes in surface topography
	TOF-SIMS, XPS	Chemical changes – oxidation
	Mechanical testing	Changes in mechanical properties
Effects of acid rain – acid catalyzation of degradation	UV-VIS-NIR τ spectroscopy	Reduction of τ_{sol} – yellowing
	FTIR-spectroscopy	Chemical changes, e.g., formation of OH^-, $C=O$
	Dehydrochlorination Measurements-ISO 182	Residual stability (pHTS)
	DSC	Physical aging
	SEM	Small changes in surface topography
	TOF-SIMS, XPS	Chemical changes – oxidation
	Mechanical testing	Changes in mechanical properties

observed in dark thermal aging tests only when the temperature was above the glass transition temperature of PVC. The best time-transformation function for the PVC degradation was the same general photooxidation time-transformation function used to model the degradation of the PC glazing (see Table 6.2-14 and 15). Acceleration of the physical aging of PVC by a suitable test was not found possible, so physical aging had to be excluded in the accelerated life-testing program for the glazing.

6.2.4.2 Accelerated Life Testing and Assessment of Expected Service Life

During life-testing, PC and PVC glazing materials were exposed in an Atlas Ci-5000 WeatherOmeter®, which has a UV intensity of about 2X compared to

Table 6.2-12. Techniques used for measurement of degradation factors in mini solar collectors and outdoors

Degradation Mechanism	Degradation factors and Measured variables	Sensors
High T degradation	T: Surface T of the collector glazing	Pt sensors glued to the inside of the cover glazing
Photooxidation	*UV-radiation and surface temperature (see above)*	UVA and UVB sensors
Hydrolysis combined with photooxidation	RH: RH of the air in the collector box	Capacitance RH sensors carefully shielded from solar radiation and thermal radiation of the ambient
Effects of acid rain	*Atmospheric corrosivity:* Measurement of corrosion mass loss rate of standard metal specimen/coupons	Metal coupons of carbon steel, zinc, and copper and evaluation of the corrosion mass loss according to ISO 9226
	Coupons are exposed both inside mini collectors and outdoors	Exposed metal coupons analyzed with respect to the sulphate content of the corrosion products by EDX

Table 6.2-13. Estimated service life of the UV-stabilized PC glazing material for an outdoor exposure with the glazing inclined 45° from horizontal and facing south

Degradation mechanism	Time-transformation function	Estimated service life with $-\Delta\tau_{sol} < 0.05$[1]
Photooxidation (PO)	$a_{PO} = \{I^P \cdot \exp[-(E/kT)]\}_{EFF,D} / \{I^P \cdot \exp[-(E/kT)]\}_{EFF,s} = PO_{EFF,D}/PO_{EFF,s}$ in which: a = acceleration factor I = intensity of Photoactive light P = material dependent constant; the value has to be determined from accelerated testing s = index for service; D = index for test EFF = effective mean value	5 years in Golden, CO 6 years in Phoenix, AZ

[1]$-\Delta\tau_{sol} = 0.05$ corresponds to a decrease in the solar system performance of 5%.

typical outdoor terrestrial irradiation intensities) and at 50X and 100X using a UV concentrator in a series of tests at varying light intensities and surface temperatures (Jorgensen *et al.*, 1999, 2002). Hemispherical transmittance measurements were made to characterize the loss in optical performance of the

Table 6.2-14. Estimated service life of the PVC glazing material for an outdoor exposure with the glazing inclined 45° from horizontal and facing south

Degradation mechanism	Time-transformation function	Estimated service life with $-\Delta\tau_{sol} < 0.05$[1]
Dehydrogenation and Photooxidation (PO)	$a_{PO} = \{I^P \cdot \exp[-(E/kT)]\}_{EFF,D} /$ $\{I^P \cdot \exp[-(E/kT)]\}_{EFF,s} = PO_{EFF,D}/PO_{EFF,s}$ in which: a = acceleration factor I = intensity of photoactive light P = material dependent constant; the value has to be determined from accelerated testing s = index for service; D = index for test EFF = effective mean value	About 0.5 years in Golden, CO About 0.75 years in Phoenix, AZ

[1] $-\Delta\tau_{sol} = 0.05$ corresponds to a decrease in the solar system performance of 5%.

Table 6.2-15. Coefficients derived from accelerated exposure for the tested polymeric glazing materials

Polymer glazing	p	E (kJ/mole)
PVC	0.669	35.33
UV-stabilized PC	1.093	28.00

glazing materials during exposures. Performance-versus-time data were thereafter used to determine the parameters of the time-transformation function (see Table 6.2-13). The results are shown in Table 6.2-15. Values of the activation energies, E, derived are reasonable for photothermal degradation mechanisms. The value of $p \sim 2/3$ for PVC indicates that the exposure to 50–100X light intensities had a net effect of only 15–25X. We conclude from the reduced net effect that some shielding or rate limiting reactions occur and do not allow all photons to participate in degradation. For the UV-stabilized PC sample, the value of $p = 1$ indicates that exposure of this material follows strict reciprocity even up to 100X. Thus, all incident photons fully contribute to the degradation reactions and proceed at twice the rate undergone at 50X exposure and 50 times the rate experienced at 2X exposure.

The predicted service life for the glazing materials satisfying the performance criteria $-\Delta\tau_{sol} < 0.05$ was estimated based on climatic data from measurements of surface temperatures and UV-light intensities. The results are shown in Tables 6.2-13 and 14.

6.2.5 VALIDATION OF METHODOLOGY

If it is assumed that the rate in transmittance change is constant if the surface temperature and the UV-light intensity are maintained at the same values during the time interval Δt_i, then the transmittance change $\Delta \tau_i$ may be expressed as

$$\Delta \tau_i = A(I_{UV})^p \Delta t_i e^{-E/kT} \tag{1}$$

using the time-transformation function shown is Table 6.2-13. The parameter A is a constant independent of surface temperature and light intensity but it is material dependent. It may be determined from the same series of aging tests as used to determine the activation energy E and the parameter p. For $\Delta \tau_i$ equal to the mean global transmittance between 400 and 500 nm, the values of A were estimated as 2892 $(MJ/m^2)^{-1}$ for PVC and 5.497 $(MJ/m^2)^{-1}$ for UV-stabilized PC. By integrating Eqn. 1, Eqn. 2 is obtained.

$$\Delta \tau_i(t) = A \int_0^t [I_{UV}(t)]^p e^{-E/kT(t)} dt \tag{2}$$

Applying Eqn. 2, the expected transmittance after different time-periods of outdoor exposure may be estimated.

Using the values of the coefficients E and p from Table 6.2-15 and the time-monitored values of sample temperature and UV irradiance, the loss in performance was predicted for both the PVC and the UV-stabilized PC as exposed outdoors in Golden, CO, and Phoenix, AZ. Predicted values were then compared with actual measured data for these materials exposed at these sites.

The results are shown in Figure 6.2-5. The time-dependent changes in the weathering variables result in the irregular shapes of the predicted curves. Excellent agreement is evident between the measured and predicted data. Thus, the phenomenological approach (see Chapter 4.6) to data analysis is validated, i.e., obtaining model coefficients from accelerated test results and then using these coefficients to predict time-variable in-service degradation.

6.2.6 CONCLUSIONS

A large amount of durability test data for both accelerated laboratory conditions and outdoor in-service conditions have been presented for PC and PVC glazing materials. Some of the accelerated exposure data were used to demonstrate how to derive damage functions that allow prediction of performance degradation. This

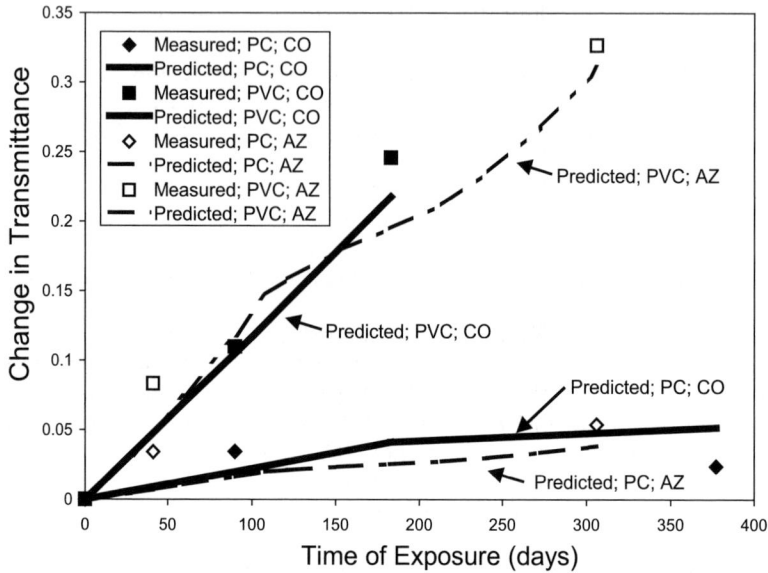

Figure 6.2-5. Comparison of the actual and predicted changes (loss) in hemispherical transmittance between 400 and 500 nm of PVC and PC for exposures of up to 380 days at Golden, CO and Phoenix, AZ.

methodology also allows the effect of multiple stress factors to be modeled. The usefulness and validity of this approach has then been confirmed by comparing predicted results with actual measured data for samples exposed to variable outdoor conditions. Consequently, highly abbreviated testing times at elevated stress conditions can be substituted for long-time exposures at lower stress levels. The procedure developed will allow much shorter development cycle times for new materials and will allow improvements to be identified and readily incorporated into new products.

REFERENCES

Carlsson, B. (Ed.): General Methodology of Accelerated Testing for Assessment of Service Life of Solar Thermal Components (2002). Final Project B1 Report, IEA Task 27, Performance of Solar Facade Components, International Energy Agency Solar Heating and Cooling Programme, November 2002.

Carlsson, B., Möller, K., Köhl, M., Heck, M., Brunold, S., Marechal, J.-C. and Jorgensen, G. (2003). Weathering of Polymer Products: Assessment of Service Life of Solar Thermal Components by Accelerated Life Testing, *Proceedings of the 1st European Weathering Symposium EWS (XXIIIrd Colloquium of Danubian Countries on Natural and Artificial*

Ageing of Polymers), Gesellschaft für Umweltsituation e.V., Prague, Czech Republic, September 25–26, 2003.

General Methodology of Accelerated Testing for Assessment of Service Life of Solar Thermal Components (2002). B. Carlsson (Ed.). Final Project B1 Report, IEA Task 27, Performance of Solar Façade Components, International Energy Agency Solar Heating and Cooling Programme, November 2002.

Jorgensen, G., Bingham, C., Netter, J., Goggin, R., Lewandowski, A. (1999). A Unique Facility for Ultra-Accelerated Natural Sunlight Exposure Testing of Materials, In: D. R. Bauer and J. W. Martin (Eds), *Service Life Prediction of Organic Coatings, A Systems Approach, ACS Symposium Series 722*, American Chemical Society, Oxford University Press: Washington, DC, 170–185.

Jorgensen, G., Bingham, C., King, D., Lewandowski, A., Netter, J., Terwilliger, K. and Adamsons, K. (2002). Use of Uniformly Distributed Concentrated Sunlight for Highly Accelerated Testing of Coatings, In: J. W. Martin and D. R. Bauer (Eds), *Service Life Prediction Methodology and Metrologies ACS Symposium Series 805*, American Chemical Society, Oxford University Press, Washington, DC, 100–118.

Jorgensen, G., Brunold, S., Carlsson, B., Heck, M.; Köhl, M. and Möller, K. (2003). Durability of polymeric glazing materials for solar applications, *Proceedings of the 1st European Weathering Symposium EWS (XXIIIrd Colloquium of Danubian Countries on Natural and Artificial Ageing of Polymers)*, Gesellschaft für Umweltsituation e.V., Prague, Czech Republic, September 25–26, 2003.

Meteonorm (1995). Meteorologische Grundlagen für die Sonnenenergienutzung; Meteotest im Auftrag des Bundesamtes für Energiewirtschaft (BEW); 1995; NovaEnergie GmbH, Schachenallee 29, CH-5000 Aarau AG.

SAE J1960, Accelerated Exposure of Automotive Exterior Materials Using a Controlled Irradiance Water-Cooled Xenon Arc Apparatus. Society of Automotive Engineers, 400 Commonwealth Drive, Warrendale, PA 15096.

Index

389